SH 329 .S53 S63 2011

Small-scale fisheries management

MHCC WITHDRAWN

D1773285

Small-scale Fisheries Management

Frameworks and Approaches for the Developing World

Small-scale Fisheries Management

Frameworks and Approaches for the Developing World

Edited by

Robert S. Pomeroy

*University of Connecticut, Connecticut Sea Grant, Connecticut, USA
and
WorldFish Center, Penang, Malaysia*

and

Neil L. Andrew

WorldFish Center, Penang, Malaysia

CABI is a trading name of CAB International

CABI Head Office
Nosworthy Way
Wallingford
Oxfordshire OX10 8DE
UK

Tel: +44 (0)1491 832111
Fax: +44 (0)1491 833508
E-mail: cabi@cabi.org
Website: www.cabi.org

CABI North American Office
875 Massachusetts Avenue
7th Floor
Cambridge, MA 02139
USA

Tel: +1 617 395 4056
Fax: +1 617 354 6875
E-mail: cabi-nao@cabi.org

© CAB International 2011. All rights reserved. No part of this publication may be reproduced in any form or by any means, electronically, mechanically, by photocopying, recording or otherwise, without the prior permission of the copyright owners.

A catalogue record for this book is available from the British Library, London, UK.

Library of Congress Cataloging-in-Publication Data

Small-scale fisheries management: frameworks and approaches for the developing world / edited by Robert S. Pomeroy and Neil L. Andrew.
　　　　 p. cm.
　Includes bibliographical references and index.
　ISBN 978-1-84593-607-5 (hardback: alk. paper)
1. Small-scale fisheries–Developing countries.　 2. Fishery management–Developing countries.　 I. Pomeroy, R. S. (Robert S.)　 II. Andrew, Neil.
III. Title.

　SH329.S53S63 2011
　338.3'727–dc22

2010033561

ISBN-13: 978 1 84593 607 5

Commissioning editor: Rachel Cutts
Production editor: Fiona Chippendale

Typeset by AMA DataSet Ltd, Preston.
Printed and bound in the UK by CPI Antony Rowe, Chippenham.

Contents

Contributors		vii
Preface		ix
1.	Under-reported and Undervalued: Small-scale Fisheries in the Developing World *David J. Mills, Lena Westlund, Gertjan de Graaf, Yumiko Kura,* *Rolf Willman and Kieran Kelleher*	1
2.	Approaches and Frameworks for Management and Research in Small-scale Fisheries *Neil L. Andrew and Louisa Evans*	16
3.	Diagnosis and the Management Constituency of Small-scale Fisheries *Louisa Evans and Neil L. Andrew*	35
4.	Human Rights and Fishery Rights in Small-scale Fisheries Management *Anthony Charles*	59
5.	Managing Overcapacity in Small-scale Fisheries *Robert S. Pomeroy*	75
6.	Adaptive Management in Small-scale Fisheries: a Practical Approach *John Parks*	93
7.	Conditions for Successful Co-management: Lessons Learned in Asia, Africa, the Pacific and the wider Caribbean *Robert S. Pomeroy, Joshua E. Cinner and Jesper Raakjær Nielsen*	115
8.	Climate Change and Other External Drivers in Small-scale Fisheries: Practical Steps for Responding *Stephen J. Hall*	132
9.	Developing Markets for Small-scale Fisheries: Utilizing the Value Chain Approach *Eusebio R. Jacinto and Robert S. Pomeroy*	160

10.	Communication *Patrick McConney and Carmel Haynes*	178
11.	**Small-scale Fisheries Compliance: Integrating Social Justice, Legitimacy and Deterrence** *Maria Hauck*	196
12.	**Poverty Reduction as a Means to Enhance Resilience in Small-scale Fisheries** *Edward H. Allison, Christophe Béné and Neil L. Andrew*	216

Index 239

Contributors

Allison, Edward H., WorldFish Center, PO Box 500 GPO, 10670 Penang, Malaysia; E-mail: e.allison@cgiar.org

Andrew, Neil L., WorldFish Center, PO Box 500 GPO, 10670 Penang, Malaysia; E-mail: n.andrew@cgiar.org

Béné, Christophe, WorldFish Center, PO Box 500 GPO, 10670 Penang, Malaysia; E-mail: c.bene@cgiar.org

Charles, Anthony, Management Science/Environmental Studies, Saint Mary's University, Halifax, Nova Scotia, Canada B3H3C3; E-mail: tony.charles@smu.ca

Cinner, Joshua E., ARC Center of Excellence for Coral Reef Studies, James Cook University, Townsville, QLD 4811, Australia; E-mail: joshua.cinner@jcu.edu.au

de Graaf, Gertjan, Fisheries and Aquaculture Department, Food and Agriculture Organization of the United Nations, via delle Terme di Caracalla, Rome 00153, Italy; E-mail: gertjan.degraaf@fao.org

Evans, Louisa, ARC Center of Excellence for Coral Reef Studies, James Cook University, Townsville, QLD 4811, Australia (formerly with WorldFish Center); E-mail: louisa.evans@jcu.edu.org

Hall, Stephen J., WorldFish Center, PO Box 500 GPO, 10670 Penang, Malaysia; E-mail: s.hall@cgiar.org

Hauck, Maria, Environmental Evaluation Unit, University of Cape Town, Private Bag X3, Rondebosch, 7701, South Africa; E-mail: maria.hauck@uct.ac.za

Haynes, Carmel, Centre for Resource Management and Environmental Studies (CERMES), University of the West Indies, Cave Hill Campus, Barbados; E-mail: carmel.haynes@cavehill.uwi.edu

Jacinto, Eusebio R., Tambuyog Development Center, Quezon City, Philippines; E-mail: ted@tambuyog.org

Kelleher, Kieran, Agriculture and Rural Development Department, World Bank, Washington, DC 20433, USA; E-mail: kkelleher@worldbank.org

Kura, Yumiko, WorldFish Center, PO Box 1135 (Wat Phnom), Phnom Penh, Cambodia; E-mail: y.kura@cgiar.org

McConney, Patrick, Centre for Resource Management and Environmental Studies (CERMES), University of the West Indies, Cave Hill Campus, Barbados; E-mail: patrick.mcconney@cavehill.uwi.edu

Mills, David J., WorldFish Center, PO Box 500 GPO, 10670 Penang, Malaysia; E-mail: d.mills@cgiar.org

Parks, John, The Nature Conservancy of Hawai'i, 923 Nu'uanu Avenue; Honolulu, Hawai'i 96817, USA; E-mail: john_parks@tnc.org

Pomeroy, Robert S., Agricultural and Resource Economics/CT Sea Grant, University of Connecticut, Connecticut 06340-6048, USA, and WorldFish Center, PO Box 500 GPO, 10670 Penang, Malaysia; E-mail: robert.pomeroy@uconn.edu

Raakjær Nielsen, Jesper, Innovative Fisheries Management (IFM), Department of Development and Planning, Aalborg University, Aalborg, Denmark; E-mail: jr@ifm.aau.dk

Westlund, Lena, 148 Pinewood Crescent, Dartmouth, NS, B2V 2P9, Canada; E-mail: lena.westlund@swipnet.se

Willman, Rolf, Fisheries and Aquaculture Department, Food and Agriculture Organization of the United Nations, via delle Terme di Caracalla, Rome 00153, Italy; E-mail: rolf.willman@fao.org

Preface

This book is about small-scale fisheries (SSF) in the developing world, that is, most of the world's fisheries. Globally, about 97% of the people directly involved in fisheries work in the developing world and they catch about half the total world catch. In the developing world, SSF account for 56% of catch and 91% of people working in fisheries. Small-scale fisheries make important but poorly quantified contributions to national and regional economies, to the food security and development of many millions of people and provide an important lever for achieving the Millennium Development Goals, particularly within rural areas.

Individually, SSF have a range of attributes that make them and the people that depend upon them vulnerable to threats operating from the largest global scale (e.g. distortions in trade and markets, and climate change) to the smallest internally derived process (e.g. overfishing and conflict). As a generalization, fishers suffer poorly defined rights, are among the poorest and most marginalized parts of society and are poorly represented in national and international policy forums. When viewed from an assessment and management perspective, these attributes are often compounded by sparse data, weak institutions within communities, overfished stocks, degraded ecosystems and lack of alternative livelihoods. Past failures to address these issues have had significant social consequences and have affected livelihoods, increased vulnerability to poverty and meant less availability of fish protein per capita.

Historically, development interventions for the fisheries sector have sought to reduce poverty through accelerated economic growth, improvements in technology and infrastructure and market-led economic policy reform. The limited results of these interventions, however, has led to a re-examination of the causes of poverty, the recognition of the significance of vulnerability and the recognition of the need for new strategies for poverty reduction. There is increasing recognition that establishing appropriate pro-poor governance for fisheries management is central to maximizing the contribution of fisheries in poverty alleviation and food security. Pro-poor strategies that include rights-based approaches, co-management regimes and fishing capacity reduction are essential for increased wealth generation from small-scale fisheries.

In order for SSF to fulfil their potential as engines of social and economic change, we need appropriate frameworks and approaches to their management and governance. It is implausible to promise sustainable SSF in the developing world within the single-species biological yield maximization research and management paradigm that has dominated fisheries since the 1950s. The last decade or so has seen fisheries research and management broaden considerably in the search for better ways of doing things.

These developments have seen new approaches, concepts and methods, such as the precautionary principle, ecosystem approaches to management, Sustainable Livelihoods Approach, participatory methods and co-management, adaptive management, and so forth. Led by The United Nations Food and Agriculture Organization, important international instruments and laws have been promoted to normalize their use.

For all this endeavour there remains no unifying set of principles nor agreed structure for attacking the particular problem of SSF in the developing world. Furthermore, the more biological approaches are sometimes set as alternatives or in opposition to the 'people-centred' or economic approaches. This is unhelpful as, trivially, we need to integrate principles and concepts from all streams of enquiry.

Fishers, resource managers, policy makers, international development agencies, non-governmental organizations and academics continue to request guidance on the practical and applied tools, methods, approaches and strategies to managing small-scale fisheries. This book is meant to address this need by providing a guide to innovative and alternative management strategies and measures for small-scale fisheries. It is meant to provide guidance on specific applications and 'how to make it work in practice' on these management strategies and measures. It serves as a companion volume to two recent books on small-scale fisheries management – the 2001 book *Managing Small-scale Fisheries: Alternative Directions and Methods*, authored by Berkes, Mahon, McConney, Pollnac and Pomeroy; and the 2006 book *Fishery Co-management: a Practical Handbook*, authored by Pomeroy and Rivera-Guieb.

The book is composed of 12 chapters from authors who have years of experience working on small-scale fisheries management all over the world. The authors represent academic, research, governmental and non-governmental institutions and organizations. The chapters range in content from adaptive management, to markets, to co-management, to communication, to enforcement and compliance, to managing overcapacity, to human rights and fishing rights.

Small-scale fisheries are full of innovation. While there is reason to be pessimistic about their future, that innovation and the resilience that comes from it give cause for optimism. We agree with the many commentators that write of a transition or revolution in the management and governance of the sector. We hope that the approaches and frameworks presented in this book contribute to that process and the hope of better lives for people who depend upon these resources.

The editors acknowledge and thank the contributors for their patience and insightful chapters. Thanks to Shwu Jiau Teoh, Sally McNeill, Suhaila Abdullah, Shabeen Ikbal and Ben Starkhouse for editorial support. Rachel Cutts from CABI was steadfast in her support throughout the extended gestation of the book – we are grateful for guidance and advice. We are grateful to Kristian Parker and the Oak Foundation, and The WorldFish Center for financial support.

The Editors

1 Under-reported and Undervalued: Small-scale Fisheries in the Developing World

David J. Mills, Lena Westlund, Gertjan de Graaf, Yumiko Kura, Rolf Willman and Kieran Kelleher

Small-scale Fisheries: a Shifting Target

Perhaps in the 1950s, the term 'small-scale' could have been used neatly to categorize a discrete set of fisheries: those operated by family or community interests, fished in coastal or inland waters from shore or small vessels, used simple technology with low levels of mechanization and fed market chains that were short and local. This is clearly no longer the case. Globally, fisheries have evolved and diversified at a breakneck pace, nowhere more so than in the small-scale sector of developing countries. Rather than discrete boundaries, we are confronted with a continuum in which the lines defining scale have been bridged in particular through integration of markets (e.g. Kurien, 1998; Ponte et al., 2007), technology transfer (e.g. Platteau, 1984; Jensen, 2007) and changing socio-economic contexts (e.g. Hapke, 2001; Warhaft, 2001; FAO, 2004).

Although encompassing considerable diversity, the term 'small-scale' clearly remains a useful descriptor for a group of fisheries unified by social, structural and institutional characteristics affecting governability as well as their vulnerability context. It is appropriate, then, that recent commentaries on the nature of small-scale fisheries (SSF), notably by Johnson (2006) and Kurien and Willman (2009), focus on the social organization and structural characteristics of the subsector while downplaying the classic technological descriptors such as vessel size or gear type. These commentaries adopt a 'broad brush' approach by highlighting commonalities in operational context. Fundamental among descriptors are the decentralized nature of fishing activities, relatively low levels of capitalization, the household or community focus of enterprise structure and the dynamism of fishing activities in space, time and technology.

In the face of increasing pressure on fisheries globally (Delgado et al., 2003; FAO, 2009; Smith et al., 2010), there is a renewed call from the international scientific and development communities for SSF governance to be reshaped according to the specificities and vulnerability context of the sector and its potential to generate economic benefits beyond the participants in the fishery (FAO-SEAFDEC, 2005; Andrew et al., 2007; Béné et al., 2007; Garcia et al., 2008; Mahon et al., 2008; World Bank/FAO, 2009). Reforms must contribute to a more sustainable fisheries sector, but in doing so seek to protect and sustain the broader benefits accruing from SSF, particularly to the poor. There exists a significant and growing compendium of case studies highlighting the food security (e.g. Thompson et al., 2002; Bell et al., 2009; Singh, 2009), economic (e.g. Neiland et al., 2000; FAO, 2006; Béné et al.,

2010a) and livelihood (e.g. Allison, 2005; Smith et al., 2005) value of SSF in the developing world. Yet these benefits are increasingly at risk, as the vulnerability among such fisheries to internal (e.g. Salayo et al., 2008; Béné, 2009) and external (e.g. Cochrane and Doulman, 2005; FAO, 2006; Mills et al., 2009) shocks is rarely considered in policy development.

A necessary dimension of effective reform is coherent policy and institutional arrangements, from practical action by fishers and stakeholders through to enabling legislation and policy support at a national level (Berkes, 2003; FAO-SEAFDEC, 2005; Hauck and Sowman, 2005; Andrew et al., 2007; Grafton et al., 2008; Mahon et al., 2008). Widespread implementation of improved systems of governance has been held back by limited progress in mainstreaming SSF in national fishery policy or national development agendas (Walmsley et al., 2006; Thorpe et al., 2007; Friend, 2009), but for recent progress see Kébé (2008a) and Reid et al. (2008). Similarly, information regarding relativities among fisheries sectors is often inaccurate or absent. This situation is set to continue while critical information on the importance, potential and vulnerability of SSF remains largely within a limited number of case study outputs in the form of scientific reports.

National- and global-level syntheses that highlight the diversity and roles of SSF will help break this impasse. The first such headline-level global comparison of the small- and large-scale sectors was published by Thomson (1980). Thomson presented global estimates for employment, annual catches and fuel consumption as well as efficiency indicators in small- and large-scale marine fisheries. Notable conclusions were that small-scale marine fisheries produced almost the same quantity of fish for human consumption as large-scale fisheries, but caught those fish using up to four times less fuel and employed well over an order of magnitude more people than the large-scale sector. While acknowledged as rough calculations, subsequent studies provided support to Thomson's estimates (e.g. Kurien and Willmann, 1982). Thomson's insightful table has been updated on several occasions since (e.g. Lindquist, 1988; Berkes et al., 2001; Pauly, 2006). It is notable, however, that with the exception of Berkes et al. (2001) none consider inland fisheries, and with the exception of Pauly (2006) none are backed up with details of data sources or synthesis methodology.

In a quest to improve data quality and accessibility, the World Bank recently supported FAO and The WorldFish Center to engage with national experts to compile and synthesize information on SSF in selected developing countries (World Bank/FAO/WorldFish, 2010). The objectives were to collate and synthesize data highlighting the relative importance of SSF in comparison with large-scale fisheries, identify critical data gaps and possible solutions for these, and provide insights into the design of a platform for future data collection. While not directly replicating the outputs of Thomson (1980), the project updated many of Thomson's indicators, and further disaggregated marine and inland fisheries. In this chapter we present an overview of outputs from this project, and in doing so also highlight current concerns and limitations in national data systems, and lessons learned in the process of synthesizing collected data. The reader is referred to World Bank/FAO/WorldFish (2010) for additional methodological details.

Missing the Target: Where Current Data Systems Fall Short

National fisheries information systems globally are taken to provide a measure of the importance of fisheries resources at a country level, and a comparative measure of reliance on fisheries at a global scale. Good systems capable of supporting effective and equitable governance and providing timely feedback on management policy effectiveness are rare (Coates, 2002; FAO, 2003; FAO-SEAFDEC, 2005; Chuenpagdee et al., 2006). In the developing world there is frequently a mismatch between the nature of fisheries and the data systems used to characterize them. Collection of production data is dominated by direct measurement techniques such as surveys of catch from selected gear types, vessels, landing areas or markets. While undoubtedly useful where centralized management is strong, where licensing systems are the norm and where catch is landed in a limited number of ports, these characteristics are not well suited to highly dispersed and informal SSF. Compounding these issues, existing data systems in developing countries are almost universally under-resourced.

Understanding fisheries' livelihoods in developing countries requires a different set of metrics from those used in the developed world. While commercial fishing in developed countries is generally a full-time occupation, SSF activities in developing countries frequently constitute a component of a diversified livelihood or subsistence strategy. Such activities include fishing regularly for part of the working day (e.g. Williams and Awoyomi, 1998), seasonal fishing dictated by weather (e.g. Teh et al., 2007) or biology of target species (e.g. Kang et al., 2009) and opportunistic fishing to fill 'livelihood gaps' created by seasonality in other activities such as agriculture (Thompson et al., 2002; Hori et al., 2006). Particularly in inland systems, those who fish but do not consider themselves primarily fishers may outnumber by several-fold full-time fishers, for example in the Mekong Delta (Sjorslev, 2001) or in Bangladesh (Hussain, 2010). National population and economic census methods used in many countries record only primary occupation and may often misrepresent the nature of rural agricultural livelihoods (Keskinen, 2003), and may miss entirely a significant component of the fishing workforce. Sample sizes of national socio-economic household surveys that allow for deeper insights into livelihood strategies are often too small to provide useful information on fishing households.

Considerable value in the fishing sector lies in its ability to generate employment in preharvest (e.g. gear construction and maintenance, port facilities support, ice supply) and postharvest (processing, marketing and distribution) occupations (McGoodwin, 2001; Béné et al., 2007; Kébé, 2008b). Postharvest occupations in particular are often part-time and may be carried out by family members of fishers. The great contribution of women to these sectors in particular has long been recognized by researchers and the development community (ICSF, 2002; Williams et al., 2002; Chao et al., 2006; Choo et al., 2006). Again, the near-complete absence of data on all but primary occupations substantially underestimates the value generated by non-harvest activities.

While less common, in many countries women are also directly involved in fishing. This involvement again tends to be part-time, and centres on harvesting methods such as gleaning, small-cast or lift-net operation etc. (see Choo et al., 2006 and references therein; Holvoet, 2009). Fishing activities by women are often important in the direct supply of protein to families, and the risk-averse nature of women's fishing can be important for stability of protein supply to the household (Chapman, 1987; Bird, 2007). Some recent studies have highlighted a shift in the traditional division of labour, especially in inland water fisheries, where the economic necessity and need for food security override social norms (ODI, 2002; FiA, 2007; CBNRM-LI and FiA, 2008). These processes and the implications of such changes in fishing patterns must be considered in the implementation of effective management.

Collectively, the nature of involvement of women in both harvest and non-harvest activities, and the inability of data systems to capture these activities, have contributed to a lack of recognition of the important role of women in fisheries. This in turn leads to lack of consideration in policy development and marginalization in decision-making. For these reasons, particular attention has been paid in the current study to estimating the number of women engaged in the sector.

Beyond Official Statistics: Country Case Studies

The project summarized here used case studies to collate disaggregated data in developing countries. Fifteen case studies were conducted across 17 countries: Bangladesh; Brazil; Cambodia; China; Ghana; India; Indonesia; Mozambique; Myanmar; Nigeria; Philippines; Senegal; Thailand; Vietnam; and around Lake Victoria (Kenya, Tanzania,

Uganda). Countries were selected based on the importance of fisheries as a source of livelihoods, income and nutrition. Inclusion was moderated by the need to achieve a reasonable global spread, and expert opinion of the study team regarding the likelihood of obtaining useful and robust data. The analysis focused on metrics for production, employment and postharvest use disaggregated among the key subsectors; namely, small- and large-scale, inland and marine fisheries. Additional data collected on efficiency indicators are not presented here (see World Bank/FAO/WorldFish, 2010).

No attempt was made to apply a generic definition for SSF for the case studies. In some case study countries where scale is expressly recognized as an important differentiator within the fishing sector, data were already disaggregated on this basis; any attempt to reclassify data based on an alternative definition would have undermined the country-specific context of the differentiation without necessarily enabling more effective among-country comparisons. While the intention was to use 'official' country-derived definitions to differentiate scale, several case study countries had no accepted definitions. In these countries we worked with national expert groups to develop a context-specific definition appropriate for this analysis.

The study presents a 'snapshot' survey of data from a single year selected at the country level based on data availability. Selected years generally ranged from 2004 to 2007, although in some cases older data were incorporated and where possible rescaled accordingly. It was not possible to prescribe a single methodology that could be applied in all case study countries. Instead, case study leaders estimated the agreed metrics using the best available means in each country. Case studies were conducted via interviews with key informants and focus group discussions as well as via detailed investigations of existing statistics, published scientific data and 'grey' literature. In a limited number of cases existing data were supplemented by primary data collection from particular fishing fleets (Bangladesh, Cambodia, China). Re-analysis of existing household production or consumption data constituted an important input into the assessment of production and use indicators for some countries such as Thailand and Vietnam.

Data availability within case study countries varied considerably, so the selected metrics were not available for each case study. Similarly, the reliability of estimates varies between countries. Where alternative national scientific or survey (household income and expenditure survey, agricultural census, etc.) datasets were available (e.g. Vietnam marine fisheries, Thailand inland fisheries), national metrics were re-estimated independently. In other studies (e.g. China, Ghana, Vietnam inland fisheries), detailed localized surveys were used to re-scale and disaggregate official national estimates. In those case studies in which no independent data sources were identified (e.g. Indian, Indonesian and Thai marine catches), efforts focused on disaggregating existing statistics between small- and large-scale sectors.

To further strengthen the analysis, ancillary data on proportion of catch from small- and large-scale fleets in 19 additional countries were incorporated into the analysis: Angola; Argentina; Chile; Ecuador; Greenland; Iran; Korea, Malaysia; Mauritania; Mexico; Morocco; Namibia; Panama; Paraguay; Peru; Taiwan; Turkey; Uruguay; and Yemen. Existing catch data disaggregated by scale were available from national statistics for these countries (obtained primarily from FAO Fishery Country Profiles and literature quoted therein). This last step makes estimates of yield from small-scale fleets more conservative because, in many instances, small-scale production is underestimated relative to large-scale production. However, the strategy also allows obvious outliers to be excluded from the extrapolation. In cases with known structural anomalies within fishing sectors additional data were sought. Chief amongst these were Chile and Peru, where massive catches of small pelagic species are almost entirely reduced to fishmeal and fish oil. For these countries, national data (OECD, 2009; IMARPE, 2010), available international datasets (IFFO, 2010) and scientific studies (Majluf et al., 2005; Fréon et al., 2010; Péron et al., 2010) were used to estimate sectoral catches and postharvest use indicators. For data-poor countries (with neither case study nor ancillary data) catches were split between large- and SSF according to proportions calculated from case studies: these were 64% and 91% small-scale for marine and inland catches, respectively.

Data from all case studies were not used in all extrapolations. In particular, data from China were not used to extrapolate catches. This case study suggested that catches were some 13% lower than national statistics for the same year. Chinese statistics have recently (2009) been revised downwards, so case study estimates were considered valid, but it was not seen as appropriate to allow the discrepancy between case study and official statistics from this case study to influence extrapolations to other developing counties. Postharvest use statistics from Thailand and China were also excluded from extrapolations, as these countries rank third and fourth (after Peru and Chile) among high fishmeal-producing countries. No other developing countries rank among the top ten fishmeal producers (Péron et al., 2010), although a significant proportion of marine catches in Vietnam is used directly for animal feed (Edwards et al., 2004).

Coverage of the study in terms of total fish production capacity among developing countries was high; among the top ten producers, eight (the exceptions being Peru and Chile) were covered by case studies. Case studies alone account for 70% of reported inland catch from developing countries, and for 54% of marine catch (calculated from unadjusted national statistics (FAO, 2008)). Ancillary data on production disaggregated by scale complete the coverage of the top 20 fish-producing developing countries, and in combination with case study countries account for 88% of marine catches and 74% of inland catches.

An additional metric of the annual catch used for domestic human consumption is included in the analysis. Clearly, this is used to highlight the direct food security function of fisheries and provide a picture of relativities among subsectors. These estimates provide a numerical basis for discussions on the direct and indirect (trade-based) food security role of fisheries, particularly noting the increasing trade with developed countries (e.g. Alder and Sumaila, 2004; Béné et al., 2010b).

Data on numbers of full- and part-time fishers and postharvest workers were collated for case study countries. Full-time workers were defined as those receiving at least 90% of their livelihood from, or spending at least 90% of their working time engaged in, fishing or postharvest activities. Part-time workers received between 30% and 90% of their livelihood from, or spent between 30% and 90% of their time engaged in, fishing or postharvest activities. Occasional fishers, those spending less than 30% of their working time fishing, were not included in extrapolations, as data tended to be localized and scarce. Despite clear definitions, it was often difficult to distinguish between categories within available data sources, and accordingly future studies may improve the precision of these estimates.

Case study data were extrapolated to data-poor countries using catch per fisher disaggregated among small- and large-scale fisheries operating in inland and marine waters. In this instance, countries were treated as data-poor if official statistics did not provide a breakdown of employment between either inland and marine, or small- and large-scale fleets. While the extrapolations from the case studies to the data-poor countries can be questioned in so far as the case study countries were not randomly selected and may not be representative, the high coverage of input data serves substantially to reduce errors associated with extrapolations. The estimates represent a step towards greater global coverage of these key indicators and provide for a level of analysis not possible where existing country statistics only are used.

People and Fisheries Revisited

In the case study countries there are close to 26 million fishers. By including ancillary data for prominent non-case study countries with large fishery sectors and extrapolating results to data-poor countries, it is conservatively estimated that there are 36 million full-time and part-time fishers in the developing world (Table 1.1). With some one million fishers

Table 1.1. Estimates of selected parameters for *developing country* fisheries extrapolated from country case study data. Lower estimates for production and domestic consumption (in parentheses) use country case study data and unadjusted national data (FAO, 2008) and are considered highly conservative. High estimates use data from country case studies supplemented by national data for other developing countries rescaled according to correction factors derived from comparing case study data with national data.

	Small-scale fisheries			Large-scale fisheries			Total
	Marine	Inland	Total	Marine	Inland	Total	
Employment							
Number of fishers (million)*	12.4	20.7	33.1	1.8	0.8	2.6	35.7
Number of postharvest jobs (million)*	34.6	40.7	75.3	6.6	0.3	6.9	82.1
Total employment (million)	47.1	59.7	107.0	8.4	1.1	9.5	117.8
Women in workforce (%)	36	55	47	67	31	63	47
Production and utilization							
Total annual catch (million t)	(27)–28	(11)–14	(39)–42	(31)–33	(0.4)–0.5	(31)–34	(70)–75
Annual catch for domestic human consumption (million t)	(19)–20	(10)–13	(29)–33	(11)–13	(0.2)–0.3	(11)–13	(40)–46
Annual catch for domestic human consumption (% of total catch)**	72	94	79	39	63	39	62
Catch per fisher (t)	2.3	0.7	1.3	18.0	0.6	12.7	2.2

Source: World Bank/FAO/WorldFish (2010) and ancillary data as described in the text. Numbers vary slightly from those in World Bank/FAO/WorldFish (2010), due largely to the inclusion of revised estimates for Indonesian inland catch.

*Full-time and part-time employment only (does not include occasional/subsistence and short-term seasonal);
**percentage values are derived from actual data, and therefore may not correspond with the calculation based on rounded figures in the table.

operating in the developed world (FAO, 2009), developing countries can claim 97% of the world's fishers.

While catching fish is clearly an important source of employment in the developing world, the bulk of fish workers are employed on land – moving, processing and trading the catch. Among case study countries, for each person employed as a fisher, two to three are employed in postharvest activities. An estimated figure of 82 million postharvest workers brings the total employment in developing countries to 118 million fish workers. Contributing substantially to these high numbers are part-time activities, notably among family members of fishers, which were detected in project case studies whereas they are most often absent from official statistics. Accepting that there are three million fish workers in the developed world (FAO, 2009) implies that 97% of all fish workers live in developing countries.

Excluded from these estimates is 'upstream' employment, i.e. employment related to input supplies such as boat building, engine and gear manufacturing and repairs, as well as support services in harbours and at landing sites. The paucity of data even from case study countries meant that meaningful extrapolations could not be made. Case study data from Ghana and Senegal indicated that upstream employment could add another 5–10% to the total employment figures. Again, we emphasize that employment figures do not include occasional fishers, defined as those spending less than 30% of productive working hours over the course of a year engaged in fishing. This is significant, as this group could include, for example, fishers who operate full-time for three months of the year. Data covering participation in occasional and subsistence fishing activities were rare. Where useable data were identified (Vietnam and Bangladesh inland fisheries) there were many more occasional fishers than full-time and part-time fishers (see World Bank/FAO/WorldFish, 2010 for further details). These countries, however, have extensive and highly productive inland systems, and estimates cannot be extrapolated to other countries.

The estimates produced by the inclusion of part-time employment in this analysis challenge some entrenched perceptions regarding the nature of employment in the fishery sector; in particular these relate to the importance of inland fisheries for livelihoods and nutrition, as well as to the importance of women in the fisheries' workforce. It is apparent from the striking case study results that information on women's role in fisheries has remained buried in the data-poor subsectors of part-time employment and postharvest activities. Case studies revealed that there are almost as many women as men employed in the fisheries sector: 47% of all fish workers in developing countries are women. Extrapolating from the case study results, close to 56 million jobs in the fisheries sector are held by women in developing countries.

A further examination of the gender data from case studies reveals some interesting dichotomies (Table 1.2). Case study data generally supported the traditional notions of division of labour, with the direct involvement of women in fish-harvesting activities being substantially lower than in the postharvest sector. However, this situation varies considerably among countries based on dominant fishing methods, cultural norms and religious beliefs. Participation rate is generally low in countries where social or religious norms limit women's participation in certain economic activities. The dichotomy between India (72% of the workforce are women) and neighbouring Bangladesh, where societal/religious norms discourage women from engaging in fishing or fish marketing (5% of the workforce are women) is particularly striking. In China, 50% of the relatively small inland fisheries workforce are women, but participation in marine fisheries is far lower (15%). None the less, due to the immense scale of marine fisheries, a combined total of over two million women work in the fishery sector in China. Similarly, over 160,000 women work in the sector in Bangladesh although they constitute only 5% of the total employment.

Large-scale fisheries employ a higher proportion of women than the small-scale sector, although total numbers of workers in the former are far lower (see Table 1.1). Given the high catch-per-fisher figure for large-scale fisheries, and the resulting high demand for postharvest labour, this is not surprising.

Inland and Marine Sectors Compared

From a total developing-country catch of 75 million t, about 61 million t (81%) comes from marine waters and 14 million t (19%) from inland fisheries. These figures represent a substantial departure from 'official' figures

Table 1.2. Women in the fisheries workforce (full-time and part-time; fishing and postharvest activities) in selected case study countries. Source: World Bank/FAO/WorldFish (2010).

Country/case study	Total workforce	Percentage women
Nigeria	6,500,000	73
India	10,316,000	72
Cambodia	1,624,000	57
Ghana	372,000	40
Senegal	129,000	32
Brazil	493,000	30
China	12,078,000	19
Bangladesh	3,253,000	5
Mozambique	265,000	4

from government sources. Comparing official statistics with case study output, marine catches were underestimated by 10% while inland catches were underestimated by 80% (note: for inland catches, World Bank/FAO/ WorldFish, 2010 used a 70% discrepancy between case study and official figures. In that study, the Indonesian case study did not revise the estimate for inland catches while here we incorporate Coates' (2002) estimate, which is approximately double the reported catch). These figures – particularly for marine catches – could still be considered conservative. The lack of independent re-estimates of catches in several case studies meant that only minor adjustments were made. Detailed reconstructions of marine catches for a range of countries (e.g. Zeller *et al.*, 2007; Wielgus *et al.*, 2010) and studies of illegal fishing (Agnew *et al.*, 2008) suggest that actual underestimates could in many cases be well above 10%. Such reconstructions are complex and beyond the scope of the current study, but clearly provide important insights into common sources of error in existing data.

Similarly for inland fisheries, data in a limited number of case studies allowed for independent re-estimates of catch. In such cases, corrections were often well above the average of 80% used for extrapolations here. Among these, in Vietnam localized consumption data were used to re-scale official national estimates, suggesting production was underestimated by a factor of 5.7. For Ghana, localized market surveys extrapolated to the national level suggest that the inland catch was under-reported by more than five times. The global inland extrapolation factor (1.8) used here is slightly more conservative than the factor of 2.0 suggested by Coates (1995).

To illustrate the effect of re-scaling reported catches, in Table 1.1 we present two estimates for catch in each sector; the more conservative of the two (in parentheses) does not include re-scaled catches, but rather presents case study data and disaggregated official national data only. The higher estimate for production metrics incorporates the re-scaling adjustments. The high proportion of total catch covered in case studies limits the discrepancy between outputs from the two methods. Total catch reported from the case studies (40.3 million t) was 12% higher than the official national statistics for the same set of countries (36.2 million t). Excluding figures from China, the difference is 22%.

The majority of small-scale fish-workers in developing countries operate in the inland sector. Given that yield from these fisheries is less than 20% of the estimated total, this may seem counter-intuitive. However, a primary consideration here is one of the relative scale of inland and marine capture operations. Accepted national definitions for small-scale marine activities encompass fishing units many times larger than most inland operations. Among case study countries, the average annual yield per small-scale inland fisher was 0.7 t (Table 1.1), while that of small-scale marine fishers was 2 t (2.3 t after ancillary data were included). As well as differences in fishing methods, livelihood strategy may also play a part here. The notion – perhaps not always backed up empirically – that specialization is more common in coastal fisheries than inland, would lead to greater yields from higher numbers of full-time fishers in marine fisheries.

A particularly striking result from the synthesis is the importance of inland fisheries for direct food security. An estimated 94% (13 million t) of small-scale inland production is consumed within the country of origin. This contrasts particularly strongly with large-scale marine fisheries which, although producing three times the catch, provide about the same amount for domestic consumption. Clearly, the large-scale marine sector is important for foreign exchange earnings, and the large quantities of fish reduced to fishmeal and fish oil further influence this figure. Few significant export commodities were identified from inland catches. Among the case study countries, Nile perch exports from Lake Victoria, freshwater fish and prawn exports from Cambodia and *kapenta* (Tanganyika sardine) exports from Mozambique were the exceptions.

Large- and Small-scale Fisheries Compared

In developing countries, SSF produce more fish than large-scale fisheries. Of an estimated 75 million t produced by capture fisheries in developing countries annually (Table 1.1), about 55% (42 million t) comes from the small-scale sector. Marine catches account for some 80% of the total, and small-scale fishing 46% of the marine total. Inland fisheries (19% of total production) are almost entirely small-scale (96% of inland production). Among the case studies, Cambodia (36% of national inland production), Mozambique (36%), Myanmar (28%) and Brazil (11%) reported significant large-scale inland fisheries.

Among an estimated 118 million fish workers in the developing world, 107 million (92%) operate in the small-scale sector. This equates to 88% of fish workers globally. Despite the dominance of small-scale fishing, the large-scale sector in developing countries employs more than three times as many workers as the large-scale sector in developed nations.

We estimate that 62% of total production in developing countries was used directly for local consumption. Marine SSF provide the greatest share of fish for local human consumption. Among case study countries, some 56% of large-scale marine catches were used for local consumption. When the industrial catches of pelagic fish (used primarily for reduction to fishmeal and oil) from Peru and Chile were included as ancillary data, the estimated figure for all developing countries dropped substantially (to 39%). Among case study countries reporting high rates of non-food use, in China 18% (some 1.5 million t) of the catch of small-scale marine fisheries is reduced to fishmeal, while in Thailand and Vietnam some 20–30% of the total fish production is destined for non-food uses.

As noted previously, the proportion of fish consumed locally is but one of several indicators that could be used in relation to the contribution of fish to food security. Regional trade in Africa in particular has been highlighted as playing an important direct role in food security (Maasdorp, 1998; ICSF, 2002; Béné, 2009), and a three-way comparison of domestic, regional and global trade would be of particular value.

Data for Pro-poor Fisheries Governance

Case studies confirmed that, with few exceptions, fisheries data systems in the developing world focus on production statistics, and typically do so in a way that paints only a partial picture of the national fishery sector. While highlighting shortcomings in resourcing, coverage, sampling regimes and level of disaggregation in particular, this type of data remains important for quantifying aspects of fisheries that can be effectively sampled through direct measurement methods. The critical step to be made at a national level is one of acknowledging that existing systems typically miss a substantial proportion of fisheries production, and that complementary methods must be employed to capture data on components of the small-scale sector.

While not providing a universal solution to existing issues, case studies of the current project have been significant in highlighting areas where substantial improvements in current data collection can be made. A number of case studies tested a revised data system providing a high level of disaggregation among fishing fleets and sectors. In yet other cases, data sources external to fishery information systems were harnessed to re-scale or complement existing production statistics. As examples, analysis of household fish consumption data from the Mekong region in Vietnam produced an estimate of inland fish production several times higher than the existing statistics gathered from fishing gear surveys. For Thailand, household fish production data collected from provincial production surveys served the same purpose. Similar household- and village-level data sources were identified in Indonesia, although comprehensive re-analysis could not be pursued within the timeframe of the case study. Market surveys from Lake Volta provided re-estimates of inland production in Ghana. In each case, official figures for inland fishery production were estimated by direct methods such as measuring and extrapolating production for particular gear-types; it is clear that these methods had missed substantial components of production.

The value of household-level data for estimating small-scale and subsistence production, particularly if collected as part of an agricultural or household census, cannot be understated. The dispersed nature of fishing activities, and the fact that a significant component of production may not pass though centralized landing sites or enter the formal economy, renders direct estimation methods ineffective. Beyond expanding the scope of existing metrics of fisheries production to capture a broader range of fishing activities, if information systems are to contribute in a meaningful way to policy development, data describing the broader contribution of fisheries to society (Allison, 2005; Thorpe et al., 2007; Béné et al., 2010a) must be incorporated. As noted in FAO (1995, p. 12):

> In order to ensure sustainable management of fisheries and to enable social and economic objectives to be achieved, sufficient knowledge of social, economic and institutional factors should be developed through data gathering, analysis and research.

Current national-level data on the socio-economic dimensions of SSF can best be described as scattered and incomplete. While project case studies were able to provide some insights into food security and livelihood functions though synthesizing data from a broad range of sources, assumptions and extrapolations feature prominently in these estimates, most notably at the global level. Even so, the value of this exercise goes well beyond updating our picture of fisheries sectors in the developing world. The diversity of data sources incorporated into case studies provides useful insights into approaches to effectively improve the availability of policy-relevant data.

Data that underscore the relationship between fisheries and poverty, the welfare function of fisheries and the role of fisheries in local food security are ultimately as important for policy development as is production information. A better understanding of who fishes, the drivers affecting fishing activities and the role that fish, fishing and fishing-related activities play at the household level are all important. While detailed research in these areas will remain the purvey of specialist institutions, the co-collection and analysis of fisheries data with general demographic, poverty, livelihood and social data provides a particularly powerful tool in understanding these relationships. A critical point here is that we are not simply advocating collection of fisheries data *and* household socio-economic data, but rather fisheries data *with* household socio-economic data. The ability directly to link the two *at the household level* provides powerful insights into the societal role of fisheries – a crucial point for pro-poor policy development.

Collection of such data falls well outside of the capabilities and mandate of most national fisheries agencies responsible for routine fisheries data collection; laying out a huge compendium of additional data for collection would be counterproductive and unsustainable. However the macro-economic value of the broader agricultural sector and the drive to develop effective poverty reduction strategies means that broader household survey systems are often relatively well developed and resourced. In accessing national household income and expenditure surveys, labour force surveys, agricultural census and production surveys through this study, the existing value in such datasets was clear and data from these sources were used effectively in a number of project case studies (see Lymer et al., 2008; World Bank/FAO/WorldFish, 2010, Figs 1 and 2).

Beyond the results presented here, in some case studies (most notably Bangladesh) existing data from household surveys were used effectively to fill gaps in understanding of various socio-economic functions of fisheries, drawing direct links between participation, production and household-level parameters such as income and education levels, or assets and expenditure. In other cases, it was clear that minor changes such as increasing the degree of data disaggregation or improving livelihood descriptors would present an unparalleled opportunity to improve our understanding of fisheries. As noted above, existing systems can be greatly enhanced by ensuring sample sizes are large enough and sampling frames are specified in a way that allows for statistically significant analyses of fishing households and the specific occupational factors relating to fishing livelihoods.

In suggesting the use of household census and sample data, we note that effective inter-agency collaboration will be critical. Cursory efforts at collecting specialized data such as nutrition and fish production information can give rise to misleading and even damaging results. In particular, issues relating to seasonality and the spatial distribution of fishing activity will require special attention. Without the need to develop and support further stand-alone systems, this represents efficient use of existing resources and offers a productive pathway towards the provision of policy-relevant information.

The ultimate consequence of the observed shortcomings in fisheries data collection and information systems is a distorted view of the fisheries sector. Case studies confirmed that SSF as a whole are undervalued at the national level, and the full weight of their contribution to food security, livelihood provision and poverty alleviation in developing countries is not recognized. Similarly, there is clearly a vacuum of country-level information on key internal drivers of SSF, adding substantially to the difficulty of developing effective local-level governance systems. While not providing a universal solution to these issues, the study summarized here provides new insights into both these areas. Not only does it provide improved estimates for fisheries production and participation in developing countries, but also highlights effective pathways to getting the types of data that can contribute to pro-poor fishery policy development.

Acknowledgements

The authors acknowledge the valuable contributions to the developing-country case study consultants and specialists: I. Adikwu (Nigeria); L.I. Braimah (Ghana); M. Démé (Senegal); L. Garces and N. Salayo (Philippines); D. Lymer, S. Funge-Smith, P. Khemakorn, S. Naruepon and S. Ubolratana (Thailand); S. Koeshendrajana, L. Adrianto, T. Trihartono and E. Anggraini (Indonesia); J. Kurien (India – marine); A.M. Menezes (Mozambique); N.V. Nghia and B.V. Hanh (Vietnam); M.G. Mustafa and M. Bose (Bangladesh); N. Thuok, P. Somany, S. Kao and D. Thomson (Cambodia); M. van der Knaap (Lake Victoria – Kenya, Tanzania and Uganda); G. Velasco Canziani (Brazil); V. Vivekanandan (India – inland); and X. Yingliang (China). Research and data analysis by Eriko Hoshino (World Bank consultant) and Pavel Salz (FAO consultant) is gratefully acknowledged. Further invaluable advice and inputs from the following people are gratefully acknowledged: David Walfoort (WorldFish), Chris Barlow (MRC), Simon Funge-Smith (FAO), Michael Akester (DANIDA/SUMA), Anders Poulsen (DANIDA/SCAFI), Kent Hortle (MRC), Francis Chopin (FAO), Blaise Kuemlangan (FAO), Susana Siar (FAO), Ben Starkhouse (WorldFish), Marie-Caroline Badjeck (WorldFish) and participants at the side-event at the Sustaining Small-scale Fisheries conference in Bangkok in October 2008 for the constructive comments on the preliminary report 'Small-scale capture fisheries: a global overview with emphasis on developing countries'. The work summarized in this chapter was funded through PROFISH – the World Bank's Global Program on Sustainable Fisheries, FAO and The WorldFish Center.

References

Agnew, D., Pearce, J., Peatman, T., Pitcher, T.J. and Pramod, G. (2008) *The Global Extent of Illegal Fishing*. MRAG, London and FERR, Fisheries Centre, University of British Columbia, Vancouver, Canada.

Alder, J. and Sumaila, U.R. (2004) Western Africa: a fish basket of Europe past and present. *Journal of Environment and Development* 13, 156–178.

Allison, E.H. (2005) The fisheries sector, livelihoods and poverty reduction in eastern and southern Africa. In: Ellis, F. and Ade Freeman, H. (eds) *Rural Livelihoods and Poverty Reduction Policies: Routledge Studies in Development Economics*. Routledge, New York, pp. 256–273.

Andrew, N.L., Béné, C., Hall, S.J., Allison, E.H., Heck, S. and Ratner, B.D. (2007) Diagnosis and management of small-scale fisheries in developing countries. *Fish and Fisheries* 8, 227–240.

Bell, J.D., Kronen, M., Vunisea, A., Nash, W.J., Keeble, G., Demmke, A., Pontifex, S. and Andrefouet, S. (2009) Planning the use of fish for food security in the Pacific. *Marine Policy* 33, 64–76.

Béné, C. (2009) Are fishers poor or vulnerable? Assessing economic vulnerability in small-scale fishing communities. *Journal of Development Studies* 45, 911–933.

Béné, C., Macfadyen, G. and Allison, E.H. (2007) Increasing the contribution of small-scale fisheries to poverty alleviation and food security. FAO Fisheries Technical paper No. 481, FAO, Rome, Italy.

Béné, C., Hersoug, B. and Allison, E.H. (2010a) Not by rent alone: analysing the pro-poor functions of small-scale fisheries in developing countries. *Development Policy Review* 28, 325–358.

Béné, C., Lawton, R. and Allison, E.H. (2010b) Trade matters in the fight against poverty: narratives, perceptions and (lack of) evidence in the case of fish trade in Africa. *World Development* 38, 933–954.

Berkes, F. (2003) Alternatives to conventional management: lessons from small-scale fisheries. *Environment* 31, 1–14.

Berkes, F., Mahon, R., McConney, P., Pollnac, R. and Pomeroy, R. (2001) *Managing Small-scale Fisheries. Alternative Directions and Methods*. International Development Research Centre (IDRC), Ottawa, Canada.

Bird, R.B. (2007) Fishing and the sexual division of labour among the Meriam. *American Anthropologist* 109, 442–451.

CBNRM-LI and FiA (2008) *Gender Implications in CBNRM: the Roles, Needs, and Aspirations of Women in Community Fisheries – six Case Studies in Cambodia*. CBNRM-LI and FiA, Phnom Penh, Cambodia.

Chao, N.H., Chen, M.H. and Chen, Y.H. (2006) Women's involvement in processing and the globalization of processing in fisheries and aquaculture in Taiwan. In: Choo, P.S., Hall, S.J. and Williams, M.J. (eds) *Global Symposium on Gender and Fisheries, Seventh Asian Fisheries Forum*, 1–2 December 2004, Penang, Malaysia. WorldFish Center, Penang, Malaysia, pp. 81–96.

Chapman, M. (1987) Women's fishing in Oceania. *Human Ecology* 15, 267–288.

Choo, P.S., Hall, S.J. and Williams, M.J. (2006) *Global Symposium on Gender and Fisheries, Seventh Asian Fisheries Forum*, 1–2 December 2004, Penang, Malaysia. WorldFish Center, Penang, Malaysia. Available at http://www.worldfishcenter.org/resource_centre/Gender%20and%20Fisheries%20Dec%202004.pdf.

Chuenpagdee, R., Liguori, L., Palomares, M.L.D. and Pauly, D. (2006) Bottom-up, global estimates of small-scale marine fisheries catches. *Fisheries Centre Research Reports* 14(8), University of British Colombia, BC, Canada.

Coates, D. (1995) Inland capture fisheries and enhancement: status, constraints and prospects for food security. Paper presented at the *International Conference on the Sustainable Contribution of Fisheries to Food Security*, Kyoto, Japan, 4–9 December 1995. FAO, Rome, Italy.

Coates, D. (2002) *Inland capture fishery statistics of Southeast Asia: current status and information needs*. RAP Publication No. 2002/11, Asia-Pacific Fishery Commission, Bangkok, Thailand.

Cochrane, K.L. and Doulman, D.J. (2005) The rising tide of fisheries instruments and the struggle to keep afloat. *Philosophical Transactions of the Royal Society B* 360, 77–94.

Delgado, C.L., Wada, N. and Rosegrant, M.W. (2003) *Fish to 2020. Supply and Demand in Changing Global Markets*. International Food Policy Research Institute, Washington, DC.

Edwards, P., Tuan, L.A. and Allan, G.L. (2004) *A survey of marine trash fish and fishmeal as aquaculture feed ingredients in Vietnam*. ACIAR Working Paper No. 57, Australian Centre for International Agricultural Research, Canberra, Australia.

FAO (1995) *Code of Conduct for Responsible Fisheries*. FAO, Rome, Italy.

FAO (2003) *New approaches for the improvement of inland capture fishery statistics in the Mekong Basin*. Regional office for Asia/Pacific publication 2003/01, Bangkok, Thailand.

FAO (2004) Report of the second session of the Working Party on Small-scale Fisheries, Bangkok, Thailand, 18–21 November 2003. FAO Fisheries Report No. 735, FAO, Rome, Italy.

FAO (2006) *Contribution of fisheries to national economies in West and Central Africa. Policies to increase the wealth generated by small-scale fisheries*. New Directions in Fisheries – a series of policy briefs on development issues No. 03. FAO, Rome, Italy.

FAO (2008) *Capture production 1950–2006. FISHSTAT Plus: universal software for fishery statistics time series. Dataset: FAO Fisheries and Aquaculture Department, Information and Statistics Service*. FAO, Rome, Italy. Available at: http://www.fao.org/fi/statist/ FISOFT/FISHPLUS (accessed 31 May 2010).

FAO (2009) *The State of World Fisheries and Aquaculture 2008*. Available at: http://www.fao.org/fishery/sofia/en (accessed 31 May 2010).

FAO-SEAFDEC (2005) Improvement of Fishery Data and Information Collection Systems in Southeast Asia. *Proceedings of the FAO/SEAFDEC Regional Workshop on the Improvement of Fishery Data and Information Collection Systems*, Bali, Indonesia, 15–18 February 2005. Report of Workshop, FishCode-STF-WP2005/1, FAO, Bangkok, Thailand.

FiA (2007) *The Gender Mainstreaming Policy and Strategy in the Fisheries Sector (GMPSFS)*. Fisheries Administration of MAFF, Phnom Penh, Cambodia.

Fréon, P., Bouchon, M., Domalain, G., Estrella, C., Iriarte, F., Lazard, J., Legendre, M., Quispe, I., Mendo, T., Moreau, Y., Nuñez, J., Sueiro, J.C., Tam, J., Tyedmers, P. and Voisin, S. (2010) Impacts of the Peruvian anchoveta supply chains: from wild fish in the water to protein on the plate. *GLOBEC International Newsletter* 16, 27–31.

Friend, R.M. (2009) Fishing for influence: fisheries science and evidence in water resource development in the Mekong basin. *Water Alternatives* 2, 167–182.

Garcia, S.J., Allison, E.H., Andrew, N.L., Béné, C., Bianchi, G., de Graaf, G., Kalikoski, G.J., Mahon, R. and Orensanz, J.M. (2008) Towards integrated assessment and advice in small-scale fisheries: principles and processes. FAO Fisheries and Aquaculture Technical Paper No. 515, FAO, Rome, Italy. Available at: http://www.fao.org/docrep/011/i0326e/i0326e00.htm (accessed 31 May 2010).

Grafton, R.Q., Hilborn, R., Ridgeway, L., Squires, D., Williams, M., Garcia, S., Groves, T., Joseph, J., Kelleher, K., Kompas, T., Libecap, G., Lundin, C.G., Makino, M., Matthiasson, T., McLoughlin, R., Parma, A., Martin, G.S., Satia, B., Schmidt, C.C., Tait, M. and Zhang, L.X. (2008) Positioning fisheries in a changing world. *Marine Policy* 32, 630–634.

Hapke, H.M. (2001) Petty traders, gender, and development in a south Indian fishery. *Economic Geography* 77, 225–249.

Hauck, M. and Sowman, M. (2005) Coastal and fisheries co-management in South Africa: is there an enabling legal environment. *South African Journal of Environmental Law and Policy* 12, 1–21.

Holvoet, K. (2009) Mainstreaming gender in fisheries. In: Westlund, L., Holvoet, K. and Kébé, M. (eds) Achieving poverty reduction through responsible fisheries: strategies and lessons from the West and Central Africa Sustainable Fisheries Livelihoods Programme. FAO Fisheries Technical Paper No. 513, FAO, Rome, Italy.

Hori, M., Ishikawa, S., Heng, P.L., Thay, S., Ly, V., Nao, T. and Kurokura, H. (2006) Role of small-scale fishing in Kompong Thom Province, Cambodia. *Fisheries Science* 72, 846–854.

Hussain, M.G. (2010) Freshwater fishes of Bangladesh: fisheries, biodiversity and habitat. *Aquatic Ecosystem Health and Management* 13, 85–93.

ICSF (2002) *Report of the study on problems and prospects of artisanal fish trade in West Africa*. International Collective in Support of Fishworkers. Available at: http://icsf.net/icsf2006/uploads/publications/reports/pdf/english/issue_1/chapter02.pdf (accessed 31 May 2010).

IFFO (2010) *The production of fishmeal and fish oil from Peruvian anchovy*. IFFO datasheet, International Fishmeal and Fish Oil Organisation. Available at: http://www.iffo.net/intranet/content/archivos/67.pdf (accessed 31 May 2010).

IMARPE (2010) *Desembarques de recursos pelágicos*. Available at: http://www.imarpe.pe/ (accessed 31 May 2010).

Jensen, R. (2007) The digital provide: information (technology), market performance, and welfare in the South Indian fisheries sector. *Quarterly Journal of Economics* 122, 879–924.

Johnson, D.S. (2006) Category, narrative and value in the governance of small-scale fisheries. *Marine Policy* 30, 747–756.

Kang, B., He, D.M., Perrett, L., Wang, H.Y., Hu, W.X., Deng, W.D. and Wu, Y.F. (2009) Fish and fisheries in the Upper Mekong: current assessment of the fish community, threats and conservation. *Reviews in Fish Biology and Fisheries* 19, 465–480.

Kébé, M. (2008a) Mainstreaming fisheries in development policy. In: Westlund, L., Holvoet, K. and Kébé, M. (eds) Achieving poverty reduction through responsible fisheries: lessons from West and Central Africa. FAO Fisheries and Aquaculture Technical Paper No. 513, FAO, Rome, Italy, pp. 55–65.

Kébé, M. (2008b) Reassessing the economic and social contribution of fisheries in developing countries. In: Westlund, L., Holvoet, K. and Kébé, M. (eds) Achieving poverty reduction through responsible fisheries: lessons from West and Central Africa. FAO Fisheries and Aquaculture Technical Paper No. 513, FAO, Rome, Italy, pp. 39–54.

Keskinen, M. (2003) Socio-economic survey of the Tonle Sap Lake, Cambodia. MSc thesis, Helsinki University of Technology, Helsinki, Finland.

Kurien, J. (1998) *Small-scale fisheries in the context of globalization*. Center for Development Studies, Thiruvananthapuram, India.

Kurien, J. and Willman, R. (1982) *Economics of artisanal and mechanized fisheries in Kerala. A study on costs and earnings of fishing units*. Regional FAO/UNDP Small-Scale Fisheries Promotion in South Asia Project RAS/77/044, Working Paper No. 34, FAO, Madras, India.

Kurien, J. and Willman, R. (2009) Special considerations for small-scale fisheries management in developing countries. In: Cochrane, K.L. and Garcia, S.M. (eds) *A Fishery Manager's Guidebook*. Wiley/Blackwell, Oxford, UK.

Lindquist, A. (1988) Thanks for using NAGA. *NAGA, ICLARM Quarterly* 11, 16–17.

Lymer, D., Funge-Smith, S., Khemakorn, P., Naruepon, S. and Ubolratana, S. (2008) *A review and synthesis of capture fisheries data in Thailand: large versus small-scale fisheries*. RAP Publication 2008/17, FAO Regional Office for Asia and the Pacific, Bangkok, Thailand.

Maasdorp, G. (1998) Regional trade and food security in SADC. *Food Policy* 23, 505–518.

Mahon, R., McConney, P. and Roy, R.N. (2008) Governing fisheries as complex adaptive systems. *Marine Policy* 32, 104–112.

Majluf, P., Barandearán, A. and Sueiro, J.C. (2005) *Evaluacion Ambiental del Sector Pesquero en el Peru World Bank Country Environmental Analysis Peru*. World Bank, Washington, DC.

McGoodwin, J.R. (2001) Understanding the cultures of fishing communities: a key to fisheries management and food security. FAO Fisheries Technical Paper No. 401, FAO, Rome, Italy.

Mills, D., Béné, C., Ovie, S., Tafida, A., Sinaba, F., Kodio, A., Russell, A., Andrew, N.L., Morand, P. and Lemoalle, J. (2009) Vulnerability in African small-scale fishing communities. *Journal of International Development*. Available at: http://dx.doi.org/10.1002/jid.1638 (accessed 31 May 2010).

Neiland, A.E., Jaffrym, S., Ladu, B.M.B., Sarch, M.T. and Madakan, S.P. (2000) Inland fisheries of North East Nigeria including the Upper River Benue, Lake Chad and the Nguru-Gashua wetlands I. Characterisation and analysis of planning suppositions. *Fisheries Research* 48, 229–243.

ODI (2002) *Inland fisheries. Key sheets for sustainable livelihoods: resource management*. Overseas Development Institute (ODI), UK. Available at: http://www.odi.org.uk/Publications/keysheets.html (accessed 31 May 2010).

OECD (2009) *An appraisal of the Chilean fisheries sector*. Organisation for Economic Co-operation and Development, Paris.

Pauly, D. (2006) Major trends in small-scale marine fisheries, with emphasis on developing countries, and some implications for the social sciences. *MAST* 4, 22–74.

Péron, G., Mittaine, J.F. and Le Gallic, B. (2010) Where do fishmeal and fish oil products come from? An analysis of the conversion ratios in the global fishmeal industry. *Marine Policy* 34, 815–820.

Platteau, J.P. (1984) The drive towards mechanization of small-scale fisheries in Kerala – a study of the transformation process of traditional village societies. *Development and Change* 15, 65–103.

Ponte, S., Raakjaer Nielsen, J. and Campling, L. (2007) Swimming upstream: market access for African fish exports in the context of WTO and EU negotiations and regulation. *Development Policy Review* 25, 113–138.

Reid, C., Thorpe, A. and Smith, S.F. (2008) Mainstreaming fisheries in development and poverty reduction strategies in the Asia-Pacific region. *Journal of the Asia Pacific Economy* 13, 518–541.

Salayo, N., Garces, L., Pido, M., Viswanathan, K., Pomeroy, R., Ahmed, M., Siason, I., Seng, K. and Masae, A. (2008) Managing excess capacity in small-scale fisheries: perspectives from stakeholders in three Southeast Asian countries. *Marine Policy* 32, 692–700.

Singh, S.K. (2009) Nutritional and food security for rural poor through multi-commodity production from a lake of eastern Uttar Pradesh. *Aquaculture Asia Magazine* 14, 23–25.

Sjorslev, J.G. (2001) *An Giang Fisheries Survey*. AMFC/MRC and RIA 2, Vientiane, Laos.

Smith, L.E.D., Nguyen Khoa, S. and Lorenzen, K. (2005) Livelihood functions of inland fisheries: policy implications in developing countries. *Water Policy* 7, 359–383.

Smith, M.D., Roheim, C.A., Crowder, L.B., Halpern, B.S., Turnipseed, M., Anderson, J.L., Asche, F., Bourillon, L., Guttormsen, A.G., Khan, A., Liguori, L.A., McNevin, A., O'Connor, M.I., Squires, D., Tyedmers, P., Brownstein, C., Carden, K., Klinger, D.H., Sagarin, R. and Selkoe, K.A. (2010) Sustainability and global seafood. *Science* 327, 784–786.

Teh, L., Zeller, D., Cabanban, A., Teh, L.C.L. and Sumaila, U.R. (2007) Seasonality and historic trends in the reef fisheries of Pulau Banggi, Sabah, Malaysia. *Coral Reefs* 26, 251–263.

Thompson, P., Roos, N., Sultana, P. and Thilsted, S.H. (2002) Changing significance of inland fisheries for livelihoods and nutrition in Bangladesh. *Journal of Crop Production* 6, 249–317.

Thomson, D. (1980) Conflict within the fishing industry. *NAGA, ICLARM Newsletter* 3, 3–4.

Thorpe, A., Andrew, N.L. and Allison, E.H. (2007) Fisheries and poverty reduction. *CAB Reviews: Perspectives in Agriculture, Veterinary Science, Nutrition and Natural Resources* 2, 1–12.

Walmsley, S., Purvis, J. and Ninnes, C. (2006) The role of small-scale fisheries management in the poverty reduction strategies in the Western Indian Ocean region. *Ocean and Coastal Management* 49, 812–833.

Warhaft, S. (2001) No parking at the Bunder: fisher people and survival in capitalist Mumbai. *Journal of South Asian Studies* 24, 213–223.

Wielgus, J., Zeller, D., Caicedo-Herrera, D. and Sumaila, R. (2010) Estimation of fisheries removals and primary economic impact of the small-scale and industrial marine fisheries in Colombia. *Marine Policy* 34, 506–513.

Williams, M.J., Chao, N.H., Choo, P.S., Matics, K., Nandeesha, M.C., Shariff, M., Siason, I., Tech, E. and Wong, J.M.C. (2002) *Global Symposium on Women in Fisheries. Proceedings of the Sixth Asian Fisheries Forum*, 29 November 2001, Kaohsiung, Taiwan. WorldFish Center, Penang, Malaysia.

Williams, S.B. and Awoyomi, B. (1998) Fish as a prime mover of the economic life of women in a fishing community. *Proceedings of the IXth International Conference of the International Institute of Fisheries Economics and Trade (IIFET)*, Tromso, Norway, 8–11 July, 1996. International Institute of Fisheries Economics and Trade, Oregon, pp. 286–292.

World Bank/FAO (2009) *The Sunken Billions: the Economic Justification for Fisheries Reform*. Agriculture and Rural Development Department, Sustainable Development Network, World Bank, Washington, DC.

World Bank/FAO/WorldFish (2010) *The Hidden Harvests: the Global Contribution of Capture Fisheries*. Agriculture and Rural Development Department, Sustainable Development Network. World Bank, Washington, DC.

Zeller, D., Booth, S., Davis, G. and Pauly, D. (2007) Re-estimation of small-scale fishery catches for U.S. flag-associated island areas in the western Pacific: the last 50 years. *Fishery Bulletin* 105, 266–277.

2 Approaches and Frameworks for Management and Research in Small-scale Fisheries

Neil L. Andrew and Louisa Evans

Introduction

There is general agreement that commonly adopted approaches to managing small-scale fisheries (SSFs) in developing countries have been less effective than they need to be to ensure sustainability (Garcia and Grainger, 1997; Mahon, 1997; Cochrane, 2000; Welcomme, 2001; FAO, 2003; Béné et al., 2004; Cochrane and Doulman, 2005). Given the importance of SSFs in the social and economic fabric of many least developed countries, it is essential that new management approaches are developed and adopted. This is complicated because SSFs present particular challenges to managers in terms of their diversity and complexity (Berkes et al., 2001; Berkes, 2003; Jentoft, 2006, 2007).

The search for innovation in SSF management is not impeded by a lack of raw material: fishery managers face an overwhelming range of approaches, frameworks, perspectives and methods for analysing fisheries and 'doing' management. As a way forward, we suggest taking a fresh look at the tools already available and synthesizing them into a coherent scheme that joins management with innovations in research. This chapter aims to clarify and make explicit overarching management, implementation and research frameworks and the choices available to managers. As a first step, it is useful to recognize four perspectives on the management of SSF and their contributions to sustainable development. Each perspective has its own emphasis, objectives, constituency and points of entry. From the smallest scale to the largest, these perspectives are:

Inside looking in. This is the classical view. Threats and solutions come from within the domain of the fishery. Managers can ensure sustainability by focusing on the fish and the fishers. Responses include size limits, total allowable catches (TACs), effort restrictions and the like. Classically, fisheries management uses these tools to optimize sustainable yield. This perspective on management may be appropriate if key threats and opportunities come from the fishery itself, and if management promotes learning and adaptability to unforeseen shocks.

Inside looking out. This view recognizes that many threats and opportunities come from outside the domain of the fishery and that in many instances intra-sectoral management alone has little prospect of success. From this perspective, management not only aims to address processes under its direct influence, but also to reduce vulnerability and increase adaptive capacity in the face of threats over which it has no control. 'Resilience' concepts and principles of natural resource management are well suited to this perspective, but other ways of thinking about the management problem are also appropriate.

Outside looking in. This view mainstreams fisheries management and governance within the broader rural (and urban) development challenge in which national issues such as governance, rule of law, literacy, use rights and health become appropriate entry points for improved fisheries. Fisheries remain the focus, but solutions are sought from a larger perspective, usually outside the sector.

Outside looking out. Fisheries per se are not important from this perspective, which arises from the perceived failure of investments at a smaller spatial scale and in SSFs themselves. Investments in such things as macro-economic reform, governance, human rights and national infrastructure are seen as the long-term path to lifting fishery-dependent people out of poverty. Benefits to fisheries will, it is thought, flow from these broader development initiatives. Implicitly, there will be 'winners' who will gain or preserve entitlements and fishery benefits that flow from the generation of taxable revenue. The role of SSFs as open-entry, open-exit social safety nets is downplayed.

In general, the focus of fisheries research and management is shifting along the spectrum from the conventional view to others that consider external disturbance and uncertainty, and wider governance dynamics. Researchers often attribute the failures of conventional fisheries management (target species and resource-oriented management) to an overemphasis on centralized organization, prescriptive design and the search for optimal use of ecosystems. Such management has largely ignored differences in the expectations of stakeholders, the complexity and non-linearity of ecosystem dynamics, and the linkages between ecological, social, political and economic subsystems. Accommodating such issues is widely considered to be essential for effective and legitimate management. More recently, researchers have also advocated for perspectives that view SSFs within a broader context that includes threats and opportunities from outside the classical intra-sectoral domain of fisheries management (see earlier citations, and Andrew *et al.*, 2007 for recent examples). This realization has led donors, governments and non-governmental organizations (NGOs) to place more emphasis on inter-sectoral approaches and on larger-scale responses, such as macro-economic reforms in which the fishery may be only a small part of a broader solution.

No single class of response at any single scale of organization or time horizon will offer a panacea for the ills facing the management and wider governance of SSFs (Ostrom, 2007; Ostrom *et al.*, 2007). Effective management requires a range of perspectives and the inclusion of different actors in the management process, as well as better engagement in wider governance within society. A range of perspectives may be taken on *management approaches, implementation frameworks* and *research approaches*; all three phases of the process are related, but separating them helps clarify a complex problem. We concentrate on fisheries from *within* the system; we do not discuss broader cross-sectoral governance issues, though it is important to recognize these different dimensions of SSF management. The governance of fisheries, particularly within the development agendas of countries, is a critical issue and a hot topic for research (see, for example, Kooiman *et al.*, 2005, Cash *et al.*, 2006; Jentoft, 2007; Mahon *et al.*, 2008).

We begin by describing three primary approaches to fisheries management: ecosystem-based management, rights-based management and management for resilience. All three are well established in the scientific literature if not in practice and are, in fact, complementary. Rights-based approaches may, for example, be used to deliver on ecosystem-based objectives. Resilience approaches are much newer innovations that remain largely untested, but offer the prospect of integrating many research concepts and methods within an overarching management approach. We then introduce a range of management implementation frameworks that can give structure (i.e. set an order in which things need to be done) to the research and management process. Implementation frameworks are partially independent of the principles and objectives underlying management and research approaches.

Finally, we outline a range of possible research approaches. Both management and research approaches provide the broader context and structure of the fisheries problem. However, the ability of the research community to evolve more rapidly than fisheries law and policy means that research approaches are leading rather than being subservient to management approaches (a case in point is resilience approaches to analysing fisheries, which are discussed below). As a result of this lag, management's assessment and advisory demands are increasingly out of step with the types of analyses considered useful/interesting by researchers (Garcia et al., 2008). Here we try to bridge this gap in a way that is appropriate for SSF management in least developed countries.

Management Approaches

Almost all countries have laws and policies that articulate the broad objectives of their fisheries sector. The approach taken to managing a fishery will largely be driven by these prevailing policies and laws, but will also be influenced by international conventions, global goals such as the Millennium Development Goals, and international and regional collaborative agreements. Conventionally, most fisheries seek to maximize production over the long term. Most often this refers to fish catch, but it is sometimes phrased in terms of employment or other societal benefits. The fact that these objectives are increasingly being adjusted to accommodate principles of democracy, human rights, decentralization, integration, empowerment, accountability and adaptability, among many others, is causing authorities to rethink their goals.

Approaches to management include: (i) ecosystem-based approaches, notably the ecosystem approach to fisheries (EAF; FAO, 2003); (ii) rights-based approaches; (iii) integrated approaches (e.g. integrated conservation and development projects (ICDPs) or integrated coastal zone (or catchment) management); and (iv) participatory or collaborative approaches (see Varjopuro et al., 2008 for an overview). It is important to note that these approaches are not mutually exclusive, and many share the underlying principles necessary for more sustainable, legitimate and holistic management. For instance, integration and participation are widely incorporated into other perspectives. In the search for practical solutions to SSF management problems, however, it is not sufficient simply to say that fisheries management should become more holistic, participatory or equitable – we must find more effective ways to achieve these things. One entry point is to understand the management implications and practical potential of alternative frames of reference. Below we focus on three related approaches that appear most suited to SSFs in the developing world.

Ecosystem approaches to fisheries management and governance

Ecology has been part of fisheries research for a very long time (see Cushing, 1975 and Welcomme, 1979) as entry points to this early literature). Explicit inclusion of ecological objectives in state-based fisheries management is a more recent phenomenon that is gaining considerable momentum (Hall, 1999; Welcomme, 2001; Degnbol, 2003; Sinclair and Valdimarsson, 2003). Christie and co-authors (2007) provide a useful summary of the evolution and differences in interpreting these management approaches. Some interpretations of ecosystem-based fisheries management remain within the natural sciences tradition, while others seek to balance societal and economic objectives within a sectoral approach (Murawski, 2000; Browman and Stergiou, 2004, 2005; Arkema et al., 2006). Principal among these broader interpretations of an ecosystem approach is the ecosystem approach to fisheries (EAF), discussed more fully below. Broader still is that class of 'ecosystem' approaches that takes a perspective outside the fisheries sector and includes large marine ecosystems, coastal zones or catchments in the system under 'management' (Grumbine, 1994; Cicin-Sain and Knecht, 1998; Sherman and Duda, 1999).

On the global stage, FAO has led the drive to reform fisheries management by promoting and mainstreaming the EAF (e.g. FAO, 1995, 2003; Sinclair and Valdimarsson, 2003). The EAF is now incorporated into many international conventions, including

Agenda 21, the Rio Declaration and the Biodiversity Treaty (CBD). While law and policy often lag a long way behind conceptual advances (Lugten and Andrew, 2008), the EAF and the associated Code of Conduct for Responsible Fisheries (FAO, 1995), have considerable legal and policy status in many jurisdictions and are now enshrined in the national laws of many countries. This legitimacy is important from a practical perspective. As a consequence, the EAF provides the most appropriate overarching approach to SSF management in the developing world. As defined by FAO (2003, p. 14), the EAF:

> ... strives to balance diverse societal objectives, by taking account of the knowledge and uncertainties about biotic, abiotic and human components of ecosystems and their interactions and applying an integrated approach to fisheries within ecologically meaningful boundaries.

The aspirations included in this definition are unarguable and sufficiently broad to be reinterpreted as advances in research and methods demand. Yet, despite the substantial normative power of the EAF (and the Code of Conduct for Responsible Fisheries), there is still much to be done to make it a reality on the ground, particularly in developing countries (Garcia and Cochrane, 2005; Christie *et al.*, 2007; De Young *et al.*, 2008; Garcia *et al.*, 2008). Interpretation and operationalization of the EAF as a practical management approach for SSFs in the developing world remains a central challenge for improved fisheries in these countries. In this respect, the EAF provides a sufficiently broad policy umbrella within which advances in research and management can be tested and refined. However, although the EAF is the most appropriate management approach, we suggest that progress is stymied by, among other things, the absence of an integrative research tradition that is capable of delivering assessment and advice appropriate to its holistic ambitions.

Rights-based approaches and co-management

Another class of management approaches, the rights-based approaches, is less explicit about the objective of sustainability and more concerned with the allocation of rights and responsibilities. Rights-based approaches to fisheries management have expanded in focus from *property rights*, to *access rights*, to *human rights*. The creation and exercise of legitimate rights offer substantial hope that the subsector will achieve sustainable economic development, but simple prescriptions for rights-based management are not sensible in light of the diversity of perspectives in SSFs. Instead, rights-based fisheries management requires a suite of political, legal and policy settings that need to evolve in ways appropriate to the diversity of rights that are often part of the fishery's societal objectives.

Property rights issues still dominate fisheries management. Even within this category there is a range of perspectives, from debate over private property rights and the roles that the market, the state, the judiciary and monitoring and evaluation play in defining, distributing and upholding these rights, to understanding the conditions conducive to managing commonly owned resources (including community-based management and co-management). Natural resource management has been heavily influenced by economic models of human behaviour and Hardin's (1968) 'tragedy of the commons' metaphor (Ostrom, 1990; Hilborn *et al.*, 2005). In fisheries, the Gordon–Schaefer model of a predictable relationship between cost, effort and benefit has dominated law, policy and research in fisheries. The relationship between these attributes is controlled in a number of ways, for example, by capping total catches, controlling effort and regulating how fish are caught. Stakeholders continue to debate how best to allocate and regulate property regimes for fisheries (see papers collected in *Marine Resource Economics*, 2007, 22(2) and *Philosophical Transactions of the Royal Society B*, 2005, 360(1453).

In developing countries, the diversity of SSFs makes such management strategies more difficult to implement. As Clay and Olson (2008) note, property rights are not the only institution that can serve as an organizing principle for collective action. Work on common property rights re-established communities as viable stewards of shared resources (Ostrom, 1990; Berkes, 1995; Ostrom et al., 2002), while work in cultural and political ecology emphasized the right of communities and local people to be involved in managing themselves and their resources (Berkes et al., 2002; Degnbol et al., 2006). Consequently, there has been a proliferation of participatory and collaborative management forms (e.g. Wilson et al., 2003; Pomeroy and Rivera-Guieb, 2006).

The terms 'co-management' and 'community-based management' are interpreted differently by many authors and are evolving into relatively complex ideas. Collaborative management, in its simplest form, refers to management processes that include entities (in addition to the state) in decision-making, usually resulting in a partnership between state and resource users, but also cooperation with other stakeholders and independent organizations (NGOs and research organizations; Pomeroy and Berkes, 1997; Pomeroy and Rivera-Guieb, 2006). However, co-management and its derivatives also aspire to embody a number of principles of 'good' governance, including democracy, transparency, accountability and sustainability (Wilson et al., 2003). These principles are necessary to ensure that co-management confers the responsibility to share power, knowledge and capacity, as well as to assign tasks. Research continues to examine the conditions suitable for effective communal management of resources (e.g. Agrawal, 2001, 2003), which is proving relatively elusive in practice (Wells and McShane, 2004; Plummer and Armitage, 2007). Property rights and broader access rights continue to be integral to how co-management and community-based management manifest in practice. Several issues complicate the practice of rights-based management:

- Property rights consist of bundles of rights, including access (right to enter), withdrawal (right to extract), management (right to regulate use), exclusion (right to deny access) and alienation (right to sell, lease or transfer), which can influence how resources are allocated (see Ribot and Peluso, 2003).
- Management consists of a variety of functions (policy, service delivery, research and monitoring, institutional design, enforcement, use), stages (planning, implementation and evaluation), levels (instructive to informative) and scales (spatial, administrative and institutional) at which participation or collaboration can occur (see Sen and Raakjær Nielsen, 1996).
- The notion of a distinct, equitable and consensual 'community' is flawed; thus management that requires a defined set of stakeholders for power-sharing must itself define this 'community'.
- Effective and sustainable collaborative management is likely to require broader political and social transitions (such as decentralization and effective legislative and judicial institutions) and integrated planning to support it (see Ribot et al., 2006).

Most resource management agreements, conventions and guidelines, including guidelines for putting the EAF into practice (FAO, 2005), stress the importance of collaborative and participatory forms of management. The resilience literature also specifies the need for collaborative management to enable and foster adaptive capacity of social-ecological systems (e.g. Olsson et al., 2004).

More recently, human rights approaches that integrate concerns for access rights, user rights, postharvest rights and human rights are coming to the fore (SAMUDRA, 2008). Broadly, such approaches espouse principles found in the Millennium Development Goals and the Universal Declaration of Human Rights, including well-being, dignity, non-discrimination and equality, as well as other, more common, principles of good governance such as participation and accountability. Awareness of rights and the capacity to demand rights and hold states accountable are central within this framework. In the fisheries sector, a move towards human rights concerns is evidenced by the mainstreaming of health and education in fisheries research and management and by campaigns such as the right to food (e.g. FAO, 2009) as a counterweight to the marginalization of local communities from resources as a consequence of conservation, tourism, development and large-scale fisheries activities.

Management for resilience

The notion of resilience has risen to prominence in the academic literature on natural resource management in the last decade. Concepts gathered under the 'resilience' banner are characterized by a focus on non-linear change, unpredictability, thresholds, adaptive management, transformation, institutional learning and vulnerability and adaptation to external drivers (Carpenter *et al.*, 2001; Walker *et al.*, 2002, 2004; Folke *et al.*, 2004; Pikitch *et al.*, 2004; Folke, 2006). As complex systems, SSFs exemplify the dynamic and unpredictable interdependencies of people and nature. People in SSFs are vulnerable to the compounding effects of stresses within fishery systems, as well as to ecological and social forces outside their domain of influence. Building the adaptive capacity of ecosystems and of people is, therefore, central to realizing the conservation, social and economic potential of SSFs in the developing world.

When integrated within the EAF's overarching legal and policy environment, resilience approaches have the potential profoundly to improve SSF management. However, while resilience has become a powerful metaphor for sustainability, advances in theory have yet to be translated into more resilient aquatic ecosystems or better lives for poor fishery-dependent people in developing countries (Carpenter *et al.*, 2001, 2005; Walker *et al.*, 2010). The real challenge now is to build bridges between the rapid advances in research and analysis and the real-world legal, policy and organizational constraints of SSF management, particularly in developing-country contexts. Poverty and vulnerability, dynamic non-equilibrial ecosystems and limited capacity and data combine to make this challenge the most important frontier for SSF research. We offer a perspective on the definitional issues that may provide a starting point for 'resilience in practice'.

Definitions of resilience may be traced to Holling's original (1973, p. 17) definition of resilience in ecological systems:

> Resilience determines the persistence of relationships within a system and is a measure of the ability of these systems to absorb changes of state variables, driving variables, and parameters, and still persist.

This definition is value-neutral, i.e. it is silent about the desirability (or otherwise) of the system configuration, and there are many examples of undesirable but persistent ecosystem configurations. Problems arise when this definition is broadened to include people as part of the system (Brand and Jax, 2007). Definitions of resilient social-ecological systems (SESs) have also been value-neutral (see below), but when it comes to people, 'resilience' is a good thing and much of the message implicit in many definitions is that resilient SESs are desirable. Building on earlier papers, notably Carpenter *et al.* (2001), Walker *et al.* (2004, p. 2), provide a widely cited definition of the resilience of a social-ecological system:

> … the capacity of a system to absorb disturbance and reorganize while undergoing change so as to still retain essentially the same function, structure, identity, and feedbacks …

As Brand and Jax (2007) note, simultaneous claims on the term challenge both conceptual clarity and practical relevance. These authors present a useful typology of definitions of resilience and conclude that a clear descriptive definition is useful in ecological science but that a more vague usage is appropriate to foster trans-disciplinary approaches to social-ecological systems (Brand and Jax, 2007). This conclusion presents problems in the search for sustainable SSFs, particularly if resilience concepts are to be incorporated into policies, laws and regulations. As noted earlier, management that defines sustainability solely in terms of ecology has largely failed in the context of the developing world. People are an integral part of these ecosystems, and their exclusion from analysis and the search for practical solutions will not provide a path to sustainability. This means that we must find an operational form of the term 'social-ecological resilience' (Carpenter *et al.*, 2005) that is appropriate for developing-country contexts (see also Vogel *et al.*, 2007). To achieve this goal, we must deal with the problem of *'value'* – who decides what a desirable configuration is, and to whom the benefits flow (Lebel *et al.*, 2006; Nadasdy, 2007).

Given an underlying motivation to reduce poverty through improved fisheries, it is possible to provide a generic definition that is compatible with democratic, participatory forms of management. A resilient SSF in the developing world may be defined as one that 'absorbs stress and reorganizes itself following disturbance, while still delivering benefits for poverty reduction'.

Within this overarching definition, there is room for the management constituency to address the political (*'value'*) dimensions of resilience approaches to management and to be specific about beneficiaries. Note that the words 'reorganize itself' are central to any generic definition. The capacity of people and institutions to learn and adapt, and to self-organize and reorganize, is critical to building resilience (Folke *et al.*, 2003; Walker *et al.*, 2004; Berkes and Seixas, 2005; Kooiman *et al.*, 2005; Folke, 2006; Mahon *et al.*, 2008 and references within). This individual and institutional capacity to organize and to respond better to surprises is especially important in an adaptive management context (McLain and Lee, 1996). Garaway and Arthur (2004) refine the familiar aphorism 'learning by doing' to 'learning as an objective of doing' to emphasize the centrality of the process. Interestingly, the emphasis placed by these and other authors on learning and empowering participants in a fishery places the conventional usage of the phrase 'capacity building' in sharper relief (see MacFadyen and Huntington, 2004 for discussion and review).

Although capacity building is fundamental for making the future less uncertain and for reducing the impact of threats as yet unknown, at the level of the fishery many issues are clear and present. As part of capacity-building initiatives, practical resilience management can be pursued by getting on with the business of addressing these threats and opportunities. Within the context of a fishery, the focus then shifts to defining resilience 'of what' and 'to what' (Carpenter *et al.*, 2001). Answers to these questions are matters of policy choice and stakeholder negotiations. For resilience-based management to be effective, stakeholders need to be involved in identifying and maintaining system attributes that make up an SSF's identity or in transforming a fishery into a new configuration that will provide more appropriate ecosystem services for social and economic benefits (Cumming *et al.*, 2005).

First, it is necessary to define the boundaries of the fishery system; this provides its identity, which has consequences for governance, the legitimacy of management institutions, the resource harvested, the nature of assessments and the appropriateness of management responses. Critically, it also makes the focal scale for management explicit (Walker *et al.*, 2004). The fishery will be influenced by processes working at both smaller and larger scales, but recognizing the primary scale of focus is a necessary step.

The resilience of an SSF may be threatened by the effects of stressors from within the fishery itself, such as fishing or debt accumulation among fishers, or by discrete disturbances such as storms or dam construction (WCD, 2000; Walker *et al.*, 2002; Béné *et al.*, 2007). Adapting the description of Walker and co-authors (2002) of the generic objectives of resilience management, we can argue that management for a resilient SSF in the developing world should prevent the fishery from failing to deliver benefits by nurturing and preserving ecological, social and institutional attributes that enable it to endure, renew and reorganize itself.

In this definition, 'benefits' refer to the ecosystem goods and services derived from the fishery. This statement of the management objective, when paired with a definition of the particular fishery being considered, would seem to provide a useful interpretation of the EAF objective and, therefore, a bridge to reconcile two largely parallel streams of thinking in fisheries. What is missing is a portfolio of case studies that are based on learning from deliberate attempts to implement resilience-based management. There are four related and practical reasons for this gap. First, resilience theory is still evolving and, in the view of many, not yet 'investment-ready'. There are few resilience 'products' available for people responsible for managing fisheries to use. Second, testing resilience-based sustainability in SSFs in developing countries using scales and timeframes appropriate to ecosystems and societies requires large changes in institutions and in the expectations of some of the poorest, most marginalized people in the world. These challenges not only present important ethical dilemmas, but also reduce the probability that failures will be adequately reported and lessons incorporated into other initiatives. Third, the time scale for building a portfolio of case histories from which to learn is a decadal one. Finally, managers and other decision-makers operate within the statutes and policies of governments. Fisheries law and policy of most developing countries frame 'sustainable exploitation' of fisheries as maximum sustainable yield (MSY), and government ministers and their agencies are unlikely to step beyond their statutory obligations to test emerging theory. In the short term, management experiments of resilience theory will have to use definitions of resilience and objectives that are reinterpretations of prevailing law and policy rather than radical departures from existing legal frameworks.

To reiterate, ecosystem-based management, rights-based approaches and resilience perspectives are compatible with each other. For instance, management for resilience, within a broad EAF framework, could apply rights-based strategies, including adaptive co-management. Clearly, there are different emphases which suggest that these different approaches, or combinations of them, may be more or less successful in different types of fisheries. A resilience approach seems highly suitable to an *'inside looking out'* perspective, but human rights approaches may be more pertinent to a perspective that sees SSF as a broader governance issue. How these management approaches shape practical management is the responsibility of the management constituency. Implementation frameworks provide a link between these conceptual concerns and choices, and actual practice.

Implementation Frameworks

Beneath the conceptual approach to managing a fishery, the framework used to implement management provides another level of organization. In particular, it describes relationships among elements of the research and management problem and suggests an order for doing things. Fisheries implementation frameworks may include many elements that, though overlapping, are distinct phases in the process. Common elements include, for example, scoping, assessment, (adaptive) management, monitoring and evaluation. In a strict sense, an implementation framework is independent of management objectives, but in practice both the management approach and the implementation framework contribute concepts and ways of thinking that guide the choices that are made.

Two recent implementation frameworks, both derived from FAO's (1995) generalized fishery management cycle but with different areas of emphasis, provide a bridge between the EAF's concepts and aspirations and its implementation in developing-world SSFs. These are presented below alongside a third implementation framework (Andrew *et al.*, 2007) designed for the diagnosis and management of SSFs in developing-country contexts. Together these frameworks can guide the diagnosis and management process from the perspective of any chosen management approach, as informed by international and national policy and legislation characterizing a particular fishery.

In 1995, FAO introduced a general Fishery Policy and Management Cycle with nested levels of activity that scale down from international laws and policies, to national governance issues, to operational management of a fishery. Each level is connected to those above by a series of feedback loops that allow finer-scale and faster-moving processes to be incorporated within the larger and slower levels above. More recently, FAO has promoted a Management Planning and Implementation Cycle (FAO, 2003) for implementing the EAF, which unpacks the management and implementation components of the original framework. The implementation cycle recognizes a series of steps in the management process, beginning with a scoping phase and running through the conventional steps of setting objectives, making rules, implementing management and monitoring and assessing outcomes (Fig. 2.1).

The Management Planning and Implementation Cycle (FAO, 2003) emphasizes assessment and advice, and the feedback loops of management and planning. The basic elements of the framework and their ordering are common to all fisheries, but their emphasis will clearly vary. The framework does not address the political and social process of deciding who is 'in the fishery' and how benefits are allocated, but the cycle does make clear the need to consult with stakeholders at all phases of the cycle. In some fisheries, managers and stakeholders are easily identified. In many other cases, they are not, which may lead to less powerful actors being marginalized and more powerful ones wrongly assumed to be central. By extension, access rights and management objectives are frequently unknown or contested, particularly in developing-country SSFs.

Garcia et al. (2008) adapt the research elements of the EAF implementation cycle to the particular circumstances of SSFs in the developing world (Fig. 2.2). The resulting integrated assessment and advisory framework, again, explicitly restricts itself to the assessment and advisory parts of the cycle. This framework advances previous versions in that, in addition to the three classical dimensions of fisheries (ecological, social and economic), the authors highlight processes outside the domain of the fishery that need to be considered (see also Andrew et al., 2007). However, the framework does not cover management issues and so is silent on how management is done, what the management objectives are, what access rights are and who enjoys them.

The Integrated Assessment and Advisory Framework is complete in the sense that it contains all the elements of earlier FAO frameworks but is also idealized. The authors make clear that investment in assessment and management must, in practice, be proportionate to the value of the fishery and appropriate to its complexity (see also Mahon et al., 2008). The capacity of SSFs in developing

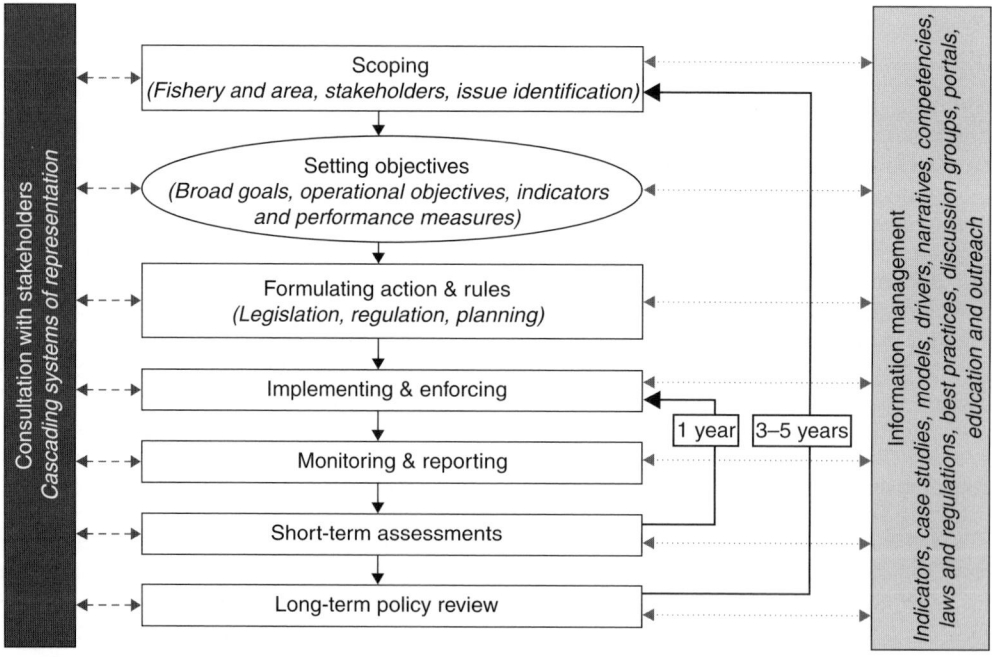

Fig. 2.1. The EAF implementation cycle (redrawn from FAO, 2003).

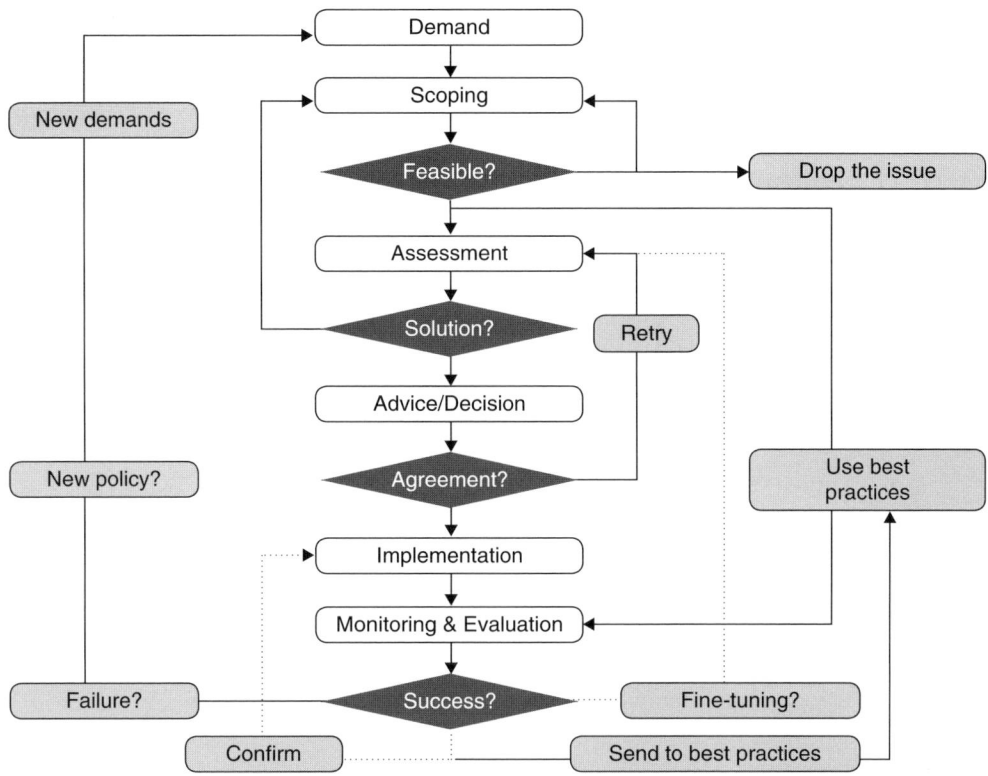

Fig. 2.2. Integrated assessment and advisory framework for SSF (redrawn from Garcia et al., 2008).

countries to conduct assessments and to monitor and evaluate outcomes will often be severely limited. As a result, it is unlikely that many SSFs, especially in the least developed countries, will have the resources to fully implement an integrated assessment process. An important part of the scoping exercise for any fishery will be to ensure that the assessment and management process is commensurate with the ecological, social and economic attributes of the fishery (Mahon et al., 2008).

Andrew and colleagues (2007) proposed an implementation framework that specifically addresses the challenges presented by SSFs in the least developed countries (Fig. 2.3). They were motivated by the need to have a flexible framework that provides the minimum set of elements in the research and management cycle. It places greater emphasis on: (i) the broader non-sectoral drivers of fisheries management performance and the opportunities and threats they present to people's livelihoods (the *'inside looking out'* perspective); and (ii) the institutions that govern fisheries, particularly the nature and legitimacy of use rights as a central and identifiably separate precursor to effective management. Underpinning this framework is the issue of defining the fishery and, therefore, making a judgment about what is within the fishery (and directly under the influence of an agreed set of actors) and what is external to it. Management should seek to make the fishery less vulnerable to those external drivers.

This framework attempts to integrate assessment and advice into the management implementation cycle of the fishery. In some respects, it is less prescriptive than EAF implementation or the integrated assessment and advisory frameworks described above, in

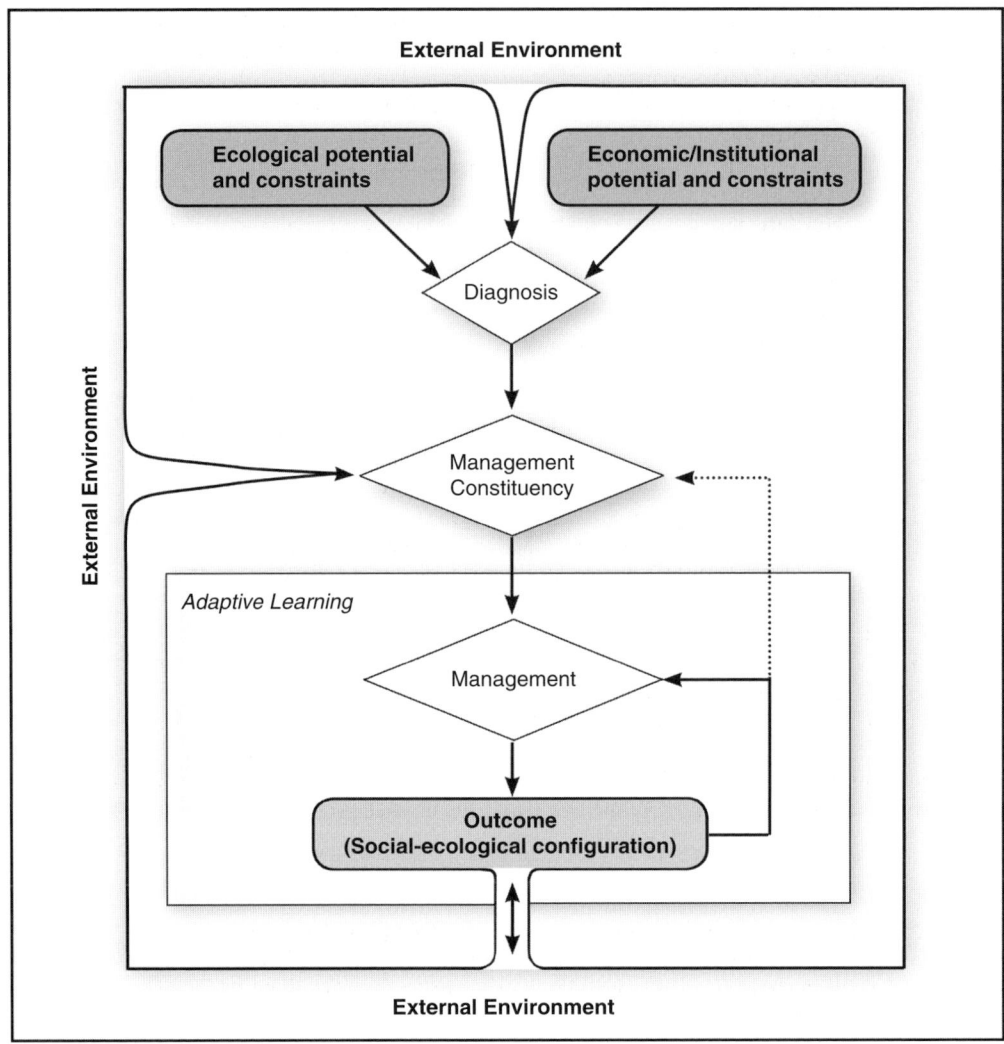

Fig. 2.3. A general framework for diagnosis and management of SSFs in the developing world (redrawn from Andrew et al., 2007).

that it places more emphasis on achieving clarity and building legitimacy of the management process and less on gathering and interpreting data. Following others (e.g. Walters and Hilborn, 1978; Charles, 2001; Arthur and Garaway, 2004; Armitage et al., 2009), Andrew and co-authors (2007) advocate an adaptive management process as the most promising way to learn about the responses of the fishery system to drivers of change.

Research Approaches

There are many different research perspectives that can be used for the diagnosis and advisory portions of the management process. The overarching management approach and the research traditions and capacities of the researchers involved will influence which research perspectives are appropriate and viable. Conventionally, research has been designed to serve the management approach of the day and, therefore, to estimate maximum sustainable yield from target species and monitor the effectiveness of management interventions (see Hilborn and Walters, 1992 as the classic text and as an entry point to this literature). In recent years, as limitations in the target resource-oriented management (TROM) approach (an '*inside looking in*' perspective) have become ever more apparent (see earlier references) and approaches such as the EAF have been promoted, other research approaches have come to prominence as researchers seek to provide policy makers and managers with more 'holistic' advice on the sustainability of fisheries. Some of these research approaches – for example, the sustainable livelihoods approach – have been adopted by large development and management organizations, such as the FAO or the World Bank, while others remain within the domain of research.

The choice of research approach has profound implications for the way the fishery is viewed, the questions asked and the methods or tools employed. Like management approaches, research approaches have different emphases. Some focus primarily on ecological components, while others are founded on rights and entitlements principles or are concerned with institutions and broader governance issues (Table 2.1). A variety of research analysis tools and data collection techniques can be used to undertake research guided by any one of these approaches. In fact, many research analysis tools and techniques are common across research frameworks. This paper focuses on overarching management choices and so does not delve into data collection techniques, although many of these are explained in the links in Table 2.1.

Ecosystem-based management perspectives, including the EAF, have implications for research. Broadly, these perspectives reiterate a multidimensional focus on ecosystems, people and livelihoods, and governance and institutions. Many applications of an EAF-based perspective continue to prioritize ecological domains over social or institutional ones.

Also originating from the natural sciences is the resilience approach. Analyses of fishery systems viewed through a resilience lens are emerging (e.g. Berkes and Seixas, 2005; Gelcich *et al.*, 2006; Marschke and Berkes, 2006; McClanahan *et al.*, 2008) and offer important insights into the dynamics of fishery systems on a local scale. Some authors focus primarily on

Table 2.1. Web-based reference material on different research approaches. The links refer to background guidelines and toolkits depending on how established the research frameworks are and whether they cross the research–implementation divide. URLs accessed 3 February 2010.

Research approaches	Further reference materials
Ecosystem-based approaches	FAO (2003): http://www.fao.org/docrep/005/Y4470E/y4470e00.htm Resilience Alliance (2007): http://www.resalliance.org and http://www.resalliance.org/3871.php
Rights-based and entitlements approaches	DFID (undated): http://www.chronicpoverty.org/toolbox/toolboxcontents.php Eldis (undated): http://www.eldis.org/go/topics/dossiers/livelihoods-connect NZAID (2007): http://nzaidtools.nzaid.govt.nz/sustainable-livelihoods-approach NZAID (undated): http://nzaidtools.nzaid.govt.nz/tools/analytical-tools University of Bath (2002–2007): http://www.bath.ac.uk/econ-dev/wellbeing/research/research.htm
Wealth-based approaches	World Bank (2004): http://siteresources.worldbank.org/INTPOVERTY/Resources/335642-1098192957114/op1_pa_guidance.pdf
Institutional assessments and governance approaches	Bavinck *et al.* (2005): http://www.fishgovnet.org/downloads/documents/bavinck_interactive.pdf Indiana University (undated, a): http://www.indiana.edu/~workshop/ Indiana University (undated, b): http://dlc.dlib.indiana.edu/dlc/ The Fisheries Governance Network (undated): http://www.fishgovnet.org/

the ecological aspects of resilience management problems, while others address resource use within social-ecological systems and look at both ecological processes and adaptive capacity. For example, Berkes and Seixas (2005) categorize factors that build social-ecological resilience in lagoon systems into four clusters: (i) learning to live with change and uncertainty; (ii) nurturing diversity for reorganization and renewal; (iii) combining different kinds of knowledge; and (iv) creating opportunities for self-organization. However, they find that only certain 'resilience surrogates' are shared among different lagoonal case studies.

Earlier, we suggested that a generic management objective for SSFs in the developing world might be to 'prevent the fishery from failing to deliver benefits by nurturing and preserving ecological, social and institutional attributes that enable it to renew and reorganize itself'. Under this definition, research needs to: (i) identify internal and external pressures and drivers that threaten the delivery of benefits; (ii) identify ecological, social and institutional attributes that are critical to the delivery of benefits; and (iii) identify opportunities and conditions for learning and self-organization. Furthermore, once management objectives have been agreed and management implemented, research must monitor and evaluate the efficacy of management.

The Resilience Alliance (2007) has published beta versions of generic workbooks for assessment and management of resilience, but a 'how-to' manual for resilience analysis for SSFs has yet to be written. The workbooks suggest questions organized around the core issues that underlie resilience thinking, such as:

- Defining and understanding the system by considering past, present and future states. Subsets of this broad issue include questions such as resilience 'of what' and 'to what', and who the people involved are, as well as institutional constraints and management opportunities.
- Developing conceptual models of current and alternative states of the system, including definition of the system under management (the focal scale), and possibly even developing alternative future scenarios to guide management decisions. A particularly challenging subset of questions focuses on recognizing thresholds of change.
- Developing preliminary management responses to maintain desirable system configurations or to transform undesirable ones.

It is unlikely that the full set of analyses canvassed in the Resilience Alliance workbooks will be possible for most SSFs in least developed countries. As a consequence, the adaptive management phase of the management cycles assumes even greater importance, given that learning is much more likely to come from managing and evaluating. We also note that, as with the FAO management cycle described above, the workbooks do not emphasize who the participants, managers and stakeholders are, and which institutions confer legitimacy. These questions are central for SSFs in the developing world, where the identity of these people and organizations is often contested.

From another angle, the 'capacities, capabilities and entitlements' research approaches (Bebbington, 1999) prioritize people-centred, rather than resource- or economics-based, perspectives for both social development and natural resource management. They also emphasize human agency (as capacity) in contrast to broader structural constraints. These approaches have raised awareness of the multiple dimensions of poverty that exist, beyond the lack of access to financial capital. Examples include the sustainable livelihoods approach (SLA), vulnerability assessments and well-being approaches (see Table 2.1). The SLA aims to understand the role and diversity of individual and household livelihoods in the context of factors that make them vulnerable. For example, Allison and Ellis (2001) used an SLA to understand the strategies of fisherfolk facing resource fluctuations. They showed that use rights can restrict the flexibility of fishers to migrate and to move in and out of the fishery in response to variability; thus, contrary to popular assumptions, use rights can undermine both livelihood strategies and ecological sustainability.

One strength of the SLA lies in the micro-level analysis of the dimensions of poverty, which is highly pertinent for SSFs in developing-country contexts. However, in some SSFs, it may be more relevant to examine the meso- and macro-level aspects of poverty; in such cases, vulnerability and poverty assessments provide a broader research focus. Such approaches have sometimes served as useful frameworks for implementation as well as for thinking about a problem. For example, the SLA's emphasis on the many cross-sectoral dimensions of poverty and vulnerability has made it useful in designing poverty-reduction projects (e.g. Ellis, 2000; Allison and Ellis, 2001; Allison and Horemans, 2006).

Another subset of research approaches within the development and environmental sciences focuses more on the institutional conditions needed for successful management and wider governance. While entitlements and rights-based approaches are primarily concerned with poverty reduction and social justice, institutional frameworks aim to develop a theory of effective collective action. Institutional approaches highlight a diverse set of multidisciplinary variables that go beyond institutions per se to consider a range of contextual factors. Examples of institutional assessment frameworks include the institutional analysis and development (IAD) framework, that for analysing social-ecological systems (Ostrom, 2007) and the interactive governance approach, which was developed specifically for fisheries (Bavinck et al., 2005; Kooiman et al., 2005). These frameworks have been used to address a variety of research questions in natural resource management and are appropriate for understanding context, the subtle conditions necessary for cooperation and collaboration between stakeholders, and the potential of different institutions successfully to coordinate management and resource use.

Conclusions

The gap between policy, legislation and the practical aspects of fisheries management, and academic research, is a persistent barrier to integration and progress in SSF management and governance. In one respect, the simplicity and apparent explanatory power of conventional models and metrics (e.g. MSY) mean they are firmly engrained in management practice. On the other hand, the relative lack of investment in operationalizing new concepts has resulted in a divergence between fisheries practice and science.

In an attempt to help bridge this gap, we have focused on themes of ecosystem, rights and resilience to show some consistency between overarching management approaches and the research perspectives available for the assessment and advisory process. Each of these approaches provides significant and unique contributions to fisheries management. More importantly, these approaches and frameworks can work together. We argue that the EAF provides the most appropriate overarching approach to management, because it is established in national and international law and policy. Within the EAF's broad objective, a resilience perspective and associated concepts of adaptive management and institutional learning can provide a way of moving beyond management forms that are based on control and optimization. As an ideal, a democratic and participatory form of management can also address the political issues of what is desirable in terms of system configuration, what are the (internal and external) threats and opportunities for SSFs, and who benefits from a particular system regime. Integrating appropriate rights-based approaches as principles (e.g. equality and accountability) and practical management strategies (e.g. property rights and co-management) can add to a comprehensive management and research approach to SSFs. In particular, rights-based perspectives can help balance the ecological – and potentially conservation-oriented – bias of ecosystem-based and resilience approaches. This is crucial for SSFs in developing-country contexts. The conceptualization of resilience outlined in this paper is consistent with human rights-based thinking.

Second, alternative management and research approaches, and the multiple principles that underpin them, are often considered somewhat restrictive in developing-country contexts. In particular, managing for resilience through adaptive management, monitoring and learning processes is seen as capacity- and resource-intense. We think this is a misconception. The EAF, resilience and rights-based management and research approaches are all underpinned by principles of participation, appreciation for multiple perspectives and knowledge, and collaborative learning and decision-making. These approaches are expected to enable more flexible management processes, in contrast to control by external experts, and investment in management can be scaled to be appropriate to the fishery.

The greatest impediment to progress has probably been the fact that, within the conventional fisheries research tradition, there has been little capacity to integrate across the many dimensions (ecological, social and economic) of a fishery system (Charles, 2001; Garcia and Charles, 2007). Furthermore, its intra-sectoral focus has meant that external threats and opportunities have not been well addressed in assessment and policy advice. A result of this has been that management has performed poorly in SSFs in the world's least developed countries. Innovation in research occurs much faster than innovation in management. Combining the broad management approach of the EAF and the innovation in the resilience approach to analysing fishery systems presents an important opportunity in fisheries management in the developing world. If management is able to prevent the fishery from failing to deliver benefits that reduce poverty by nurturing and preserving ecological, social and institutional attributes that enable it to renew and reorganize itself, then these approaches to SSF are more likely to succeed than conventional approaches. Nevertheless, there remains considerable work to be done to crystallize appealing theory into well-grounded and tested approaches and frameworks for analysis and policy advice. Building such a portfolio of practice in alternative SSF management is the principal challenge to reconciling EAF, resilience and rights-based SSF approaches.

Acknowledgements

We are grateful to Edward Allison, Chris Béné, Stephen Hall, David Mills and Bob Pomeroy for discussion and comments on the manuscript. Insightful comments from Robin Mahon clarified some muddles. This work was funded by PROFISH, the World Bank's global programme on fisheries management, the Challenge Programme for Water and Food and the WorldFish Center.

References

Agrawal, A. (2001) Common property institutions and sustainable governance of resources. *World Development* 29, 1649–1672.

Agrawal, A. (2003) Sustainable governance of common-pool resources: context, methods and politics. *Annual Review of Anthropology* 32, 243–262.

Allison, E.H. and Ellis, F. (2001) The livelihoods approach and management of small-scale fisheries. *Marine Policy* 25, 377–388.

Allison, E.H. and Horemans, B. (2006) Putting the principles of the sustainable livelihoods approach into fisheries development policy and practice. *Marine Policy* 30, 757–766.

Andrew, N.L., Béné, C., Hall, S.J., Allison, E.H., Heck, S. and Ratner, B.D. (2007) Diagnosis and management of small-scale fisheries in developing countries. *Fish and Fisheries* 8, 227–240.

Arkema, K.K., Abramson, S.C. and Dewsbury, B.M. (2006) Marine ecosystem-based management: from characterization to implementation. *Frontiers in Ecology and Environment* 4, 525–532.

Armitage, D.R., Plummer, R., Berkes, F., Arthur, R.I., Charles, A.T., Davidson-Hunt, I.J., Diduck, A.P., Doubleday, N.C., Johnson, D.S., Marschke, M., McConney, P., Pinkerton, E.W. and Wollenberg, E.K. (2009) Adaptive co-management for social–ecological complexity. *Frontiers in Ecology and Environment* 7, 95–102.

Arthur, R.I. and Garaway, C.J. (2004) Creating understanding and ownership of collaborative research results through 'learning-by-doing'. *STREAM Journal* 3, 1–2.

Bavinck, M., Chuenpagdee, R., Diallo, M., van der Heijden, P., Kooiman, J., Mahon, R. and Williams, S. (2005) *Interactive Fisheries Governance: a Guide to Better Practice*. Eburon Academic Publishers, Delft, The Netherlands. Available at: http://www.fishgovnet.org/downloads/documents/bavinck_interactive.pdf (accessed 3 February 2010).

Bebbington, A. (1999) Capital and capabilities: a framework for analyzing peasant viability, rural livelihoods and poverty. *World Development* 27, 2021–2044.

Béné, C., Bennett, L. and Neiland, A.E. (2004) The challenge of managing small-scale fisheries with reference to poverty alleviation. In: Neiland, A. and Béné, C. (eds) *Poverty and Small-Scale Fisheries in West Africa*. Kluwer Academic Publishers and Food and Agriculture Organization, Amsterdam, pp. 83–102.

Béné, C., Macfadyen, G. and Allison, E.H. (2007) Increasing the contribution of small-scale fisheries to poverty alleviation and food security. FAO Fisheries Technical paper No. 481, FAO, Rome, Italy. Available at: www.fao.org/docrep/009/a0965e/a0965e00.htm (accessed 3 February 2010).

Berkes, F. (1995) Community-based management of common property resources. *Encyclopaedia of Environmental Biology* 1, 371–378.

Berkes, F. (2003) Alternatives to conventional management: lessons from small-scale fisheries. *Environment* 31, 1–14.

Berkes, F. and Seixas, C.S. (2005) Building resilience in lagoon social-ecological systems: a local-level perspective. *Ecosystems* 8, 967–974.

Berkes, F., Mahon, R., McConney, P. and Pomeroy, R.S. (2001) *Managing Small-scale Fisheries: Alternative Directions and Methods*. IDRC, Ottawa, Canada.

Berkes, F., Colding, J. and Folke, C. (2002) *Navigating Social-ecological Systems: Building Resilience for Complexity and Change*. Cambridge University Press, Cambridge, UK.

Brand, F.S. and Jax, K. (2007) Focusing the meaning of resilience: resilience as a descriptive concept and a boundary object. *Ecology and Society* 12(1), art. 23. Available at: http://www.ecologyandsociety.org/vol12/iss1/art23/ (accessed 3 February 2010).

Browman, H.I. and Stergiou, K.I. (eds) (2004) Perspectives on ecosystem-based approaches to the management of marine resources. *Marine Ecology Progress Series* 274, 269–303.

Browman, H.I. and Stergiou, K.I. (eds) (2005) Politics and social-economics of ecosystem-based management of marine resources. *Marine Ecology Progress Series* 300, 241–296.

Carpenter, S., Walker, B., Anderies, J.M. and Abel, N. (2001) From metaphor to measurement: resilience of what to what? *Ecosystems* 4, 765–781.

Carpenter, S., Westley, R.F. and Turner, M.G. (2005) Surrogates for resilience of social-ecological systems. *Ecosystems* 8, 941–944.

Cash, D.W., Adger, W., Berkes, F., Garden, P., Lebel, L., Olsson, P., Pritchard, L. and Young, O. (2006) Scale and cross-scale dynamics: governance and information in a multilevel world. *Ecology and Society* 11(2), art. 8. Available at: www.ecologyandsociety.org/vol11/iss2/art8/ (accessed 3 February 2010).

Charles, A.T. (2001) *Sustainable Fishery Systems*. Blackwell Science Ltd., London.

Christie, P., Fulharty, D.L., White, A.T., Eisma-Osorio, L. and Jatulan, W. (2007) Assessing the feasibility of ecosystem-based fisheries management in tropical contexts. *Marine Policy* 31, 239–250.

Cicin-Sain, B. and Knecht, R.W. (1998) *Integrated Coastal and Ocean Management*. Island Press, Washington, DC, USA.

Clay, P. and Olson, J. (2008) Defining fishing communities: vulnerability and the Magnuson–Stevens Fishery Conservation and Management Act. *Human Ecology Review* 15, 143–160.

Cochrane, K.L. (2000) Reconciling sustainability, economic efficiency and equity in fisheries: the one that got away. *Fish and Fisheries* 1, 3–12.

Cochrane, K.L. and Doulman, D.J. (2005) The rising tide of fisheries instruments and the struggle to keep afloat. *Philosophical Transactions of the Royal Society Biological Sciences* 360, 77–94.

Cumming, G.S., Barnes, G., Perz, S., Schmink, M., Sieving, K.E., Southworth, J., Binford, M., Holt, R.D., Stickler, C. and Van Holt, T. (2005) An exploratory framework for the empirical measurement of resilience. *Ecosystems* 8, 975–987.

Cushing, D. (1975) *Marine Ecology and Fisheries*. Cambridge University Press, Cambridge, UK.

De Young, C., Charles, A. and Hjort, A. (2008) Human dimensions of the ecosystem approach to fisheries: an overview of context, concepts, tools and methods. FAO Fisheries Technical Paper No. 489, FAO, Rome, Italy.

Degnbol, P. (2003) Science and the user perspective: the scale gap and the need for co-management. In: Wilson, D.C., Nielsen, J.R. and Degnbol, P. (eds) *The Fisheries Co-management Experience. An Evaluation Management Experience: Accomplishments, Challenges and Prospects*. Kluwer, Amsterdam, pp. 31–49.

Degnbol, P., Gislason, H., Hanna, S., Jentoft, S., Raakjaer Nielsen, J., Sverdrup-Jensen, S. and Wilson, D.C. (2006) Painting the floor with a hammer: technical fixes in fisheries management. *Marine Policy* 30, 534–543.

DFID (undated) *Chronic Poverty Research Centre: Toolbox for SLA*. Available at: http://www.chronicpoverty.org/toolbox/toolboxcontents.php (accessed 3 February 2010).

Eldis (undated) *Livelihoods Connect*. Available at: http://www.eldis.org/go/topics/dossiers/livelihoods-connect (accessed 3 February 2010).

Ellis, F. (2000) The determinants of rural livelihood diversification in developing countries. *Journal of Agricultural Economics* 51, 289–302.

FAO (1995) *Code of Conduct for Responsible Fisheries*. FAO, Rome, Italy. Available at: www.fao.org/DOCREP/005/v9878e/v9878e00.htm (accessed 3 February 2010).

FAO (2003) *Fisheries Management: the Ecosystem Approach*. Technical Guidelines for Responsible Fisheries Suppl. 2, FAO, Rome, Italy. Available at: http://www.fao.org/docrep/005/y4470e/y4470e00.htm (accessed 3 February 2010).

FAO (2005) *Putting into Practice the Ecosystem Approach to Fisheries*. FAO, Rome, Italy. Available at: www.fao.org/docrep/009/a0191e/A0191E00.htm (accessed 3 February 2010).

FAO (2009) *The Right to Food*. FAO, Rome, Italy. Available at: http://www.fao.org/righttofood/ (accessed 3 February 2010).

Folke, C. (2006) Resilience: the emergence of a perspective for social-ecological systems analyses. *Global Environmental Change* 16, 253–267.

Folke, C., Colding, J. and Berkes, F. (2003) Synthesis: building resilience and adaptive capacity in social-ecological systems. In: Berkes, F., Colding, J. and Folke, C. (eds) *Navigating Social-ecological Systems: Building Resilience for Complexity and Change*. Cambridge University Press, Cambridge, UK, pp. 352–387.

Folke, C., Carpenter, S., Walker, B., Scheffer, M., Elmqvist, T., Gunderson, L. and Holling, C.S. (2004) Regime shifts, resilience and biodiversity in ecosystem management. *Annual Review of Ecology Evolution and Systematics* 35, 557–581.

Garaway, C.J. and Arthur, R.I. (2004) *Adaptive Learning: a Practical Framework for the Implementation of Adaptive Co-management – Lessons from Selected Experiences in South and Southeast Asia*. Marine Resources Assessment Group, London.

Garcia, S.M. and Charles, A.T. (2007) Fishery systems and linkages: from clockwork to soft watches. *ICES Journal of Marine Sciences* 64, 580–587.

Garcia, S.M. and Cochrane, K.L. (2005) Ecosystem approach to fisheries: a review of implementation guidelines. *ICES Journal of Marine Science* 62, 311–318.

Garcia, S.M. and Grainger, R. (1997) Fisheries management and sustainability: a new perspective of an old problem? In: Hancock, D.A., Smith, D.C., Grant, A. and Beumer, J.P. (eds) Developing and sustaining world fisheries resources. The state of science and management, *Second World Fisheries Congress*, Brisbane, Australia, 1996. CSIRO Publishing, Melbourne, Australia, pp. 175–236.

Garcia, S.J., Allison, E.H., Andrew, N.L., Béné, C., Bianchi, G., de Graaf, G., Kalikoski, G.J., Mahon, R. and Orensanz, J.M. (2008) Towards integrated assessment and advice in small-scale fisheries: principles and processes. FAO Fisheries and Aquaculture Technical Paper No. 515, FAO, Rome, Italy. Available at: http://www.fao.org/docrep/011/i0326e/i0326e00.htm (accessed 3 February 2010).

Gelcich, S., Edwards-Jones, G., Kaiser, M.J. and Castilla, J.C. (2006) Co-management policy can reduce resilience in traditionally managed marine ecosystems. *Ecosystems* 9, 951–966.

Grumbine, R.E. (1994) What is ecosystem management? *Conservation Biology* 8, 27–38.

Hall, S.J. (1999) *The Effects of Fishing on Marine Ecosystems and Communities*. Blackwell Science, Oxford, UK.

Hardin, G. (1968) Tragedy of the commons. *Science* 162, 1243–1248.

Hilborn, R. and Walters, C.J. (1992) *Quantitative Fisheries Stock Assessment: Choice, Dynamics and Uncertainty*. Chapman and Hall, New York.

Hilborn, R., Parrish, J.K. and Little, K. (2005) Fishing rights or fishing wrongs? *Reviews in Fish Biology and Fisheries* 15, 191–199.

Holling, C.S. (1973) Resilience and stability of ecological systems. *Annual Review of Ecology and Systematics* 4, 1–23.

Indiana University (undated, a) *Workshop in Political Theory and Policy Analysis*. Indiana University, Bloomington, Indiana. Available at: http://www.indiana.edu/~workshop/ (accessed 3 February 2010).

Indiana University (undated, b) *Digital Library of the Commons*. Indiana University, Bloomington, Indiana. Available at: http://dlc.dlib.indiana.edu/dlc/ (accessed 3 February 2010).

Jentoft, S. (2006) Beyond fisheries management: the phronetic dimension. *Marine Policy* 30, 671–680.

Jentoft, S. (2007) Limits of governability: institutional implications for fisheries and coastal governance. *Marine Policy* 31, 360–370.

Kooiman, J., Bavinck, M., Jentoft, S. and Pullin, R. (eds) (2005) *Fish for Life: Interactive Governance for Fisheries*. Amsterdam University Press, Amsterdam.

Lebel, L., Anderies, J.M., Campbell, B., Folke, C., Hatfield-Dodds, S., Hughes, T.P. and Wilson, J. (2006) Governance and the capacity to manage resilience in regional social-ecological systems. *Ecology and Society* 11(1), art. 19. Available at: http://www.ecologyandsociety.org/vol11/iss1/art19/ (accessed 3 February 2010).

Lugten, G. and Andrew, N.L. (2008) Maximum sustainable yield of marine capture fisheries in developing archipelagic states – balancing law, science, politics and practice. *The International Journal of Marine and Coastal Law* 23, 1–37.

MacFadyen, G. and Huntington, T. (2004) Human capacity development in fisheries. FAO Fisheries Circular No. 1003, FAO, Rome, Italy. Available at: http://www.fao.org/docrep/007/y5613e/y5613e00.HTM (accessed 3 February 2010).

Mahon, R. (1997) Does fisheries science serve the needs of managers of small stocks in developing countries. *Canadian Journal of Fisheries and Aquatic Sciences* 54, 2207–2213.

Mahon, R., McConney, P. and Roy, R.N. (2008) Governing fisheries as complex adaptive systems. *Marine Policy* 32, 104–112.

Marschke, M.J. and Berkes, F. (2006) Exploring strategies that build livelihood resilience: a case from Cambodia. *Ecology and Society* 11(1), art. 42. Available at: http://www.ecologyandsociety.org/vol11/iss1/art42/ (accessed 3 February 2010).

McClanahan, T.R., Cinner, J.E., Maina, J., Graham, N.A.J., Daw, T.M., Stead, S.M., Wamukota, A., Brown, K., Ateweberham, M., Venus, V. and Polunin, N.V.C. (2008) Conservation action in a changing climate. *Conservation Letters* 1, 53–59.

McLain, R.J. and Lee, R.G. (1996) Adaptive management: promises and pitfalls. *Environmental Management* 20, 437–448.

Murawski, S.J. (2000) Definitions of overfishing from an ecosystem perspective. *ICES Journal of Marine Science* 57, 649–658.

Nadasdy, P. (2007) Adaptive co-management and the gospel of resilience. In: Armitage, D., Berkes, F. and Doubleday, N. (eds) *Adaptive Co-management: Collaboration, Learning and Multi-level Governance*. UBC Press, Vancouver, Canada, pp. 208–226.

NZAID (2007) *Sustainable Livelihoods Approach Guidelines: Analytical tools*. Available at: http://nzaidtools.nzaid.govt.nz/sustainable-livelihoods-approach (accessed 3 February 2010).

NZAID (undated) *Tool Clusters: Analytical tools*. Available at: http://nzaidtools.nzaid.govt.nz/tools/analytical-tools (accessed 3 February 2010).

Olsson, P., Folke, C. and Berkes, F. (2004) Adaptive co-management for building resilience in social-ecological systems. *Environmental Management* 34, 75–90.

Ostrom, E. (1990) *Governing the Commons: the Evolution of Institutions for Collective Action*. Cambridge University Press, Cambridge, UK.

Ostrom, E. (2007) A diagnostic approach for going beyond panaceas. *Proceedings of the National Academy of Sciences* 104, 15181–15187.

Ostrom, E., Dietz, T., Dolsak, N., Stern, P.C., Stovich, S. and Weber, E.U. (eds) (2002) *The Drama of the Commons*. National Academy Press, Washington, DC.

Ostrom, E., Janssen, M.A. and Anderies, J.M. (2007) Going beyond panaceas. *Proceedings of the National Academy of Sciences of the United States of America* 104, 15176–15178.

Pikitch, E.K., Santora, C., Babcock, E.A., Bajun, A., Bonfil, R., Conover, D.O., Dayton, P., Doukaki, P., Fluharty, D., Heneman, B., Houde, E.D., Link, J., Livingston, P.A., Mangel, M., McAllister, M.K., Pope, J. and Sainsbury, K.J. (2004) Ecosystem-based fishery management. *Science* 305, 346–347.

Plummer, R. and Armitage, D. (2007) A resilience-based framework for evaluating adaptive co-management: linking ecology, economics and society in a complex world. *Ecological Economics* 61, 62–74.

Pomeroy, R.S. and Berkes, F. (1997) Two to tango: the role of government in fisheries co-management. *Marine Policy* 21, 465–480.

Pomeroy, R.S. and Rivera-Guieb, R. (2006) *Fishery Co-management: A Practical Handbook*. CABI Publishing with IDRC, Wallingford, UK.

Resilience Alliance (2007) *Assessing and Managing Resilience in Social-ecological Systems: a Practitioner's Workbook* (vol. 1, version 1.0). Available at: http://www.resalliance.org/3871.php (accessed 3 February 2010).

Ribot, J.C. and Peluso, N.L. (2003) A theory of access. *Rural Sociology* 68, 153–181.

Ribot, J.C., Agrawal, A. and Larson, A.M. (2006) Recentralising while decentralising. How national governments re-appropriate forest resources. *World Development* 34, 1864–1886.

SAMUDRA (2008) Human rights of fishers. Triannual report of the international collective in support of fishworkers No. 51, November 2008. Chandrika Sharma for International Collective in Support of Fishworkers, Chennai, India. Available at: http://icsf.net/icsf2006/jspFiles/icsfMain/ (accessed 3 February 2010).

Sen, S. and Raakjær Nielsen, J. (1996) Fisheries co-management: a comparative analysis. *Marine Policy* 20, 405–418.

Sherman, K. and Duda, A.M. (1999) An ecosystem approach to global assessment and management of coastal waters. *Marine Ecology Progress Series* 190, 271–287.

Sinclair, M. and Valdimarsson, G. (2003) *Responsible Fisheries in the Marine Ecosystem*. FAO, Rome and CABI Publishing, Wallingford, UK.

The Fisheries Governance Network (undated) Hosted by the Centre for Maritime Research (MARE) in Amsterdam. Available at: http://www.fishgovnet.org/ (accessed 3 February 2010).

University of Bath (2002–2007) *Wellbeing in Developing Countries Research Group*. Center for Development studies. Available at: http://www.bath.ac.uk/econ-dev/wellbeing/research/research.htm (accessed 3 February 2010).

Varjopuro, R., Gray, T., Hatchard, J., Rauschmayer, F. and Wittmer, H. (2008) Introduction: interaction between environment and fisheries and the role of stakeholder participation. *Marine Policy* 32, 158–168.

Vogel, C., Moser, S.C., Kasperson, R.E. and Dabelko, G.D. (2007) Linking vulnerability, adaptation, and resilience science to practice: pathways, players, and partnerships. *Global Environmental Change* 17, 349–364.

Walker, B., Carpenter, S., Anderies, J., Abel, N., Cumming, G., Janssen, M., Lebel, L., Norberg, J., Peterson, G.D. and Pritchard, R. (2002) Resilience management in social-ecological systems: a working hypothesis for a participatory approach. *Conservation Ecology* 6(1), art. 14. Available at: http://www.consecol.org/vol6/iss1/art14/ (accessed 3 February 2010).

Walker, B., Holling, C.S., Carpenter, S.R. and Kinzig, A. (2004) Resilience, adaptability and transformability in social-ecological systems. *Ecology and Society* 9(2), art. 5. Available at: http://www.ecologyandsociety.org/vol9/iss2/art5/ (accessed 3 February 2010).

Walker, B.D., Sayer, J., Andrew, N.L. and Campbell, B.D. (2010) Resilience in practice: challenges and opportunities for natural resource management in the developing world. *Crop Science* 50, 10–19.

Walters, C.J. and Hilborn, R. (1978) Ecological optimisation and adaptive management. *Annual Review of Ecology and Systematics* 9, 157–188.

WCD (2000) *Dams and Development. Report of the World Commission on Dams*. Earthscan, London.

Welcomme, R.L. (1979) *Fisheries Ecology of Floodplain Rivers*. Longman, London.

Welcomme, R.L. (2001) *Inland Fisheries Ecology and Management*. Fishing News Books, Oxford, UK.

Wells, M.P. and McShane, T.O. (2004) Integrated protected areas management with local needs and aspirations. *AMBIO: a Journal of the Human Environment* 33, 513–519.

Wilson, D.C., Nielsen, J.R. and Degnbol, P. (2003) *The Fisheries Co-management Experience: Accomplishments, Challenges and Prospects*. Kluwer Academic Publishers, Amsterdam.

World Bank (2004) *Guidance Note on Poverty Assessments*. World Bank, Washington, DC, pp. 1–7. Available at: http://siteresources.worldbank.org/INTPOVERTY/Resources/335642-1098192957114/op1_pa_guidance.pdf (accessed 3 February 2010).

3 Diagnosis and the Management Constituency of Small-scale Fisheries

Louisa Evans and Neil L. Andrew

Introduction

Small-scale fisheries (SSF) provide essential services to more than 180 million people living in developing country contexts characterized by poverty and food insecurity (Delgado *et al.*, 2003; FAO, 2004a, 2008; Pauly, 2006; Béné *et al.*, 2007; Zeller *et al.*, 2007). Management is widely regarded to have failed to deliver fisheries that contribute fully to economic and social development (FAO, 2003a, 2004a; Cochrane and Doulman, 2005). Small-scale fisheries present particular challenges for management in that they are diverse, in terms of participants, resources and ecosystem services, gears and contexts, and complex in their connectivity to other livelihoods, other ecological systems and across multiple scales (Berkes *et al.*, 2001; Berkes, 2003). Small-scale fisheries are also vulnerable to drivers of change external to the fishery domain, but these factors have often been neglected in classical fisheries management (Andrew *et al.*, 2007). Innovations in management that include wider system dynamics and enhance the ability to better cope with and adapt to both external drivers of change and internal sources of uncertainty are needed to facilitate a broader management focus.

Small-scale fisheries are diverse and not easily categorized. The constraints and opportunities they face demand a focus on getting the basics of management right, rather than on seeking to optimize benefits, as traditionally and narrowly defined in terms of yield. Widely recognized constraints include a lack of research and management capacity in government agencies, political marginalization of fishery participants, lack of quantitative data on trends in fish stocks and vulnerability to factors outside the fishery (Allison and Ellis, 2001; Charles, 2001; Wilson *et al.*, 2003; Pomeroy and Rivera Guieb, 2006). Opportunities arise from the dynamic ecological and social environment of these fisheries, such as the capacity of the fishery system (including the people integral to it) to self-organize and adapt, and to change harvest patterns to suit fluctuating resources. Blueprint solutions or panaceas are inappropriate for the fisheries management problem and, instead, diagnostic approaches that seek to contextualize fisheries and seek appropriate entry points are proposed (Andrew *et al.*, 2007; Ostrom *et al.*, 2007; McClanahan *et al.*, 2008b; Berkes, 2009). In this chapter we explore elements of one such framework, the Participatory Diagnosis and Adaptive Management (PDAM) framework (Andrew *et al.*, 2007; Fig. 3.1), which provides a flexible basis for implementation.

There is little in the PDAM framework that is, of itself, novel (see Walters and

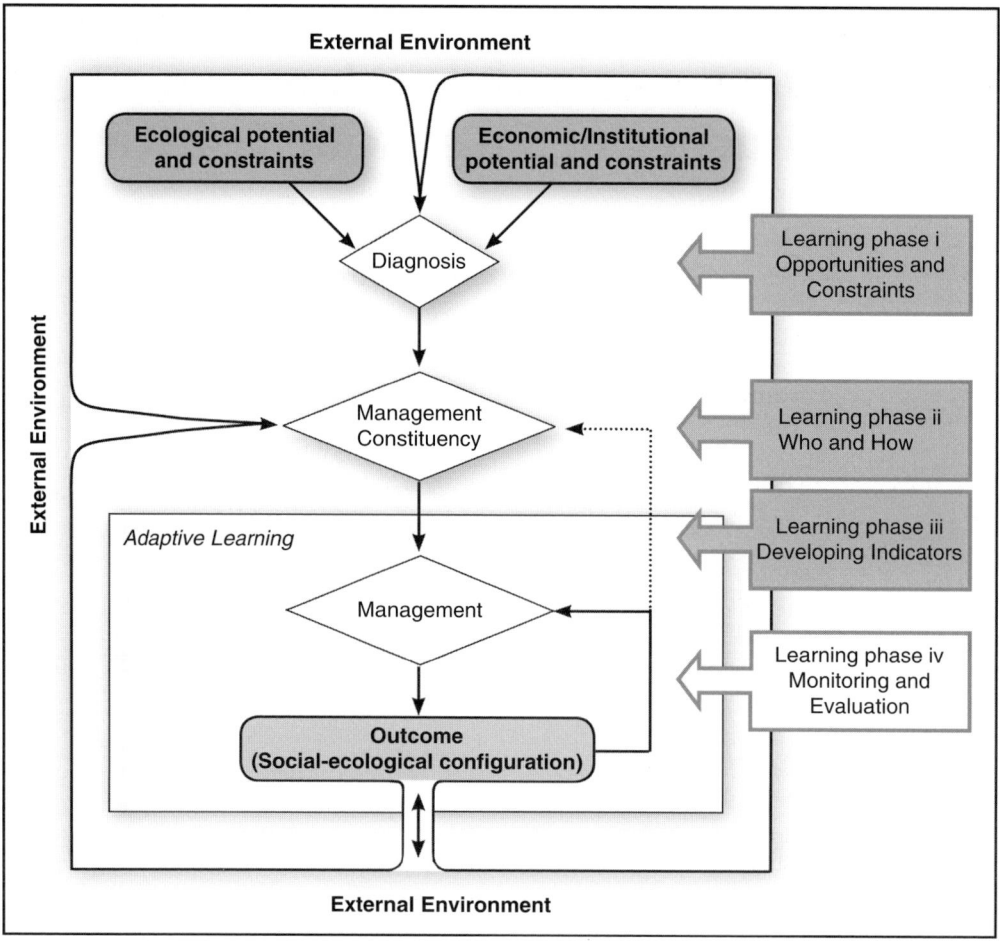

Fig. 3.1. A general framework for diagnosis and management of small-scale fisheries in the developing world. Learning phases i to iii are discussed in this chapter (adapted from Andrew et al., 2007).

Hilborn, 1978; Allison and Ellis, 2001; Berkes et al., 2001; Charles, 2001; Garaway and Arthur, 2004 among others for antecedent work). However, the framework emphasizes different aspects of the fisheries management problem. First, it emphasizes the factors arising from outside the fishery domain that may offer opportunities and act as constraints on the fishery system itself, so driving fishery change and influencing management performance and the livelihoods and well-being of fishery stakeholders. Second, the framework emphasizes the institutions that govern fisheries and, particularly, the nature and legitimacy of access rights as a central and distinct precursor to effective management. Third, it focuses on the potential of adaptive management as the primary vehicle for addressing uncertainty and sustainability.

The PDAM lays out four distinct opportunities for learning and acquiring the right information for better management, which include: (i) scoping threats and opportunities for management (diagnosis); (ii) clarifying the management constituency (fishery beneficiaries and wider stakeholders) and how the constituents wish to manage their fishery; (iii) developing management indicators to enable reflection and learning for adaptive management, phase i; and (iv) monitoring and evaluation (adaptive management, phase ii). Although these opportunities for learning are laid out as sequential and progressive steps, the overall diagnosis and management process should be iterative and flexible, not linear. How effective this is will depend largely on the capacities and will of the fishery participants involved in the process.

Here, we focus on the diagnosis and management constituency phases of the PDAM framework (learning phases i and ii). In the sections below, we briefly discuss issues and questions, and the tools that might be used to address these. We do not discuss management itself in this chapter, except briefly to link objectives with the indicators that can be used in adaptive management phases to reflect upon performance. Our discussion, particularly with respect to the diagnosis or scoping phases, draws on the comprehensive overview of integrated assessment and advice provided in Garcia *et al.* (2008).

Assessment in Small-scale Fisheries

Before we break down the diagnosis and management constituency phases of the PDAM framework, it is worth taking a step back to consider the implications of assessment processes for SSF in developing countries. Traditional fisheries science and sophisticated management measures that require extensive monitoring and enforcement are unsuitable in developing-country contexts that are generally characterized by data scarcity and relatively low capacity for intensive management (Johannes, 1998; Berkes *et al.*, 2001; McClanahan *et al.*, 2008b). Investment in diagnosis processes needs to be commensurate with the value of the fishery (Garcia *et al.*, 2008).

The value of many small-scale fisheries, particularly those defined by small stocks and/or resources of low economic value, will not support large investments in assessment and management intervention. However, evaluating the value of a fishery to determine how much investment is warranted is complicated. The values of SSF are multidimensional and do not necessarily manifest as only or primarily economic value. These values then benefit different stakeholders to different degrees, influencing who is willing and able to participate in fishery assessment. Finally, the value of a fishery may not be fully realized in contexts where fisheries management is failing and the provision of ecosystem services is undermined. At the same time, while the potential value of the fishery may be high, capacity for diagnosis and management may be limited by the costs of mitigating current threats to the fishery and of reversing existing degradation, or by the funds available for investment.

At the very least, diagnosis and management of SSF in developing countries requires adaptive and collaborative forms of management whereby multiple stakeholders are involved in learning and action. Such management can benefit from assessment processes founded on multiple sources of knowledge and relatively cheap and accessible data and information.

Rapid and/or participatory assessment techniques, or Participatory Rural Appraisal (PRAs), are research tools specifically designed to elucidate the perspectives, knowledge and values of local peoples affected by management decisions. Rapid assessments use intensive team interaction *in situ* for data collection and interpretation to gain a preliminary understanding of the context, from the stakeholders' perspective. Participatory methods aim to legitimize and increase the relevance of assessments and subsequent management. There are many approaches to PRA, but all are founded on key principles: understanding multiple perspectives, encouraging group learning processes and enabling self-mobilization and context-specific change (Pretty *et al.*, 1995; Campbell and Salagrama, 2001). Methods designed specifically for participatory research include, among many others, transect walks, seasonal calendars, mapping (e.g. ecological processes) and ranking (e.g. wealth) exercises (Pretty *et al.*, 1995; Chambers, 1997, 2002).

Participatory techniques are well suited to assessment of SSF, for example to answer questions of how fisheries stakeholders behave and why, what system change looks like and what is thought to drive it. This is important in fisheries contexts characterized by poor availability of formal data and information. Pido *et al.* (1996, 1997) provide early examples of how rapid participatory assessments can be used in SSF. Participatory techniques have typically been applied in understanding the social and socio-economic dimensions of resource use systems, but can be used to gain local knowledge of fish and ecosystem trends, as well as to support local collection of both ecological and catch data. The costs of participation in collating such information are less than those incurred in understanding these processes independently of the extraction process (McClanahan *et al.*, 2008a). Participatory techniques are suitable for clarifying and ranking issues from all the different domains of a fishery (human, ecological, institutional, external drivers), and a wide range of tools and methods are available (Table 3.1).

While PRA is an improvement over more top-down, expert-driven approaches, the constraints on true participation have been well documented (e.g. Cooke and Kothari, 2001).

Table 3.1. Web-based reference material on different analysis techniques. All websites accessed 2 February 2010.

Analysis technique	Reference material
Rapid assessment	FAO (1997)
	http://www.fao.org/docrep/W3241E/w3241e09.htm#rra%20definition
	FAO (1999a)
	http://www.fao.org/DOCREP/005/X4175E/X4175E00.HTM
	University of British Columbia (undated)
	http://www.fisheries.ubc.ca/archive/projects/rapfish.php
	World Agroforestry Center (undated)
	www.worldagroforestry.org/SEA/Publications/files/book/BK0010-04.pdf
	WorldFish Center (2005)
	http://www.nrsp.org.uk/database/documents/2372.pdf
Participatory assessment	Eldis (undated)
	http://www.eldis.org/go/topics/resource-guides/manuals-and-toolkits/participation-manuals
	FAO (1996)
	http://www.fao.org/docrep/006/W2352E/W2352E00.HTM
	FAO (1999b)
	http://www.fao.org/participation/
	http://www.fao.org/participation/tools/PRA.html
	FAO (2001)
	http://www.fao.org/docrep/007/y1127e/y1127e00.htm
	International Institute for Environment and Development (undated, a)
	http://www.planotes.org/index.html
	OneFish (undated)
	http://www.onefish.org/global/index.jsp
	Social Analysis Systems (undated)
	http://www.sas2.net/
	World Bank (2005)
	http://go.worldbank.org/L84QLQN2V0
Other socio-economic tools	FAO (2004b)
	http://www.fao.org/docrep/006/y5055e/y5055e00.htm
	Global Coral Reef Monitoring Network (undated)
	http://www.gcrmn.org/publications.aspx
	Overseas Development Institute (undated)
	http://www.odi.org.uk/Rapid/Tools/Toolkits/

How participatory approaches are used and combined with other sources of knowledge is important. A broad view that accounts for the multiple scales of ecological and social interaction needs to draw on perspectives from a number of resource-user groups to build up a balanced view of SSF dynamics, as well as on other research techniques and knowledge systems. A comprehensive diagnostic process should integrate different knowledge systems (research-based, local and state knowledge) and knowledge of different dimensions of the fishery (ecological, social and institutional) at different scales.

Diagnosis

We see the key tasks within this learning phase as: (i) defining the domain of the fishery; (ii) understanding the historical and current context of the fishery; and (iii) projecting the future direction of the fishery. This latter task merges the diagnosis into the management constituency phase and ties in closely with understanding how adaptive fisheries management will be enabled and who should be involved. Most management implementation frameworks include a phase for scoping, assessment or diagnosis, but differ in the extent to which this initial phase is reliant on data and expertise.

What is 'the fishery'?

To understand the 'system' under management, to clarify and prioritize issues both within and outside the fishery domain, and to develop a constituency and set of rights and institutions that 'fit' the fishery (*sensu* Young, 2002), we need to be clear on a fishery's *identity*. Identity refers to a system's structure, function and feedbacks (Walker *et al.*, 2006). Management should seek to enhance the ecological, human and institutional attributes that enable a fishery to absorb stress and reorganize following disturbance in order to retain its essential identity. How we define a fishery's identity is, therefore, important from both a technical and political point of view. For example, it can influence the effectiveness of management in terms of the fit between the ecosystem, the institutions developed to manage it and the indicators designed to monitor it. If a fishery's identity is poorly defined or evolves without consequent changes in institutions, power relations and indicators, then management is more likely to fail. It can also influence the legitimacy of actually trying to maintain existing system identity. Understanding who benefits from the current fishery configuration is an important consideration in fisheries management. In general, defining the identity of the fishery begins explicitly to address the 'of what', 'to what' and 'for whom' questions that are raised in the more politically aware discussions of resilience (Carpenter *et al.*, 2001; Lebel *et al.*, 2006; Nadasdy, 2007; Clay and Olson, 2008).

Defining the boundaries of the fishery is an essential, but often neglected, first step in outlining a fishery's structure. These boundaries have implications for the scope and scale of management. Historically, fisheries have been defined by many criteria, including management or administrative unit, harvested species ('the tuna fishery'), ecosystem ('the floodplain fishery'), gear type ('the trawl fishery') and by the people who harvest the fish. All of these categorizations are valid, but none is sufficient by itself to fully describe the fishery. Charles (2001, p. 3) integrates the many dimensions of a fishery to describe a 'fishery system' as a web of: 'inter-related, interacting ecological, biophysical, economic, social and cultural components'.

We use the term 'fishery' as shorthand for Charles' 'fishery system'. Implicit in this general definition is a sense of place and a continuity of connections among different components (see also Cumming and Collier, 2005). Attributes of scale are also central to any definition and may range from a small reservoir to a river basin, or even larger. To paraphrase Cumming and Collier's (2005) working specification of a complex system, the definition of a fishery should contain/describe: (i) an outline of system components; (ii) the relationships between those components; (iii) the location and spatial scale of the fishery, and the degree of constancy of this scale over time; and (iv) the temporal outlook of the fishery.

For some fisheries, the boundaries of the system are obvious and there is a clear relationship between the natural resource and the people who fish it. Others are considerably more complex: they may encompass much larger scales, the people who fish may be difficult to determine or are constantly changing, and in some, the fishery is only a small part of a diversified livelihood system, meaning that 'fisheries management' is not sufficient to improve the lives of people associated with the fishery. To give an example from each end of this spectrum:

- The sea cucumber fishery in Kia community, Isabel Province, Solomon Islands. This fishery is based on the holothurian resource and the people of Kia community who harvest it. The Kia community extends from the Bahana Fisheries Center in the north to Kesoa Primary School in the south, but excludes settlements on Barora Fa Island. This community is unified under a House of Chiefs, which is responsible for its well-being and for managing the fishery. The fishery, an important source of cash in a largely subsistence local economy, has supplied benefits to the community for decades. This fishery provides a useful example of a clearly bounded and defined system.
- The Lake Chilwa fishery, southern Malawi. This is a diverse lake fishery in which fishers use a range of gear (including traps, fine-mesh seines and long lines from dugout canoes and, increasingly, planked boats) to target a large number of species, primarily *Barbus*, *Clarius* and *Oreochromis* spp. There are as many as 5000 specialist and part-time fishers who also derive their income from farming and petty trading, and who enter and leave the fishery as catches and economic opportunities rise and fall. The fishery is co-managed by the Fisheries Department and the Lake Chilwa Fisheries Management Association, composed of 43 Beach Village Committees. Management focuses on controlling access through the issuing of licences and enforcement of fines for violating the closed season or for using inappropriate gear. Chilwa is an endorheic lake that recedes and expands with rainfall patterns in the basin (it last dried up completely in 1995). Catches fluctuate with lake level and in good years account for almost half the total fish production in Malawi. The integrity of the lake system is dependent on the extensive wetlands that surround the lake and on ecological processes in the catchment. Lake Chilwa is an example of a more complex fishery because of the presence of migrant fishers and because it is strongly influenced by external drivers of change in the watershed. The fish themselves migrate up the rivers to spawn, thus further enlarging the scale of the fishery and the scope of management.

While defining boundaries of the focal scale of management will be partially arbitrary due to the multi-scale nature of any fishery, the process is necessary for devising appropriate management responses. The need to match management institutions to the ecosystems they manage is now widely recognized (Young, 2002; Dietz *et al.*, 2003). Bohensky and Lynam (2005), in a study of multi-scale governance of water in southern Africa, suggest that management responses are most effective when awareness of an impact, and the power to act or influence responses, match the scales at which impact occurs (on whom, what and for how long). If fishery managers are not aware of key threats and opportunities because their perspective is too broad to understand local natural history or societal relations, or too narrow to appreciate global drivers of change, then management responses are likely to be less effective. In complex, multi-scale fisheries, such as those on the floodplains of the Mekong or the Ganges Rivers, clarity in the definition of 'the fishery' may lead to the conclusion that new governance institutions are required to get the congruence between fishery outcomes or impacts and management responses right.

A fishery's identity is dynamic and likely to change over time. Nevertheless, once a fishery's identity has been defined, the diagnosis process can: (i) clarify and prioritize the key threats and opportunities that characterize a fishery; and (ii) outline the desired future trajectory of the fishery, at both the focal scale and at levels below and above this, if appropriate.

The fishery context: clarifying and prioritizing threats and opportunities

Once the fishery is defined, assessment can focus on clarifying the ecological, social and political context of the fishery and the constraints and opportunities it faces. To fully understand the fishery, contextualize risk and identify opportunity, we have to consider not just its present characteristics but also its history and potential future as well (Johnson, 2004; Resilience Alliance, 2007; Walker et al., 2009). One means of visualizing fisheries as dynamic systems is to discuss and assess fishery issues along time lines in order to incorporate past influences on current management. Part of this includes paying particular attention to some of the more covert political and socio-cultural processes that have and do underlie fisheries management and influence outcomes, including property rights, vulnerability and conflict.

In developing countries, the wider context of a fishery is often very different from that experienced in developed countries where mainstream fisheries science originates (e.g. North America and Europe). Research increasingly recognizes the extent to which cultural beliefs, traditional practices and even religion can influence the behaviour of managers and other stakeholders in such contexts. For instance, some management strategies are more consistent with Islamic concepts of ownership and use of aquatic resources than others; younger fishers in parts of Africa are reluctant to challenge the authority of elder fishers and so do not put themselves forward for leadership roles; and women are often marginalized from decision-making as a result of cultural norms. A participatory diagnostic process that includes a diversity of local stakeholders is likely to have a better chance of elucidating some of these context-specific dynamics.

Understanding the historical and current context of the fishery will help clarify the threats and opportunities that characterize the fishery. Prioritizing these, in turn, provides a basis for developing management objectives and performance indices to track progress in reducing risks and capitalizing on opportunities. This is a critical step in preparing for management. In many SSF, unsustainable fishing is the greatest threat to the resource and the people dependent on it. In others, particularly inland fisheries, fishing may be relatively unimportant to ecosystems in which resources wax and wane and fishing is part of a diversified livelihood that people enter and leave as appropriate (Sarch and Allison, 2000; Jul-Larsen et al., 2003; Morand et al., 2005; Welcomme and Marmulla, 2008).

Not enough is known about most SSF to reliably assume the threats that characterize them, although typically, sedentary invertebrates, such as trochus, sea cucumbers and clams, as well as spawning aggregations of long-lived, slow-growing fish such as groupers, are more vulnerable to overfishing than many small pelagic fish or mobile invertebrates (e.g. Orensanz et al., 2005; Sadovy and Domeier, 2005; Rhodes and Tupper, 2007). On the other hand, infrastructure development, such as dams, irrigation schemes or roads, appears to be a far more important threat for many river and floodplain fisheries than typically small-scale, non-capital-intensive fisheries (Allen et al., 2005; Welcomme, 2008; Welcomme and Marmulla, 2008). For many of these river and floodplain fisheries, the external drivers can often overwhelm the capacity of fishery stakeholders and management structures to preserve the internal processes necessary for sustainability, renewal and reorganization, and adaptation (see also Cumming and Collier, 2005; Kolding et al., 2008).

It is important, therefore, to assess to what extent threats to the fishery arise from within or outside its boundaries. Clearly, investments in management institutions that focus on the dynamics of fish and fishing will be inappropriate in some instances (Jul-Larsen et al., 2003; Kolding et al. 2008). In such cases, management responses might be better focused on conserving underlying adaptive capacity than on attempting to create institutions to limit harvests. In many other fisheries, however, reducing the fishing effort and changing fishing practices would clearly be the best route to improved management outcomes. Recognizing this, fisheries science and management has broadened its focus to include a wider range of drivers and, subsequently, the need for a more integrated approach to assessment has become clear (Garcia *et al.*, 2008). This means that the range of issues to be addressed, the means of addressing them and the indicators used to track progress need to encompass more dimensions of the fishery.

Various frameworks that integrate different dimensions of these systems are available. Charles (2001) recognizes three basic dimensions to fisheries: ecological, social and economic. The sustainable livelihoods approach (SLA) more broadly analyses fishery-related livelihoods in terms of five 'capitals' (natural, physical, human, social and financial) and seeks to understand how they are influenced by processes, policies, institutions and external 'shocks' (Allison and Ellis, 2001; Pretty and Ward, 2001). An extension of the SLA, CRiSTAL, provides a multidimensional framework specifically to assess threats related to climate change and the opportunities for adaptive capacity of communities (IUCN *et al.*, undated). More recently, Garcia and colleagues (2008) categorized issues according to four domains: livelihoods and people, the natural system, institutions and governance, and external threats and opportunities (Fig. 3.2). This latter categorization emphasizes external processes to a greater degree. In the context of this discussion, these categorizations serve only to organize types of issues and act as an *aide-mémoire* to ensure that a broad sweep of issues is canvassed in fisheries diagnosis. The bullet points indicated beside each domain in Fig. 3.2 suggest a range of issues to be covered, but are only examples and will not satisfy the needs of all types of fisheries.

These frameworks help us to think about multidimensional sources of risk and opportunity. The data required to clarify the issues characterizing a particular fishery can be gleaned relatively quickly from a variety of sources using data collection methods suitable for SSF in the developing world. These include secondary data from online databases (e.g. FishBase, undated and ReefBase, undated), published research, independent assessments and grey literature (policy, legislation, management plans), participatory assessment and research, as well as more traditional but straightforward research techniques including questionnaires, key-informant

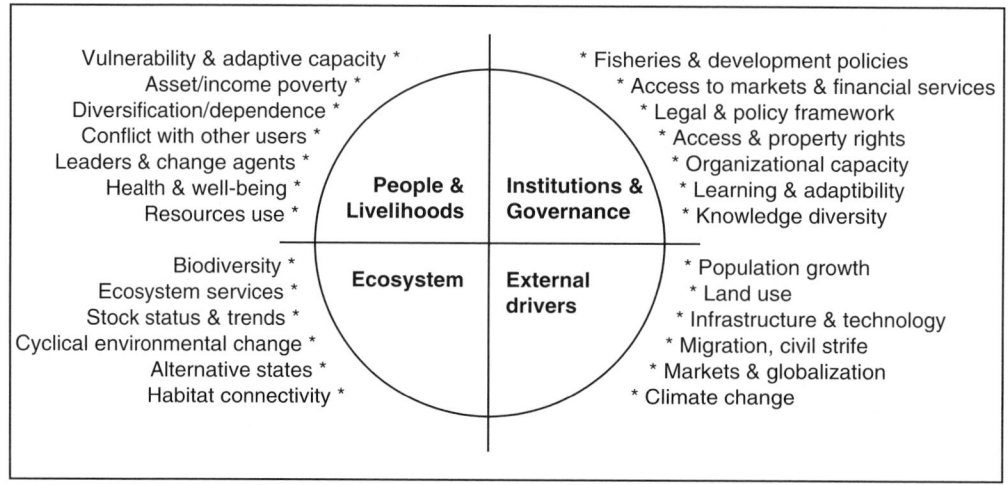

Fig. 3.2. Four domains for identifying issues in SSF, with examples of the types of issues in each category (adapted from Garcia *et al.*, 2008).

interviews, focus groups or group interviews, and so on.

Ecological risk assessments are increasingly seen as a productive cluster of techniques for guiding fishery assessments, and have been championed by influential bodies such as the Australian CSIRO, the Marine Stewardship Council and FAO. These methods range from very simple participatory techniques (e.g. Astles et al., 2006; Marine Stewardship Council, 2010) to methods that require quantitative data and simulation (e.g. Bentley and Stokes, 2009; Dichmont and Brown, 2010). Much of the innovation in this field has come from Australian researchers in response to statutory obligations to implement an ecosystem approach to fisheries (EAF) (Stobutzki et al., 2001; Fletcher et al., 2002, 2005; Fletcher, 2005; Astles et al., 2006; Hobday et al., 2007; Smith et al., 2007). These approaches provide an integrated and tested framework for identifying and prioritizing issues, which can be tailored to the context of each fishery (Fletcher et al., 2003; Cochrane et al., 2007). Central to the assessment is identifying the range of potential issues and a procedure that provides a qualitative assessment of risks and consequences. The broadening of ecological risk assessment techniques to social and economic realms, and their integration into the management process, is an active area of development that promises many advances in the assessment and management of SSF.

Once issues are identified, they need to be ranked for management purposes. Again, there are many frameworks available for developing lists of priorities, management objectives and indicators (see Garcia and Staples, 2000; Fletcher et al., 2002; Garcia and Cochrane, 2005; Reed et al., 2005; Rice and Rochet, 2005; Potts, 2006; Garcia et al., 2008; Pascoe et al., 2009 for examples and review). Most of them share common elements, which include: (i) clarifying issues through participatory methods; (ii) developing lists of candidate indicators; and (iii) prioritizing and choosing indicators. The suitability of these frameworks depends on the fishery, the degree to which the process is driven by external experts or through participatory 'bottom-up' methods, and the ecological basis of the resource (Reed et al., 2005; Fraser et al., 2006).

Ranking and prioritization of threats and opportunities from the different domains of the fishery often involve trade-offs and possible conflict. Decision-support tools are available to assist fishery stakeholders in this phase of diagnosis. These methods are broadly categorized as multi-criteria approaches (see Mardle and Pascoe, 1999 for a review in a fisheries context and De Young et al., 2008 for an FAO review). Many are highly analytical and require a lot of data, so are of limited application in the context of SSF. Others are, however, highly relevant to our context. For instance, one set of approaches uses risk ranking of broadly defined issues to identify management priorities (e.g. Fletcher et al., 2002, 2005; Fletcher, 2005; Cochrane et al., 2007). Another promising procedure is the Analytical Hierarchy Process (AHP) (Saaty, 1980) that uses a series of independent, pair-wise comparisons of indicators to rank objectives. Proponents of the AHP cite its analytical simplicity and ease of use in the field. Fishery-related applications of this tool may be found in Wattage and Mardle (2005), Himes (2007) and McClanahan et al. (2008a).

Whichever diagnostic tools are used, the objective is to identify key threats to the fishery and opportunities to sustain provision of ecosystem services and enable renewal, reorganization and adaptation in response to change. As the prioritized list of issues should guide management responses, it is important that it have legitimacy and be 'owned' by those people carrying and managing risk (see also Freebairn and King, 2003). Leadbitter and Ward (2007) suggest criteria for evaluating assessment processes. They refer to: (i) comprehensiveness: the process must evaluate a range of issues and include a diversity of stakeholders; (ii) transparency and accountability: stakeholders must agree on the legitimacy of the diagnostic process and its outcomes; and (iii) nature, use and quality of the data: there must be integration of different sources of knowledge.

Projecting the fishery's future: scenarios and objectives

Without stakeholders developing clear and agreed upon objectives for management, the fishery is unlikely to move beyond repeated failures, no matter how clearly understood the threats and potential options for the fishery are (Charles, 2001; Degnbol, 2003; FAO, 2003a; Degnbol and Jarre, 2004). Objectives for individual fisheries need to be developed within the context of broader international, regional and national policy and law. Selecting the small number of indicators required to track management progress is as much a political process as a technical one, and requires a clearly defined and empowered group of stakeholders to reach durable decisions about management objectives and the indicators used to track performance. As with other steps in the diagnostic process, inclusion of stakeholders in this phase of diagnosis is necessary to legitimize and ensure ownership of decisions related, in this case, to the future trajectory of the fishery.

There are some fundamental questions that need to be asked of stakeholders at this stage (the 'hard choices' described by Bailey and Jentoft, 1990). Is the fishery to be managed primarily for human development or for conservation? Is the fishery to be managed for its role as a social safety net or as a national income generator? Is the cost of managing the fishery commensurate with expected benefits, or should the fishery be transformed into an alternative system? Should management focus on a future ideal or should objectives be more concerned with what to avoid and defend against (Jentoft and Buanes, 2005)? How can fishery stakeholders ensure and monitor the quality of management processes in terms of legitimacy, participation, degree of precaution, cross-scale networking, accountability and so on?

One way of experimenting with different management options is through the use of scenarios or storylines. Scenarios are imagined alternative futures (optimistic and problematic), which present the likely outcomes of different development paths (see Wollenberg et al., 2000 and Evans et al., 2006 for methods). Scenario planning has been used for a number of global and regional assessments, including the Millennium Ecosystem Assessment (2005), the International Food Policy Research Institute's Fish Supply and Demand to 2020 study (Delgado et al., 2003) and the Alternative Future Scenarios for Marine Ecosystems study (Pinnegar et al., 2006). Typically, scenarios describe two to four different trajectories over timescales ranging from 5 to 10 years. Scenario planning can involve complicated quantitative modelling techniques or more simple qualitative storylines.

For small-scale fisheries management in developing country contexts, simple storylines that can be ranked by stakeholders could be developed to compare, for example: (i) management of the fishery for national or local development; (ii) various forms of governance: self-governance, co-governance or hierarchical governance and their likely outcomes (Bavinck et al., 2005); (iii) protection of traditional authority and management structures versus modernization of fishing technology and governance structures; and (iv) the outcomes if managing for different sets of drivers (internal/external). Building scenarios that map out divergent perspectives within a system and allow open and honest debate and learning are increasingly seen as more appropriate than consensus-based processes, which focus on changing the opinion of a particular person or group (Frame and Brown, 2008). Discussion of potential small-scale fishery trajectories within the context of global scenarios, such as those developed by the IPCC (2000), could add extra dimensions to the debate.

Scenario planning usually involves consultation with a panel of experts. Visioning by fisheries stakeholders affected by management decisions is, therefore, also necessary. A network of international organizations (led by IMM Ltd) developed a set of guidelines for understanding people's visions of their preferred future livelihood strategies (Cattermoul et al., 2008). The method focuses on capacities, strengths and past successes, and outlines a simple process for scaling up individual and household visions to community level (and beyond). This involves: (i) identifying the strengths and potential of individuals and households; (ii) articulating these as visions for common-interest groups (e.g. female traders, young fishers, net fishers); and (iii) developing the visions of common-interest groups into community visions. Participatory tools such as vision trees can support the scaling up of visions for fisheries and associated communities. This process is appropriate for SSF management to complement (or replace) scenario planning.

Once fishery stakeholders agree on a trajectory for a fishery, they can develop objectives and indicators. Translating normative principles and the newer, more innovative management approaches into practical management objectives can, however, be a major challenge. For example, to date there is little guidance on how to apply resilience as a concrete management aim. One considerable challenge is that the loss of resilience may only be recognized when the fishery has moved into an unsustainable (read undesirable) state (Scheffer et al., 2001; Carpenter et al., 2005). Few surrogates for resilience have been found to use as indicators to assess performance. Nevertheless, Folke et al. (2003) and, following them, Berkes and Seixas (2005) recognized four groups of factors that promote resilience:

- learning to live with change and uncertainty;
- nurturing various types of ecological, social and political diversity for increasing options and reducing risks;
- increasing the range of knowledge for learning and problem-solving; and
- creating opportunities for self-organization, including strengthening local institutions and building cross-scale linkages and problem-solving networks.

Many of the variables within these four clusters are concerned with building human and institutional capacity – through leadership, innovation, collaboration and learning – both to self-organize and reorganize. These factors do not address biophysical, technical problems but rather social-political ones, highlighting the importance of including objectives for SSF management that capture the need to develop and nurture the capacity of fisheries stakeholders, and the institutions they form, to learn and reorganize. Innovations in developing objectives (and indicators to monitor progress) that result in resilient small-scale fisheries are still needed. Some useful suggestions may, however, be drawn from fisheries-related examples, including Marschke and Berkes (2006) and McClanahan et al. (2008a).

Performance indicators

Developing performance indicators, the third phase of learning in the PDAM, occurs once the diagnosis and management constituency phases are complete. However, in some instances (for example, when the participants in the fishery are clear from the outset and the diagnostic process is participatory), it may be possible to begin the process of moving from threats and opportunities to management objectives and then to candidate indices early in the process. Performance indicators measure progress against the broader objectives of the fishery, and management revised accordingly. Sustainability indicators for fisheries management have been used for many years (e.g. Caddy, 1999; Garcia and Staples, 2000; FAO, 2002; Degnbol and Jarre, 2004). Indicators are needed because a predictive understanding of the dynamics of a fishery system is rarely possible. As Garcia and Staples (2000, p. 400) state:

> Indicators are needed to simplify, quantify and communicate information, to structure and standardize reporting, and to facilitate integration of economic and social dimensions. They assist decision-making in problem identification, objective setting, identification of gaps in research and data, monitoring, and performance assessment.

The frameworks used to develop indicators and track management performance are essentially the same as those listed above for issue identification and prioritization. For indicators to be durable and useful to adaptive management, they should reflect the experience of those affected by management decisions and system change, as well as mirror broader management approaches and overarching international, regional and national policy and law, where possible (Freebairn and King, 2003; Fraser et al., 2006). Yet for SSF in developing-country contexts, an ideal set of indicators accompanied by detailed monitoring data is unlikely to be feasible. More appropriate are indicators that can: trace simple trajectories away from reference points (improving, stable, degrading) (desirable, undesirable, critical) (see also Berkes et al., 2001 for a discussion on reference directions); (ii) reflect perceptions of system change based on local ecological knowledge and participatory science; and (iii) indicators that reflect socially relevant impacts of change (e.g. income level related to fish catch).

Management Constituency

The management constituency of a fishery refers to the *people*, *interactions* and *structures* that will influence management outcomes. How these are aligned in a particular fishery will determine how adaptive, collaborative and legitimate a management system will be. The overarching management approach taken will have implications for which organizations, institutions and people will manage the fishery. For instance, compared with conventional target resource-oriented management, the EAF (FAO, 2003a) results in a larger pool of relevant stakeholders and may require more investment in institution building. In practice, when the management constituency does not appropriately reflect and enable the management approach and the fishery-specific objectives developed under that approach, the potential of alternative management options will fall short (Christie et al., 2007). Constructing an appropriate management constituency for a particular fishery is a vital link between diagnosis and effective adaptive management. For this reason, the PDAM framework places the issue of developing an appropriate management constituency at the centre of the management process.

People and organizations within a defined fishery system can be identified as stakeholders. Those with an interest in a defined fishery, who may actually sit outside the system in the wider governance context, for example donors, environmental groups, community organizations and tourist operators, may also be important stakeholders. Interactions refer to the relationships and networks among different stakeholders within and outside the SSF. Networks may reinforce the status quo, limit adaptation or be enabling and empowering. Structures refer to institutional, political and economic guidelines that influence human behaviour, and may constitute constraints or opportunities. Institutional structures will already exist, power relations will already be in play and past events will have left their legacy. This second phase of learning is, therefore, more about adjusting and aligning existing interactions and structures to better suit the fishery than about designing new ones.

People

It is widely recognized that exclusive, centralized forms of management have failed, on the whole, to deliver sustainable and equitable fisheries (Charles, 2001; Berkes, 2003; Garcia, 2005; Varjopuro et al., 2008). Inclusion of a diverse, but appropriate, set of stakeholders is advocated by proponents of integrated, collaborative and adaptive forms of management (Wells and McShane, 2004; Brown, 2006), as well as by organizations such as the FAO. Benefits are expected to include better problem definition and ownership; a more diverse knowledge base for decision-making; greater legitimacy and, therefore, better compliance and commitment to agreed-on courses of action; and conflict resolution (Jentoft, 2000; Bryan, 2004; Rockloff and Lockie, 2006). The question is how to identify the appropriate set of stakeholders.

Small-scale fisheries management, particularly if grounded in ecosystem-based approaches, sits at the nexus between fisheries, environmental protection and development. Some would argue that because of this, wider society has a right to participate in decisions regarding SSF and their associated ecosystems as global commons (see Gray and Hatchard, 2008). Identifying appropriate stakeholders is made difficult by the fact that roles are likely to be scale-dependent and to change over time. Nevertheless, once the boundary of the fishery is spatially and administratively defined, managers can begin to consider a range of questions to guide their stakeholder identification process.

A stakeholder analysis and various data collection techniques can support this process. Questions that can facilitate stakeholder analysis include, for example: (i) which individuals and groups are involved in the SSF system at the different spatial and administrative scales included within the fishery boundary?; (ii) who is affected by management decisions and from a social justice perspective?; (iii) who can influence management outcomes and should be included from a strategic perspective, particularly in order to work towards resilience of the SSF to disturbance (including those that emerge from outside the fishery system)?; and (iv) what types of relationships do different stakeholders have?

Stakeholder analysis systematically identifies different stakeholders along a continuum of spatial scales and aims to categorize their relationships. Various typologies are available for classifying and differentiating stakeholders. For example, Brown and colleagues (2002) classify stakeholders according to attributes of influence and importance (adapted from Grimble et al., 1995), whereas Mikalsen and Jentoft (2001) prefer to use attributes of urgency, legitimacy and power (adapted from Mitchell et al., 1997). A stakeholder analysis can be conducted through document analysis, survey techniques and/or key-informant interviews or through more participatory approaches including focus groups and participatory assessment techniques (e.g. PRA). Many international development agencies use stakeholder analysis as a preliminary scoping tool for research and development (e.g. World Bank, 2003). In general, the participation of people and groups offering diverse perspectives on the SSF problem can enhance stakeholder identification and classification. Once stakeholders are identified, they need to be meaningfully included and engaged in the management process.

Interactions

Individuals and groups need to interact to make decisions about their behaviour in relation to others. Interactions such as joint decision-making opportunities and forums, relationships and networks can promote or hinder legitimate and effective management. To understand how interactions influence management, we can consider the existence of opportunities and channels for interaction (amount), the quality of interactions (type) (Mahon et al., 2005) and their outcomes (consensus/conflict, compliance/resistance). In simple terms, asking questions about whether different stakeholders are included in decision-making processes, and how and to what extent their knowledge and perspectives inform decisions, can illustrate how collaborative or exclusionary the current management approach is. Alternatively, tools such as social network analysis can be used to map both formal and informal networks between stakeholders.

Adjusting and building productive interactions and useful networks is challenging. Several factors complicate processes of participation and collaboration in management. For instance, there are a variety of functions (policy, service delivery, research and monitoring, institutional design, enforcement), stages (planning, implementation, evaluation), levels (instructive to informative) and scales (spatial and administrative) at which stakeholder participation can occur (see Sen and Raakjær Nielsen, 1996 for a detailed scale of participation). Exactly how and when participation should occur remains a highly debated topic in natural resource management and development intervention. Who is included, in what type of interaction, and to what extent, as well as the level of capacity building and institutional support they receive, will influence the quality of interactions and outcomes.

Participation is recognized as particularly problematic in developing-country contexts due to inequalities in experience, capacity and power between different groups and individuals (Jentoft, 2005; Rockloff and Lockie, 2006; Varjopuro et al., 2008). In such contexts, decision-making processes should include attention to cultural sensitivities, insecurities and language. In parallel, there needs to be capacity building of local participants so that they are better equipped to self-organize, solve problems, communicate and defend their rights when interacting in multi-scale governance processes. Outcomes of knowledge sharing, conflict resolution, consensus building and so on should not be assumed, since achieving these may require different approaches and often particular investments of time, resources and facilitation. At higher levels of decision-making, stakeholders need to focus more on issues of representation and accountability (downward) and how these can be supported (e.g. Agrawal and Ribot, 1999; Blaikie, 2006).

The stakeholder team (identified through stakeholder analysis) is best placed to negotiate the most appropriate ways of organizing in order to address potential threats to the SSF and take advantage of opportunities. Several questions can be posited:

- At which spatial and temporal scales is it useful and necessary to involve different stakeholders?
- In which management functions and stages is it useful and necessary to involve different stakeholders?
- Are the costs of participation commensurate with the value of the fishery?
- Which stakeholders need support to participate meaningfully?
- Is it appropriate and viable to weight local voices to ensure they are not diluted by more vocal, powerful and experienced stakeholders?
- Are the different types of decision-making forums achieving the expected outcomes? If not, how can they be redesigned?
- Are different knowledge systems incorporated and taken into account in management decisions? If not, how can this be facilitated?

An analytical framework (IBEFish) has recently been developed to evaluate participatory management (see Varjopuro et al., 2008 and the special issue of *Marine Policy* 32(2) (2008). Unfortunately, most of the participation-in-fisheries literature is dominated by developed-country examples. In developing-country contexts, issues of power inequality, differences in knowledge and value systems, transparency and representation are more acute.

The stakeholder team also needs to consider the networks that characterize the SSF. There is increasing interest in networks and network analysis within many strands of governance literature (Adger et al., 2005; Janssen et al., 2006; Bodin and Crona, 2009; Berkes, 2010; Cummingham et al., 2010). Networks provide critical mechanisms for dealing with the dynamism and multi-scale nature of complex systems like SSF. In the ecological subsystem, networks provide flows of energy (e.g. food web dynamics), information (environmental cues) and ecological memory (coral and fish larvae) and are used to understand and explain many ecological interactions (Cummingham et al., 2010). Social networks, on the other hand, mobilize knowledge (Crona and Bodin, 2006), innovation (Johnson, 1986), resources and capacities, and can drive consensus building, conflict resolution and power-sharing (Bodin and Crona, 2009). Networks can exist within groups, reflecting bonding social capital, or between groups, reflecting bridging social capital (Putnam, 2000). The former can foster coherence and strength to both positive and negative ends; the latter can open up new opportunities (and new threats). Networks may contribute to trust, reciprocity and the empowerment of local resource-users (Berkes and Seixas, 2005), to broader development aims (Bodin and Crona, 2009), and to the resilience of the system as a whole (Adger et al., 2005). Conversely, they may also constrain and distort governance processes; whether they contribute or constrain progress depends on the strength and structure of the networks and the motivations and capacities of the key stakeholders and agencies that are embedded within them.

Social network analysis is often used visually to map relationships, portraying people and groups as nodes, and relationships as flows between nodes. A review of empirical evidence by Bodin and Crona (2009) highlights some key characteristics of networks that are important for natural resources management. First, stakeholders with a higher number of linkages to other nodes and who play a vital role in connecting other nodes that would otherwise remain isolated have greater influence and access to resources, but also greater responsibility. Whether higher connectedness of stakeholders works in favour of SSF management or not depends on the motivations, capacities and mandates of these stakeholders. Second, increasingly connected systems have greater access to sources of novelty (e.g. technology, information, personalities) and are better able to coordinate collective action, up to a point at which high connectivity begins to restrain innovation and lead to homogenization.

More cohesive networks or those with fewer subgroups are more likely to be able to coordinate and act collectively (Bodin and Crona, 2009). Yet, the relative autonomy of subgroups facilitates diversity and independent thinking, which is important in governance of complex systems. Boundary-spanning networks and organizations are important for coordinating action and building trust where multiple subgroups exist. In mobilizing and promoting governance networks, Bodin and Crona (2009) highlight that there is no optimal network configuration, that sustainable and effective networks are best facilitated voluntarily rather than coercively and that networks and what is transferred through networks are constantly evolving. Therefore, rather than precise design, encouraging broad participation, informal and shadow networks and supporting facilitators and bridging organizations are advocated.

Structures

Management structures refer to the institutions, rights, power relations and incentives (economic and moral) that mediate human action. All of these can motivate or block collective action that may or may not facilitate more appropriate SSF management.

Institutions are the rules, norms and shared strategies that mediate human behaviour (Ostrom, 2005; Scott, 2008). They range from legislated property rights and gear enforcement rules to community expectations for appropriate fisher behaviour. They can be restrictive and regulative or enabling, for instance, in supporting collaboration, experimentation and learning. The design, functionality (Young, 2002) and enforcement characteristics (Scott, 2008) of institutions can determine their performance. Design refers to how well institutions 'fit' the ecological system they are expected to govern and the social dynamics of the system and to what extent different institutions overlap within and across spatial and administrative scales. This becomes increasingly important for larger-scale, more complex, or trans-boundary fisheries. For instance, a fishery may be defined as a small-scale, multi-species fishery where fishers from nearby communities fish near landing sites using many types of gear. Institutions that govern this fishery may, however, need to account for migrating pelagic species that come inshore and make up a substantial proportion of the catch, and for part-time or foreign fishers who seasonally access the fishing grounds.

Functionality generally refers to the effectiveness of an institution in terms of strength, compliance, resilience and adaptability (Young, 2002). Robust institutions are usually identified as a necessary feature of a successful management system (Berkes and Seixas, 2005), and yet institutions that are too rigid (requiring a significant change in fishery users' behaviour) may experience low levels of compliance. Importantly, enforcement of institutions has ethical and cognitive dimensions in addition to regulatory ones (Scott, 2008). Compliance and self-enforcement of gear regulations by local fishers can occur when fishers understand and agree with the purpose of the regulations (cognitive) and perceive them to be legitimate (ethical). More often, fishers disregard legislated rules, even when they are aware of them and understand the regulative sanctions imposed on defectors; when these rules are not considered legitimate; and when they contradict the perceived rights of local fishers.

Institutions can be created to promote a shift towards ecosystem-based management and/or co-management, if these forms are appropriate for the focal SSF. Institutions can also foster integration of systems across multiple sectors and spatial boundaries better to address issues of upstream development in floodplain fisheries or of coastal zone management. Empirical research continues to assess the types of institutions that work towards these ends in certain contexts. For example, a recent review examined the preconditions for co-management to try to understand what should happen before implementation to improve the chances that suitable institutions emerge for successful co-management (Chuenpagdee and Jentoft, 2007; see also Pomeroy et al., Chapter 7, this volume).

In fisheries where the management constituency is relatively clear and can be quickly defined, stakeholders may consider questions to guide assessment of management structures. For example:

- Do the institutions fit the ecological and social dynamics of the SSF? If not, how can they be modified?
- What types of institutions (formal/informal, rules/norms) are likely to work best in the context of this fishery?
- Is regulative enforcement adequate, fair, appropriate? What levels of (graduated) sanction are appropriate for rule-breaking?
- How can other forms of enforcement be encouraged?
- How do different stakeholders perceive the legitimacy of current institutions?
- How can management benefit from linkages between institutions at different geographical (landing site, ecosystem, watershed) and administrative (district, province, national) scales?
- How can institutional and political structures and networks foster safe experimentation and learning?
- Are there conflicts, power struggles or manipulation and domination among stakeholders? How can these be mediated?
- How can financial mechanisms support meaningful capacity building of different stakeholders, from managers to resource-users?
- How is commitment and accountability fostered?
- How can stakeholders, from resource-users to managers, be motivated and incentivized to behave in an appropriate, collective fashion?

Research tools are available to assess management structures. To understand management institutions, the Institutional Analysis and Development (IAD) framework (Ostrom, 2005) and the 'diagnosis framework for social-ecological systems' (Ostrom, 2007) provide approaches that have been used in many natural resource contexts (e.g. Yandle, 2008). The FAO (2003b) also has a guide to institutional analysis in the context of local livelihoods. More general frameworks, such as the SLA, would also be appropriate for understanding the processes, policies and institutions that influence fishers' livelihoods, as well as the power relations and incentives that drive collective action. Finally, the International Institute for Environment and Development (undated, b) has suggested a range of tools designed specifically to assess power and its influence in natural resource management. However, most of these are research frameworks. What are missing are operational frameworks, design principles and guidelines for intervention that can suggest appropriate ways of building, adjusting and aligning institutions to scales of impact (Mahon et al., 2008).

Despite stakeholders' best intentions to mobilize an appropriate management constituency for their fishery, unanticipated outcomes can be expected in complex systems such as SSF (Mosse, 1997; Cleaver, 2000; Lewins, 2007). An adaptive learning approach has the highest potential to help stakeholders cope with the uncertainty that characterizes such systems (Mahon et al., 2005). However, adaptive management itself requires deliberate planning, design and facilitation. Capacity to adapt, reorganize and learn does not automatically result from integrated, participatory or precautionary approaches. DFID UK suggests a set of strategies to enhance learning in fisheries programmes (Research4Development, undated; Fisheries Management Science Programme, undated).

Besides fostering learning about the fishery and, in particular, the integration of different knowledge systems, adaptive management also promotes the use of trial and error learning-by-doing (which can include 'safe-to-fail' experimentation and subsequent reflection) (Lebel et al., 2006). 'Safe-to-fail' experimentation is used in the adaptive management literature to suggest experimentation where even negative outcomes can be absorbed by the system without significant detrimental effect to human or ecological well-being. Stakeholders can learn by experimenting with different regulations (spatial closures, different gear restrictions in different areas), different technological investments or market chains and different alternative livelihoods. This creates options and opens up debate about their potential. Other disturbances, such as climate change, cannot be easily or 'safely' replicated, but large-scale comparisons of past events can illuminate the different responses of regions (McClanahan *et al.*, 2008a). In the longer term, funding, policy and legislative mechanisms may need to be altered to accommodate a more experimental approach to management.

Conclusions

Small-scale fishery management in the developing world is, above all, about getting the basics right while retaining the flexibility to change course if circumstances change. Recent efforts at reform are based on the assumption that fisheries management, as is, is not working, and that there is a general desire to update approaches and adopt others that account for ecosystem dynamics, function more democratically and look beyond intra-sectoral factors to account for drivers external to the system. As is clear from this chapter, many of the fundamentals of fisheries management are still relevant, including assessment, setting objectives, designing performance indicators and monitoring and evaluation (not all of which are discussed here).

If a systems perspective is taken, the social and natural dimensions of a fishery cannot be separated. The diagnostic process may be used to canvass threats and opportunities from all fishery domains (natural system, people and livelihoods, governance and institutions, and external drivers). Key to a successful systems approach is the definition of 'system' boundaries. This allows us to distinguish and choose between perspectives that focus on management within the fishery and those that emphasize building resilience against external threats, including climate variability and infrastructure development. It also allows us to pay attention to the multiple scales at which fisheries management functions, and to clarify stakeholders, networks and institutions within and beyond the focal fishery, which aids decision-making.

It is widely recognized that fishery systems and SSF in particular have non-linear and unpredictable dynamics. Given this complexity, 'management' must move beyond the control and manipulation of resources for productivity and stability, and beyond blueprint approaches. Instead, a diagnosis approach facilitates context-specific, tailor-made management of distinct small-scale fisheries. Threats and opportunities are clarified and prioritized, and a management constituency brings in stakeholders most concerned with issues specific to that fishery, so that the scales of impact are better matched to the scale at which action can be taken.

In addition to diagnosis, the PDAM advocates an adaptive approach to fisheries management. Three distinct learning phases are emphasized before the more formal management phase of monitoring and evaluation. Bringing management processes closer to the realities of a particular fishery system and including a diversity of relevant stakeholders will improve the speed and sensitivity of both environmental and social feedback mechanisms, a core property of adaptive management. Finally, the process of forming a management constituency includes the purposeful facilitation of interactions, supported by structures, which facilitate knowledge exchange, networking, learning and innovation. The inclusion of the appropriate set of stakeholders will contribute to the legitimacy and durability of management decisions and to overall social capital (trust). If nothing else, it puts some power back into the hands of local stakeholders to determine their own future. Rather than simply advocate for participatory and collaborative structures, the PDAM sets aside a specific phase for learning how best to achieve this, according to SSF requirements.

Implementation frameworks tailored to the special demands of SSF in the developing world can address some of the constraints outlined above, such as low research capacity, stakeholder marginalization and vulnerability to external threats. The challenge is to support the opportunities that characterize these systems by enhancing flexibility, diversity and sensitivity of feedback and learning. In many cases, the process will lead to incremental improvements in fisheries management when it enables more adaptive and legitimate management of fishery-specific risk (for instance, when fisheries extraction is the critical threat). In other cases, the diagnostic process will trigger transformation of fisheries management to focus more on reducing internal vulnerability and buffering against external threats that are beyond the control of fisheries managers.

Acknowledgements

We are grateful to Robin Mahon, Katherine Snyder and Tim Daw for discussion and comment. This chapter was funded by PROFISH, the World Bank's global programme on fisheries management, the Challenge Programme for Water and Food and the WorldFish Center.

References

Adger, N., Brown, K. and Tompkins, E.L. (2005) The political economy of cross-scale networks in resource co-management. *Ecology and Society* 10 (2), art. 9. Available at: http://www.ecologyandsociety.org/vol10/iss2/art9/ (accessed 2 February 2010).

Agrawal, A. and Ribot, J. (1999) Accountability in decentralisation – a framework with South Asian and West African cases. *Journal of Developing Areas* 33, 473–502.

Allen, J.D., Abell, R., Hogan, Z., Revenga, C., Taylor, B.W., Welcomme, R.L. and Winemiller, K. (2005) Overfishing inland waters. *BioScience* 55, 1041–1051.

Allison, E.H. and Ellis, F. (2001) The livelihoods approach and management of small-scale fisheries. *Marine Policy* 25, 377–388.

Andrew, N.L., Béné, C., Hall, S.J., Allison, E.H., Heck, S. and Ratner, B.D. (2007) Diagnosis and management of small-scale fisheries in developing countries. *Fish and Fisheries* 8, 227–240.

Astles, K.L., Holloway, M.G., Steffe, A., Green, M., Gannassin, C. and Gibbs, P.J. (2006) An ecological method for qualitative risk assessment and its use in the management of fisheries in New South Wales, Australia. *Fisheries Research* 82, 290–303.

Bailey, C. and Jentoft, S. (1990) Hard choices in fisheries development. *Marine Policy* 14, 333–344.

Bavinck, M., Chuenpagdee, R., Diallo, M., van der Heijden, P., Kooiman, J., Mahon, R. and Williams, S. (2005) *Interactive Fisheries Governance, a Guide to Better Practice*. Eburon, Delft, Netherlands.

Béné, C., Macfadyen, G. and Allison, E.H. (2007) Increasing the contribution of small-scale fisheries to poverty alleviation and food security. FAO Fisheries Technical paper 481, FAO, Rome, Italy. Available at: www.fao.org/docrep/009/a0965e/a0965e00.htm (accessed 3 February 2010).

Bentley, N. and Stokes, K. (2009) Contrasting paradigms for fisheries management decision making: how well do they serve data-poor fisheries? *Marine and Coastal Fisheries: Dynamics, Management, and Ecosystem Science* 1, 391–401.

Berkes, F. (2003) Alternatives to conventional management: lessons from small-scale fisheries. *Environments* 31, 1–14.

Berkes, F. (2009) Evolution of co-management: Role of knowledge generation, bridging organizations and social learning. *Journal of Environmental Management* 90, 1692–1702.

Berkes, F. (2010) Linkages and multi-level systems for matching governance and ecology: lessons from roving bandits. *Bulletin of Marine Science* 86(2), 235–250.

Berkes, F. and Seixas, C.S. (2005) Building resilience in lagoon social-ecological systems: a local-level perspective. *Ecosystems* 8, 967–974.

Berkes, F., Mahon, R., McConney, P. and Pomeroy, R.S. (2001) *Managing Small-scale Fisheries: Alternative Directions and Methods*. IDRC, Ottawa, Canada.

Blaikie, P. (2006) Is small really beautiful? Community-based natural resource management in Malawi and Botswana. *World Development* 34, 1942–1957.

Bodin, Ö. and Crona, B.I. (2009) The role of social networks in natural resource governance: what relational patterns make a difference? *Global Environmental Change* 19, 366–374.

Bohensky, E. and Lynam, T. (2005) Evaluating responses in complex adaptive systems: insights on water management from the Southern African Millennium Ecosystem Assessment (SAfMA). *Ecology and Society* 10(1), art. 11. Available at: http://www.ecologyandsociety.org/vol10/iss1/art11/ (accessed 2 February 2010).

Brown, K. (2006) Adaptive institutions for coral reef conservation. In: Coté, I.M. and Reynolds, J.D. (eds) *Coral Reef Conservation: Conservation Biology*. Cambridge University Press, Cambridge, UK, pp. 455–477.

Brown, K. Tompkins, E.L. and Adger, W.N. (2002) *Making Waves: Integrating Coastal Conservation and Development*. Earthscan Publications Ltd, London.

Bryan, T.A. (2004) Tragedy averted: the promise of collaboration. *Society and Natural Resources* 17, 881–896.

Caddy, J.F. (1999) Fisheries management in the twenty-first century: will new paradigms apply? *Reviews in Fish Biology and Fisheries* 9, 1–43.

Campbell, J. and Salagrama, V. (2001) New approaches to participation in fisheries. FAO Fisheries Technical Paper No. 956, FAO, Rome, Italy. Available at: www.fao.org/docrep/007/y1127e/y1127e00.htm (accessed 2 February 2010).

Carpenter, S., Walker, B., Anderies, J.M. and Abel, N. (2001) From metaphor to measurement: resilience of what to what? *Ecosystems* 4, 765–781.

Carpenter, S.R., Westley, F. and Turner, M.G. (2005) Surrogates for resilience of social-ecological systems. *Ecosystems* 8, 941–944.

Cattermoul, B., Townsley, P. and Campbell, J. (2008) *Sustainable Livelihoods Enhancement and Diversification (SLED): a Manual for Practitioners*. IMM Ltd, UK. Available at: www.imm.uk.com/PS/Main.aspx?projectid=22cbb689-4e26-41f2-9077-48ee89ef4de3 (accessed 2 February 2010).

Chambers, R. (1997) *Whose Reality Counts? Putting the First Last*. Intermediate Technology, London.

Chambers, R. (2002) *Participatory Workshops: a Sourcebook of 21 Sets of Ideas and Activities*. Earthscan, London.

Charles, A.T. (2001) *Sustainable Fishery Systems*. Blackwell Science Ltd, London.

Christie, P., Fulharty, D.L., White, A.T., Eisma-Osorio, L. and Jatulan, W. (2007) Assessing the feasibility of ecosystem-based fisheries management in tropical contexts. *Marine Policy* 31, 239–250.

Chuenpagdee, R. and Jentoft, S. (2007) Step zero for fisheries co-management: what precedes implementation. *Marine Policy* 31, 657-668.

Clay, P. and Olson, J. (2008) Defining fishing communities: vulnerability and the Magnuson–Stevens Fishery Conservation and Management Act. *Human Ecology Reviews* 15, 143–160.

Cleaver, F. (2000) Moral ecological rationality, institutions and the management of common property resources. *Development and Change* 31, 361–383.

Cochrane, K.L. and Doulman, D.J. (2005) The rising tide of fisheries instruments and the struggle to keep afloat. *Philosophical Transactions of the Royal Society B* 360, 77–94.

Cochrane, K.L., Augustyn, C.J., Bianchi, G., de Barros, P., Fairweather, T., Iitembu, J., Japp, D., Kanandjembo, A., Kilongo, K., Moroff, N., Nel, D., Roux, J.P., Shannon, L.J., van Zyl, B. and Vaz Velho, F. (2007) Ecosystem approaches for fisheries management in the Benguela Current Large Marine Ecosystem. FAO Fisheries Circular No. 1026, FAO, Rome, Italy.

Cooke, B. and Kothari, U. (2001) *Participation: the New Tyranny?* Zed Books Limited, New York.

Crona, B. and Bodin, Ö. (2006) What you know is who you know? Communication patterns among resource users as a prerequisite for co-management. *Ecology and Society* 11(2), art. 7. Available at: http://www.ecologyandsociety.org/vol11/iss2/art7/ (accessed 2 February 2010).

Cumming, G.S. and Collier, J. (2005) Change and identity in complex systems. *Ecology and Society* 10(1), art. 29. Available at: http://www.ecologyandsociety.org/vol10/iss1/art29/ (accessed 2 February 2010).

Cummingham, G.S., Bodin, Ö., Ernston, H. and Elmqvist, T., (2010) Network analysis in conservation biogeography: challenges and opportunities. *Diversity and Distributions* 16, 414–425.

De Young, C., Charles, A. and Hjort, A. (2008) Human dimensions of the ecosystem approach to fisheries: an overview of context, concepts, tools and methods. FAO Fisheries Technical Paper No. 489, FAO, Rome, Italy.

Degnbol, P. (2003) Science and the user perspective: the scale gap and the need for co-management. In: Wilson, D.C., Nielsen, J.R. and Degnbol, P. (eds) *The Fisheries Co-management Experience: Accomplishments, Challenges and Prospects*. Kluwer, Amsterdam, pp. 31–49.

Degnbol, P. and Jarre, A. (2004) Review of indicators in fisheries management: a development perspective. *African Journal of Marine Science* 26, 303–326.

Delgado, C.L., Nikola, W., Rosegrant, M.W., Siet, M. and Mahfuzuddin, A. (2003) *Fish to 2020: Supply and Demand in Changing Global Markets*. The International Food Policy Research Institute (IFPRI), Washington, DC and the WorldFish Center, Penang, Malaysia. Available at: http://www.ifpri.org/pubs/books/fish2020/oc44.pdf (accessed 2 February 2010).

Dichmont, C.M. and Brown, I.W. (2010) A case study in successful management of a data-poor fishery using simple decision rules: the Queensland spanner crab fishery. *Marine and Coastal Fisheries: Dynamics, Management and Ecosystem Science* 1, 1–13.

Dietz, T., Ostrom, E. and Stern, P.C. (2003) The struggle to govern the commons. *Science* 302, 1907–1912.

Eldis (undated) *Participation Manuals*. Available at: http://www.eldis.org/go/topics/resource-guides/manuals-and-toolkits/participation-manuals (accessed 2 February 2010).

Evans, K., Velarde, S.J., Prieto, R., Rao, S.N., Sertzen, S., Dávila, K., Cronkleton P. and de Jong, W. (2006) *Field Guide to the Future: Four Ways for Communities to Think Ahead*. Center for International Forestry Research (CIFOR), ASB, World Agroforestry Centre, Nairobi. Available at: http://www.asb.cgiar.org/ma/scenarios (accessed 2 February 2010).

FAO (1996) Rapid rural appraisal, participatory rural appraisal and aquaculture. FAO Fisheries Technical Paper No. 358, FAO, Rome, Italy. Available at: http://www.fao.org/docrep/006/W2352E/W2352E00.HTM (accessed on 2 February 2010).

FAO (1997) Marketing research and information systems. Marketing and agribusiness texts 4, FAO, Rome, Italy. Available at: http://www.fao.org/docrep/W3241E/W3241E00.htm (accessed 2 February 2010).

FAO (1999a) RAPFISH, a rapid appraisal technique for fisheries and its application to the Code of Conduct for Responsible Fisheries. FAO Fisheries Circular No. 947, FAO, Rome, Italy. Available at: http://www.fao.org/DOCREP/005/X4175E/X4175E00.HTM (accessed 2 February 2010).

FAO (1999b) Participation website. Established by the Informal Working Group on Participatory Approaches and Methods to Support Sustainable Livelihoods and Food Security (IWG-PA). Available at: http://www.fao.org/participation/ and http://www.fao.org/participation/tools/PRA.html (accessed 2 February 2010).

FAO (2001) New approaches to participation in fisheries research. FAO Fisheries Circular No. 965, FAO, Rome, Italy. Available at: http://www.fao.org/docrep/007/y1127e/y1127e00.htm (accessed 2 February 2010).

FAO (2002) Indicators for sustainable development of fisheries. Paper presented at the *2nd World Fisheries Congress, Workshop on Fisheries Sustainability Indicators*, Brisbane, Australia, August 1996. Available at: www.fao.org/docrep/W4745E/w4745e0f.htm (accessed 2 February 2010).

FAO (2003a) Fisheries management: the ecosystem approach to fisheries. FAO Technical Guidelines for Responsible Fisheries 4, Suppl. 2, FAO, Rome, Italy.

FAO (2003b) Local institutions and livelihoods: guidelines for analysis, FAO, Rome, Italy. Available at: http://www.fao.org/docrep/006/y5084E/y5084e00.HTM (accessed 2 February 2010).

FAO (2004a) A research agenda for small-scale fisheries. FAO Regional Office for Asia and the Pacific, Bangkok, Thailand. RAP Publication No. 2004/21 and FIPL/C 10009, FAO, Rome, Italy.

FAO (2004b) Guidelines on the collection of demographic and socio-economic information on fishing communities for use in coastal and aquatic resources management. FAO Fisheries technical paper No. 439, FAO, Rome, Italy. Available at: http://www.fao.org/docrep/006/y5055e/y5055e00.htm (accessed 2 February 2010).

FAO (2008) *The State of World Fisheries and Aquaculture*. FAO, Rome, Italy. Available at: ftp.fao.org/docrep/fao/011/i0250e/i0250e.pdf (accessed 2 February 2010).

FishBase (undated) *A Global Information System on Fishes*. Available at: www.fishbase.org (accessed 2 February 2010).

Fisheries Management Science Programme (undated) Managed by MRAG and funded by DFID. Available at: http://www.fmsp.org.uk/Home.htm (accessed 2 February 2010).

Fletcher, W.J. (2005) Application of qualitative risk assessment methodology to prioritise issues for fisheries management. *ICES Journal of Marine Science* 62, 1576–1587.

Fletcher, W.J., Chesson, J., Fisher, M., Sainsbury, K.J., Hundloe, T., Smith, A.D.M. and Whitworth, B. (2002) *National ESD Reporting Framework for Australian Fisheries: the 'How To' Guide for Wild Capture Fisheries*. FRDC Project 2000/145, FRDC, Canberra, Australia. Available at: http://www.fisheries-esd.com/c/pubs/index.cfm (accessed 2 February 2010).

Fletcher, W.J., Chesson, J., Sainsbury, K.J., Hundloe, T. and Fisher, M. (2003) *National ESD Reporting Framework for Australian Fisheries: the ESD Assessment Manual for Wild Capture Fisheries*. FRDC Project 2002/086,

FRDC, Canberra, Australia. Available at: http://www.fisheries-esd.com/a/pdf/AssessmentManualV1_0.pdf (accessed 2 February 2010).

Fletcher, W.J., Chesson, J., Sainsbury, K.J., Fisher, M. and Hundloe, T. (2005) A flexible and practical framework for reporting on ecologically sustainable development for wild capture fisheries. *Fisheries Research* 71, 175–183.

Folke, C., Colding, J. and Berkes, F. (2003) Synthesis: building resilience and adaptive capacity in social-ecological systems. In: Berkes, F., Colding, J. and Folke, C. (eds) *Navigating Social-ecological Systems: Building Resilience for Complexity and Change.* Cambridge University Press, Cambridge, UK, pp. 352–387.

Frame, B. and Brown, J. (2008) Developing post-normal technologies for sustainability. *Ecological Economics* 65, 225–241.

Fraser, E.D.G., Dougill, A.J., Mabee, W.E., Reed, M.S. and McAlpine, P. (2006) Bottom up and top down: analysis of participatory processes for sustainability indicator identification as a pathway to community empowerment and sustainable environmental management. *Journal of Environmental Management* 78, 114–127.

Freebairn, D.M. and King, C.A. (2003) Reflections on collectively working toward sustainability: indicators for indicators. *Australian Journal of Experimental Agriculture* 43, 223–238.

Garaway, C.J. and Arthur, R.I. (2004) *Adaptive Learning: a Practical Framework for the Implementation of Adaptive Co-management – Lessons from Selected Experiences in South and Southeast Asia.* Marine Resources Assessment Group, London. Available at: www.worldlakes.org/uploads/Adaptive_Learning_Guidelines.PDF (accessed 2 February 2010).

Garcia, S.M. (2005) Fishery science and decision-making: dire straits to sustainability. *Bulletin of Marine Science* 76, 171–196.

Garcia, S.M. and Cochrane, K.L. (2005) Ecosystem approach to fisheries: a review of implementation guidelines. *ICES Journal of Marine Science* 62, 311–318.

Garcia, S.M. and Staples, D. (2000) Sustainability reference systems and indicators for responsible marine capture fisheries: a review of concepts and elements for a set of guidelines. *Marine and Freshwater Research* 51, 385–426.

Garcia, S.M., Allison, E.H., Andrew, N.L., Béné, C., Bianchi, G., de Graaf, G., Kalikoski, G.J., Mahon, R. and Orensanz, J.M. (2008) Towards integrated assessment and advice in small-scale fisheries: principles and processes. FAO Fisheries and Aquaculture Technical Paper No. 515, FAO, Rome, Italy.

Global Coral Reef Monitoring Network (undated) *Monitoring Guidelines and Protocols: Various.* Available at: http://www.gcrmn.org/publications.aspx (accessed 2 February 2010).

Gray, T. and Hatchard, J. (2008) A complicated relationship: stakeholder participation and the ecosystem-based approach to fisheries management. *Marine Policy* 32, 158–168.

Grimble, R., Chan, M.K., Aglionby, J. and Quan, J. (1995) *Trees and Trade-offs: a Stakeholder Approach to Natural Resource Management.* IIED Sustainable Agriculture Gatekeeper Series No. SA52, London International Institute for Environment and Development, London.

Himes, A.H. (2007) Performance indicator importance in MPA management using a multi-criteria approach. *Coastal Management Journal* 35, 601–618.

Hobday, A.J., Smith, A., Webb, H., Daley, R., Wayte, S., Bulman, C., Dowdney, J., Williams, A., Sporcic, M., Dambacher, J., Fuller, M. and Walker, T. (2007) Ecological risk assessment for the effects of fishing: methodology. Report R04/1072 for the Australian Fisheries Management Authority, Canberra, Australia.

International Institute for Environment and Development (undated) Participatory learning and action series. Available at: http://www.planotes.org/index.html (accessed 2 February 2010).

International Institute for Environment and Development (undated) Power tools for policy influence in natural resource management. Available at: http://www.policy-powertools.org/ (accessed 2 February 2010).

IPCC (2000) Special Report on Emissions Scenarios. Nakicenovic, N. and Swart, R. (eds) Cambridge University Press, Cambridge, UK. Available at: http://www.ipcc.ch/ipccreports/sres/emission/index.php?idp=0 (accessed 2 February 2010).

IUCN, IISD, SEI-US and Intercooperation (undated) *Community-based Risk Screening Tool – Adaptation and Livelihoods (CRiSTAL).* Available at: http://www.cristaltool.org/content/about.aspx (accessed 2 February 2010).

Janssen, M.A., Bodin, O., Anderies, J.M., Elmqvist, T., Ernstson, H., McAllister, R.R.J., Olsson, P. and Ryan, P. (2006) A network perspective on the resilience of social-ecological systems. *Ecology and Society* 11(1), art. 15. Available at: http://www.ecologyandsociety.org/vol11/iss1/art15/ (accessed 26 June 2010).

Jentoft, S. (2000) Legitimacy and disappointment in fisheries management. *Marine Policy* 24, 141–148.

Jentoft, S. (2005) Fisheries co-management as empowerment. *Marine Policy* 29, 1–7.
Jentoft, S. and Buanes, A. (2005) Challenges and myths in Norwegian coastal zone management. *Coastal Management* 33, 151–165.
Johannes, R.E. (1998) The case for data-less marine resource management: examples from tropical nearshore fin-fisheries. *Trends in Ecology and Evolution* 13, 243–246.
Johnson, C. (2004) Uncommon ground: the 'poverty of history' in common property discourse. *Development and Change* 35, 407–433.
Johnson, J.C. (1986) Social networks and innovation adoption: a look at Burth's use of structural equivalence. *Social Networks* 8, 343–364.
Jul-Larsen, E., Kolding, J., Overå, R., Raakjær Nielsen, J. and van Zwieten, P.A.M. (2003) Management, co-management or no management? Major dilemmas in Southern African freshwater fisheries. 1. Synthesis report. FAO Fisheries Technical Paper No. 426/1, FAO, Rome, Italy.
Kolding, J., van Sweiten, P., Mkumbo, O., Silsbe, G. and Hecky, R. (2008) Are Lake Victoria fisheries threatened by exploitation or eutrophication? Towards an ecosystem-based approach to management. In: Bianchi, G. and Skjoldal, H.R. (eds) *The Ecosystem Approach to Fisheries*. FAO, Rome, Italy and CABI, Wallingford, UK, pp. 309–354.
Leadbitter, D. and Ward, T.J. (2007) An evaluation of systems for the integrated assessment of capture fisheries. *Marine Policy* 31, 458–469.
Lebel, L., Anderies, J.M., Campbell, B., Folke, C., Hatfield-Dodds, S., Hughes, T.P. and Wilson, J. (2006) Governance and the capacity to manage resilience in regional social-ecological systems. *Ecology and Society* 11(1), art. 19. Available at: http://www.ecologyandsociety.org/vol11/iss1/art19/ (accessed 2 February 2010).
Lewins, R. (2007) Acknowledging the informal institutional setting of natural resource management: consequences for policy-makers and practitioners. *Progress in Development Studies* 7, 201–215.
Mahon, R., Bavinck, M. and Roy, R. (2005) Fisheries governance in action. In: Kooiman, J., Bavinck, M., Jentoft, S. and Pullin, R. (eds). *Fish for Life: Interactive Governance for Fisheries*. MARE Publication Series No. 3, University of Amsterdam Press, Amsterdam, pp. 353–378.
Mahon, R., McConney, P. and Roy, R.N. (2008) Governing fisheries as complex adaptive systems. *Marine Policy* 32, 104–112.
Mardle, S. and Pascoe, S. (1999) A review of applications of multiple criteria decision making techniques to fisheries. *Marine Resource Economics* 14, 41–63.
Marine Stewardship Council (2010) *Fisheries Assessment Methodology and Guidance to Certification Bodies*. Available at: http://www.msc.org/documents/scheme-documents/methodologies/Fisheries_Assessment_Methodology.pdf/view (accessed 25 June 2010).
Marschke, M.J. and Berkes, F. (2006) Exploring strategies that build livelihood resilience: a case from Cambodia. *Ecology and Society* 11(1), art. 42. Available at: http://www.ecologyandsociety.org/vol11/iss1/art42/ (accessed 2 February 2010).
McClanahan, T.R., Cinner, J.E., Maina, J., Graham, N.A.J., Daw, T.M., Stead, S.M., Wamukota, A., Brown, K., Ateweberham, M., Venus, V. and Polunin, N.V.C. (2008a) Conservation action in a changing climate. *Conservation Letters* 1, 53–59.
McClanahan, T.R., Castilla, J.C., White, A.T. and Defeo, O. (2008b) Healing small-scale fisheries by facilitating complex socio-ecological systems. *Reviews in Fish Biology and Fisheries* 19, 33–47.
Mikalsen, K.H. and Jentoft, S. (2001) From user-groups to stakeholders? The public interest in fisheries management. *Marine Policy* 25, 281–292.
Millennium Ecosystem Assessment (2005) *Ecosystems and Human Wellbeing: Scenarios*. Available at: http://www.millenniumassessment.org/en/Scenarios.aspx (accessed 2 February 2010).
Mitchell, R.K, Agle, B.R. and Wood, D.J. (1997) Toward a theory of stakeholder identification and salience: defining the principle of who and what really counts. *The Academy of Management Review* 22, 853–886.
Morand, P., Sy, O.H. and Breuil, C. (2005) Fishing livelihoods: successful diversification, or sinking into poverty? In: Wisner, B., Toulmin, C. and Chitiga, R. (eds) *Towards a New Map of Africa*. Earthscan, London, pp. 71–92.
Mosse, D. (1997) The symbolic making of a common property resource: history, ecology and locality in a tank-irrigated landscape in South India. *Development and Change* 28, 467–504.
Nadasdy, P. (2007) Adaptive co-management and the gospel of resilience. In: Armitage, D., Berkes, F. and Doubleday, N. (eds) *Adaptive Co-management: Collaboration, Learning and Multi-level Governance*. UBC Press, Vancouver, Canada, pp. 208–226.

OneFish (undated) *A Fishery Projects Portal and Participatory Resource Gateway*. Available at: http://www.onefish.org/global/index.jsp (accessed 2 February 2010).

Orensanz, J.M., Parma, A.M. and Jerez, G. (2005) What are the key elements for the sustainability of 'S-fisheries': insights from South Africa. *Bulletin of Marine Science* 76, 527–556.

Ostrom, E. (2005) *Understanding Institutional Diversity*. Princeton University Press, Princeton, New Jersey.

Ostrom, E. (2007) A diagnostic approach for going beyond panaceas. *Proceedings of the National Academy of Sciences* 104, 15181–15187.

Ostrom, E., Janssen, M.A. and Anderies, J.M. (2007) Going beyond panaceas. *Proceedings of the National Academy of Sciences* 104, 15176–15178.

Overseas Development Institute (undated) *Research and Policy in Development: Toolkits*. Available at: http://www.odi.org.uk/Rapid/Tools/Toolkits/ (accessed 2 February 2010).

Pascoe, E., Proctor, W., Wilcox, C., Innes, J., Rochester, W. and Dowling, N. (2009) Stakeholder objective preferences in Australian Commonwealth managed fisheries. *Marine Policy* 33, 750–758.

Pauly, D. (2006) Major trends in small-scale marine fisheries, with emphasis on developing countries and some implications for the social sciences. *Maritime Studies (MAST)* 4, 7–22.

Pido, M.D., Pomeroy, R.S., Carlos, M.B. and Garces, L.R. (1996) *A Handbook for Rapid Appraisal of Fisheries Management Systems*. ICLARM Education Series 16, ICLARM, Manila, Philippines.

Pido, M.D., Pomeroy, R.S., Garces, L.R. and Carlos, M.B. (1997) A rapid appraisal approach to evaluation of community-level fisheries management systems: framework and field application at selected coastal fishing villages in the Philippines and Indonesia. *Coastal Management* 25, 183–204.

Pinnegar, J.K., Viner, D., Hadley, D., Dye, S., Harris, M., Berkout, F. and Simpson, M. (2006) Alternative future scenarios for marine ecosystems: technical report. Available at: http://www.cefas.co.uk/Publications/techrep/afmec_techrep.pdf (accessed 2 February 2010).

Pretty, J. and Ward, H. (2001) Social capital and the environment. *World Development* 29, 209–227.

Pretty, J.N., Guijt, I., Thompson, J. and Scoones, I. (1995) *Participatory Learning and Action – a Trainers' Guide*. IIED, London.

Pomeroy, R.S. and Rivera-Guieb, R. (2006) *Fishery Co-management: a Practical Handbook*. CABI Publishing with IDRC, Wallingford, UK.

Potts, T. (2006) A framework for the analysis of sustainability indicator systems in fisheries. *Ocean and Coastal Management* 49, 259–280.

Putnam, R. (2000) *Bowling Alone: the Collapse and Revival of American Community*. Simon and Schuster, New York.

Research4 Development (undated) *A Portal to DFID-funded Research*. Available at: http://www.research4development.info/ (accessed 2 February 2010).

Reed, M., Fraser, E.D.G., Morse, S. and Dougill, A.J. (2005) Integrating methods for developing sustainability indicators to facilitate learning and action. *Ecology and Society* 10(1), res. 3. Available at: http://www.ecologyandsociety.org/vol10/iss1/resp3/ (accessed 2 February 2010).

ReefBase (undated) *A Global Information System for Coral Reefs*. Available at: www.reefbase.org (accessed 2 February 2010).

Resilience Alliance (2007) *Assessing and Managing Resilience in Social-ecological Systems: a Practitioner's Workbook* (vol. 1, version 1.0.). Available at: http://www.resalliance.org/3871.php (accessed 3 February 2010).

Rhodes, K.L. and Tupper, M. (2007) The vulnerability of reproductively active squaretail coral grouper (*Plectropomus areolatus*) to fishing. *Fishery Bulletin* 106, 194–203.

Rice, J.C. and Rochet, M.J. (2005) A framework for selecting a suite of indicators for fisheries management. *ICES Journal of Marine Science* 62, 516–527.

Rockloff, S.F. and Lockie, S. (2006) Democratisation of coastal zone decision-making for indigenous Australians: insights from stakeholder analysis. *Coastal Management* 34, 251–266.

Saaty, T.L. (1980) *The Analytic Hierarchy Process*. McGraw-Hill, New York.

Sadovy, Y. and Domeier, M. (2005) Are aggregation-fisheries sustainable? Reef fish fisheries as a case study. *Coral Reefs* 24, 253–262.

Sarch, M.T. and Allison, E.A. (2000) *Fluctuating Fisheries in Africa's Inland Waters: Well-adapted Livelihoods, Maladapted Management*. Available at: www.oregonstate.edu/dept/IIFET/2000/papers/sarch.pdf (accessed 2 February 2010).

Scheffer, M., Carpenter, S., Foley, J.A., Folke, C. and Walker, B. (2001) Catastrophic shifts in ecosystems. *Nature* 413, 591–596.

Scott, R.W. (2008) *Institutions and Organizations: Ideas and Interests*, 3rd edn. Sage Publications Inc., Los Angeles, California.

Sen, S. and Raakjær Nielsen, J. (1996) Fisheries co-management: a comparative analysis. *Marine Policy* 20, 405–418.

Smith, A.D.M., Fulton, E.J., Hobday, A.J., Smith, D.C. and Shoulder, P. (2007) Scientific tools to support the practical implementation of ecosystem-based fisheries management. *ICES Journal of Marine Science* 64, 633–639.

Social Analysis Systems (undated) Available at: http://www.sas2.net/ (accessed 2 February 2010).

Stobutzki, I., Miller, M. and Brewer, D. (2001) Sustainability of fishery bycatch: a process for assessing highly diverse and numerous bycatch. *Environmental Conservation* 28, 167–181.

University of British Columbia (undated). *The Rapid Appraisal of the Status of Fisheries*. Available at: http://www.fisheries.ubc.ca/archive/projects/rapfish.php (accessed 2 February 2010).

Varjopuro, R., Gray, T., Hatchard, J., Rauschmayer, F. and Wittmer, H. (2008) Introduction: interaction between environment and fisheries and the role of stakeholder participation. *Marine Policy* 32, 158–168.

Walker, B., Gunderson, L., Kinzig, A., Folke, C., Carpenter, S. and Schultz, L. (2006) A handful of heuristics and some propositions for understanding resilience in social-ecological systems. *Ecology and Society* 11(1), art. 13. Available at: http://www.ecologyandsociety.org/vol11/iss1/art13/ (accessed 2 February 2010).

Walker, B.H.N., Abel, N., Anderies, J.M. and Ryan, P. (2009) Resilience, adaptability and transformability in the Goulburn-Broken Catchment, Australia. *Ecology and Society* 14(1), art. 12. Available at: www.ecologyandsociety.org/vol14/iss1/art12/ (accessed 2 February 2010).

Walters, C.J. and Hilborn, R. (1978) Ecological optimisation and adaptive management. *Annual Review of Ecology and Systematics* 9, 157–188.

Wattage, P. and Mardle, S. (2005) Stakeholder preferences towards conservation versus development for a wetland in Sri Lanka. *Journal of Environmental Management* 77, 122–132.

Welcomme, R.L. (2008) Conservation of fish and fisheries in large river systems. In: Nielsen, J.L., Dodson, J.J., Friedland, K., Hamon, T.R., Musick, J., Verspoor, E. (eds) *Reconciling Fisheries with Conservation; Proceedings of the Fourth World Fisheries Congress*, Vancouver, BC, Canada, 2–6 May 2004. American Fisheries Society, Bethesda, Maryland, pp. 1703–1715.

Welcomme, R.L. and Marmulla, G. (2008) Hydropower, flood control and water abstraction: implications for fisheries and fisheries: preface. *Hydrobiologia* 609, 1–7.

Wells, M.P. and McShane, T.O. (2004) Integrated protected areas management with local needs and aspirations. *Ambio* 33, 513–519.

Wilson, D.C., Nielson, J.R. and Degnbol, P. (2003) *The Fisheries Co-management Experience: Accomplishments, Challenges and Prospects*. Kluwer Academic Publishers, Amsterdam.

Wollenberg, E., Edmunds, D. and Buck, L. (2000) *Anticipating Change: Scenarios as a Tool for Adaptive Forest Management. A Guide*. Center for International Forestry Research, Bogor, Indonesia. Available at: http://www.cifor.cgiar.org/acm/methods/fs.html#top (accessed 2 February 2010).

World Agroforestry Center (undated) *Characterisation, Diagnosis and Design: Training Exercise Book*. Available at: www.worldagroforestry.org/SEA/Publications/files/book/BK0010-04.pdf (accessed 2 February 2010).

World Bank (2003) *A Users' Guide to Poverty and Social Impact Analysis*. World Bank, Washington, DC. Available at: http://go.worldbank.org/TDD9JAXK60 (accessed 2 February 2010).

World Bank (2005) *Tips for Institutional, Political and Social Analysis. A Sourcebook for Development Practitioners*. Available at: http://go.worldbank.org/L84QLQN2V0 (accessed 2 February 2010).

WorldFish Center (2005) Enabling better management of fisheries conflicts. Final Technical Report for project R8294, WorldFish Center, Dhaka, Bangladesh. Available at: http://www.nrsp.org.uk/database/documents/2372.pdf (accessed 2 February 2010).

Yandle, T. (2008) The promise and perils of building a co-management regime: an institutional assessment of New Zealand fisheries management between 1999 and 2005. *Marine Policy* 32, 132–141.

Young, R.O. (2002) *The Institutional Dimensions of Environmental Change: Fit, Interplay and Scale*. MIT Press, London.

Zeller, D., Booth, S. and Pauly, D. (2007) Fisheries contributions to the gross domestic product: underestimating small-scale fisheries in the Pacific. *Marine Resource Economics* 21, 355–374.

4 Human Rights and Fishery Rights in Small-scale Fisheries Management

Anthony Charles

Introduction

When the word 'rights' is used in fisheries discussions, two very different ideas come to mind, depending on one's perspective. First, from the perspective of the people and communities engaged in fishing or otherwise dependent on the fishery, there are human, social and economic rights that can be reinforced, or negatively impacted, by actions taken in the fishery. Second, from the perspective of fishery management, there are 'fishery rights' that define who can go fishing and who can be involved in managing the fishery. This form of rights arises in what is referred to as rights-based fishery management (Neher *et al.*, 1989), focusing on the rights (together with the responsibilities) held by individuals, communities, companies and/or governments specifically in relation to fishery management.

These two categories of rights have typically been treated separately, but there is now an emerging focus on linking human rights and fishery rights (e.g. Civil Society Preparatory Workshop, 2008). This chapter seeks to expand upon and reinforce the links between them, in the context of small-scale fisheries and their management. The following section introduces aspects of human rights and fishery rights, and summarizes current thinking on the practical links between these in the context of small-scale fisherfolk and fishing communities. This is followed by a section with more detailed discussions of fishery rights, including access rights, effort rights, harvest rights and management rights, as well as the particular importance of community fishery rights. A range of implementation issues are then examined; these arise when existing rights are being recognized or when a new rights system is being put in place, and cover questions of who can receive rights, how long the rights last, whether they can be transferred and how to choose among specific forms of rights. Finally, a set of conclusions is presented, along with potential directions forward in reinforcing or creating rights frameworks that provide better integration and balance than in many past approaches.

Human Rights, Fishery Rights and Their Interaction

When most people think of rights, it may well be *human rights* that come to mind. The United Nations has defined the overall nature of human rights, as well as accompanying social and economic rights (United Nations, 1948). Recently, efforts have been under way to examine the specific manifestations of such rights in fisheries and fishing communities – with attention to this highlighted particularly by international bodies of fishers and fishworkers, and with coverage in legal and policy debates. Some efforts in this direction will be discussed below.

At the same time, within fisheries management circles, attention is focused on so-called fishery rights – the rights of specific individual fishers, fishing communities or companies to have access to the fishery, to be able to exert a certain amount of fishing effort or to catch a certain amount of fish and/or to be involved in managing the fishery (e.g. Shotton, 2000). These rights are typically discussed in the context of achieving more effective management, both by specifying who is involved in the fishery (and how much) and by bringing fishers and others more actively and supportively into the management process. In a small-scale fishery setting, such rights may also have impacts on the well-being and security of fishers and fishing communities; the effects can be positive, given suitable recognition, design and implementation of rights, but can alternatively be negative (Charles, 2001, 2009; Béné et al., 2010). The various forms of fishery rights will be explored in detail later in this chapter.

There are clear and important relationships between human rights and fishery rights (Charles, 2009). The former reflect imperatives in terms of the relationships among people, specifically fishers, and between people and society. The latter govern who can go fishing and who can be involved in decisions relating to the fishery. The FAO (2007, p. 6) connects these together in addressing small-scale fisheries, noting that:

> A rights-based approach, in defining and allocating rights to fish, would also address the broader human rights of fishers to an adequate livelihood and would therefore include poverty-reduction criteria as a key component of decisions over equitable allocation of rights, including in decisions over inclusion and exclusion, and the protection of small-scale fishworkers' access to resources and markets.

Certainly, if there are some aspects of human rights that can be maintained and enhanced through fishing activities, then this provides a strong link to fishery rights, and a context within which decisions concerning who should hold those rights, how they should be managed and so on, can be made. This connecting of fishery-specific rights and human rights has been neglected in much of the literature on rights in fisheries management, but will become increasingly important to take into account in fishery policy development, at scales from the local to the international. Such linkages are important in particular in addressing the challenge of poverty in fishing communities.

In considering these linkages, one analyst (Kearney, 2007) has developed a list of five 'fishing rights' that apply specifically to fisheries but reflect more general statements found in the Universal Declaration of Human Rights. These five 'fishing rights' Kearney notes are as follows:

- the right to fish for food;
- the right to fish for livelihood;
- the right to healthy households, communities and cultures;
- the right to live and work in a healthy ecosystem that will support future generations of fishers; and
- the right to participate in the decisions affecting fishing.

The adoption of this human rights-based approach in fisheries has been advocated by two major international fisherfolk organizations, the World Forum of Fisher People (WFFP) and the International Collective in Support of Fishworkers (ICSF). In a briefing note prepared for the FAO Committee on Fisheries in 2009, these organizations state (ICSF-WFFP, 2009, p. 3) that such an approach:

> … recognizes that development efforts in fisheries should contribute to securing the freedom, well-being and dignity of all fisher people everywhere. Given the international consensus on achieving human rights, committed action to realizing the human rights of fishing communities, as indeed of all vital, yet marginalized groups and communities, is an obligation.

The organizations highlight two reasons for a human-rights approach in fisheries: 'The adoption of a human rights approach has an intrinsic rationale as achieving human rights of all citizens is an end in itself. Adopting this approach also has an instrumental rationale in that it is likely to lead to better and more sustainable human development outcomes'.

The ICSF and WFFP, together with many other civil society organizations, defined a human rights approach in fisheries within the 'Bangkok Statement' (Civil Society Preparatory Workshop, 2008) that was presented at the FAO-organized Global Conference on Small-Scale Fisheries (FAO, 2008). These organizations note (ICSF-WFFP, 2009, p.3) that the Statement 'expands on what a human-rights based approach to fisheries and fishing communities means, from the perspective of small-scale fishworkers and their communities'.

The approach of ICSF-WFFP (2009, p. 3) builds on the above list of Kearney (2007) to include the rights of fishing communities:

> ... (a) to their cultural identities, dignity and traditional rights, and to recognition of their traditional and indigenous knowledge systems; (b) to access territories, lands and waters on which they have traditionally depended for their life and livelihoods; (c) to use, restore, protect and manage local aquatic and coastal ecosystems; (d) to participate in fisheries and coastal management decision-making; (e) to basic services such as safe drinking water, education, sanitation, health and HIV/AIDS prevention and treatment services; and (f) of all fish workers to social security and safe and decent working and living conditions.

Furthermore, the ICSF and WFFP (2009, p. 3) specifically note the rights of women to:

> ... participate fully in all aspects of small-scale fisheries; to have access to fish resources for processing, trading, and food, particularly through protecting the diversified and decentralized nature of small-scale and indigenous fisheries; and to utilize fish markets, particularly through provision of credit, appropriate technology and infrastructure at landing sites and markets.

In considering these rights, it is important to keep in mind that the well-being of a small-scale fishery is often closely interrelated to that of the corresponding coastal communities. In particular, holding fishery rights over the use of fishery resources and the management of the fishery can empower communities, while loss of those rights (e.g. through their transfer to outside players) can lead to a loss of social cohesion in the community. This can be reflected in reduced local involvement in the fishery, reduced employment and a corresponding increase in the proportion of 'outsiders' fishing on what had been locally controlled resources. All of these impacts can run counter to the human rights of the community and its residents. Thus, attention to livelihoods and poverty reduction in the context of small-scale fisheries is directly related to fishery rights.

As Allison and Horemans (2006, p. 760) note: 'Livelihoods approaches are evolving and merging with rights-based approaches and community-development...' Indeed, these authors argue (Allison and Horemans, 2006, p. 760) that how fishery rights are dealt with is critical to the well-being and human rights of the people:

> It is policies and institutions that determine access to assets, set the vulnerability context and determine peoples' livelihood options, reactions and strategies, and ultimately, the outcomes of those strategies in terms of their ability to make a living and willingness to invest in helping to conserve the natural resource base. Addressing governance therefore remains the key challenge for both poverty reduction and responsible fisheries.

The connection of artisanal and subsistence fishing to food security and livelihoods is an important element in considering human rights and fishery rights. As Schumann and Macinko (2007, p. 716) suggest: 'Reverence for cultural concerns and anxiety over food security can both be justifiable grounds for subsistence priorities, warranting precedence over other uses of fishery resources when not all uses can be sustained...'. Consider, for example, the case of South Africa. In 1999, the South African government, reviewing management of the subsistence fishing sector, formed a Subsistence Fisheries Task Group (SFTG), which revised the definition of 'subsistence' to be more restrictive (Sowman, 2006, p. 66): 'The SFTG resource recommendations have resulted in no subsistence fishers being recognized along the west and south coasts of South Africa. This is of grave concern given the high levels of food insecurity found in fisher households in these regions'.

Along similar lines, Jaffer (2006, pp. 22–23) reports on a legal battle between the artisanal fishing sector and the government in South Africa. The artisanal fishers argued that legislation relating to fisheries management, the Marine Living Resources Act, 'deprived them of their right to choose their trade or occupation' under Section 22 of the South African Constitution. They claimed further that 'the current legislative framework violates a number of other basic socio-economic rights, most notably, the right of access to sufficient food', but also the 'right to healthcare, housing and education, and the rights of the child to basic nutrition'.

Furthermore, in certain situations, human rights and fishery rights are closely linked to aboriginal rights. This arises, for example, in some small-scale fisheries of the Asia-Pacific region, where 'Traditional management systems ... are based on property rights and associated regimes which reflect local culture, economic conditions, and structures of power and social organization' (Pomeroy, 2001, p. 121). The links are also important in the aboriginal fisheries of northern countries, such as Canada and Norway, where fishing is crucial to community food security, health and livelihoods.

Finally, the connection of fishery rights and human rights can be usefully related to recent debates over the desired focus of small-scale fishery policy – debates between a so-called 'wealth-based' approach, with an emphasis on rent maximization (Cunningham *et al.*, 2009) and a multi-objective approach that highlights the 'welfare functions' of small-scale fisheries, and the dual goals of poverty reduction and poverty prevention (Béné *et al.*, 2010).

A Focus on Fishery Rights

The discussion in the previous section emphasized the links between human rights and fishery rights in broad terms, but there remains a need to examine the various forms of fishery rights in more detail, and particularly to assess their implications in small-scale fisheries. This section will address such matters, adapting and extending the discussions in Charles (2002, 2009) to explore a range of fishery rights relevant to small-scale fisheries, from rights over fish in the sea to access rights and other rights over use of the resource, and finally to management rights. At the end of the section, a focus is placed on an approach of particular relevance in small-scale fisheries, namely implementing fishery rights at the community level.

Rights over fish in the sea

While coastal fishers have significant social, economic and human rights that relate to fisheries, they do not typically own the fish swimming in the sea, until those fish are landed on a fishing boat or on shore. Who, then, does own fish in the sea? With small-scale fisheries generally located within national exclusive economic zones (EEZ), perhaps the most common arrangement is that fish are under the jurisdiction of the particular nation in whose waters they are located. If it is possible in such a situation to speak of ownership, the fish could be thought of as the property of that nation's citizens – typically until the time at which they are caught by fishers.

Another common scenario in small-scale fisheries, particularly traditional ones, occurs when the fish in the sea are 'owned in common' by a certain identifiable group of people – e.g. the set of citizens within a specific local jurisdiction, such as a coastal community, or the members of a native tribe, as opposed to a whole nation, or a single private individual or company. In such cases, the fish, as a common-pool resource, are managed under a *common property* regime (Dolsak and Ostrom, 2003), a situation that will be explored in more detail below.

Whoever are considered the 'owners' of the fish hold certain property rights, such as the right to decide how the fish are to be used and by whom. Fishers may or may not be seen to hold those specific rights, but are likely to hold other 'fishery rights', namely access rights, harvesting 'use rights' and management rights. These rights, which are the focus of what is often referred to as 'rights-based fisheries management', are discussed in turn below.

Access rights

Whenever a fishery is managed by restricting who can have access to the fishery, those with such entitlements are said to hold access rights (Charles, 2001, 2002, 2004) – simply the right to 'use' the fishery. This right is recognized, or assigned, by the relevant management authority, whether formal or informal. For example, in a tribal fishery, it may be the chief deciding who is to have access to the resource, while in another situation a governmental fisheries authority may designate the holders of fishing licences. There is often a territorial aspect to the rights, in that those outside the community or region often lack access rights and are thus excluded from the fishery.

Access rights may be suitable where there is a recognized need for and desirability of restriction of use of fishery resources. This can be for a variety of reasons – food and livelihood security, sustainability of the resources, conflict reduction, manageability, etc. Access rights are widely accepted within fishery management, seen as a remedy to the problems of open access – unrestricted access to fishery resources. Indeed, the FAO Code of Conduct for Responsible Fisheries (FAO, 1995, para. 10.1.3) makes reference to access rights, not only within fisheries but pertaining to coastal resources in general: 'States should develop, as appropriate, institutional and legal frameworks in order to determine the possible uses of coastal resources and to govern access to them taking into account the rights of coastal fishing communities …'.

Specifying access rights is helpful to the fishery manager, both in resolving open-access problems and helping to clarify who is being affected by management. An access rights system resolves the uncertainty over who are the users of the fishery (i.e. who holds access rights and who does not). However, this only becomes clear once rights are established. Thus in any fishery, a key issue arises: who *should* hold access rights?

The above-noted Code of Conduct (para. 6.18) has addressed one aspect of this question in a clear way, stating that: 'States should appropriately protect the rights of fishers and fishworkers, particularly those engaged in subsistence, small-scale and artisanal fisheries, to a secure and just livelihood, as well as preferential access, where appropriate, to traditional fishing grounds and resources in the waters under national jurisdiction'.

However, the situation is often complicated. First, the fishers in a given location are not necessarily homogenous. For example, Pomeroy (2001) notes that in addition to full-time fishers, there are also often part-time or seasonal fishers, including those who come from their inland homes to fish on the coast. Indeed, the latter point reinforces the reality that those who have traditionally had access to a local fishery may not be limited to community residents. Allison and Ellis (2001) argue that some small-scale community-based fisheries may allow for 'reciprocal access' between differing locations, to boost sustainable livelihoods in both places: 'Outsiders can access village-based fishing territories in times of their need, or when there are local surpluses, often in exchange for an access fee' (p. 380). In such situations, they state (p. 387): 'Institutions to regulate access to resources are still important, it is just that they do not necessarily take the form of fixed fishing territories and fixed license numbers …'.

In addition to challenges in determining who should have access to a given fishery, there are also issues with making access rights in small-scale fisheries effective. Indeed, Pomeroy (2001, p. 122) has stated that many '… coastal fisheries in developing countries are in effect de facto open access …' even though access rights may be specified in these fisheries. He argues (p. 122) that: '… the ability to enforce these laws and regulations is practically non-existent due to the fact that fisheries department and enforcement agencies do not have sufficient resources. In addition, the political will is often not in place to enforce these laws and regulations due to the influence of power elites'.

Therefore, while informal and traditional access rights have existed for centuries in a wide variety of fishery jurisdictions, and such rights are being implemented with increasing frequency even where direct government regulation dominates, there are nevertheless likely to be difficulties in making access rights fully effective in many small-scale fisheries.

Access rights can be defined spatially, in terms of rights to a specific fishing ground or in terms of entry ('access') into the fishery as a whole. These two options are described in turn below.

Spatial access rights

First, in terms of spatial access rights, two key concepts are customary marine tenure (CMT) and territorial use rights in fishing (TURFs). These have long been applied by fishing communities in determining, for each fisher or household, the location where they can access fishery resources. Both approaches are inherently spatial management mechanisms, assigning rights to individuals and/or groups to fish in certain locations (thus the term 'territorial' in TURF), generally, although not necessarily, based on long-standing tradition ('customary tenure'). A classic reference on TURFs is that of Christy (1982, p. 1), who noted that: 'As more and more study is given to the culture and organisation of fishing communities, there are indications that some forms of TURFs are more pervasive than previously thought to be the case, in both modern and traditional marine fisheries'.

Indeed, TURFs have a particularly long history in traditional, small-scale/artisanal and indigenous fisheries. Two particularly well-known examples are the long-standing arrangement in coastal Japan, where traditional institutions are incorporated in modern resource management, and the small-scale lobster fisheries on the north-eastern coast of North America, where fishers in many locations have been able to maintain informal but effective community control on entry, i.e. demonstrating the capability to exclude others.

Some CMT and TURF systems have gone through periods when they lacked support in policy and thus suffered declines over time. However, there are now moves to maintain or restore many such systems. For example, in the fisheries of Oceania, traditional CMT/TURF systems declined as fisheries were 'modernized', but as recognition of the efficiency of such systems grew, there have been initiatives in some nations (notably in the South Pacific) to re-establish them. As Johannes (2002, p. 317) noted: 'Factors contributing to the upsurge include a growing perception of scarcity, the restrengthening of traditional village-based authority, and marine tenure by means of legal recognition and government support, better conservation education, and increasingly effective assistance, and advice from regional and national governments and NGOs'.

For example, Veitayaki (1998) reported on the case of Fiji, where customary marine tenure over traditional fishing grounds was historically the principal marine resource management practice, but had been in a significant state of decline. However, it was suggested that recent initiatives to formally register the boundaries related to CMT could be an important step in helping to restore community ownership over these areas.

As with any management mechanism, CMT and TURFs are not suitable in all cases. For example, Allison and Ellis (2001, p. 385) point out that:

> Creating TURFS associated with individual fishing villages is a currently fashionable form of institution building in fisheries development; however temporary migration to places where fish are available is a prevalent feature of artisanal fisheries worldwide, and one that does not sit comfortably with the notion of territorial rights being based on resident populations in shoreline villages.

While caution is thus necessary, there is a broad sense that for appropriate cases, these 'traditional sea tenure systems' can hold considerable potential to provide efficient and relatively stable socially supported fishery management, particularly if implemented within the framework of existing social institutions and livelihood approaches (Ruddle, 1989).

Limited entry access rights

The second key form of access rights is found in the form of a fishing licence, reflecting the 'limited entry' approach that is common in modern state management of fisheries. Indeed, this form of management is often expressed as a regulatory tool to control the activities of fishers and fishing communities, in which the government (typically) issues a limited number of licences to fish. Each licence conveys a 'right' on a fisher, a fishing group or a community to access the fishery (to go fishing); some will thus have this right to 'use' the fishery, while all others will not. In this way, limited entry seeks to prevent the expansion of the number of fishing boats and/or fishers, with the aim of controlling potential fishing effort (fleet capacity), thereby helping to conserve the resource and generating higher incomes for the licence holders (i.e. those holding the access right).

Limiting access is also common in small-scale fisheries. Indeed, Berkes *et al.* (2001, p.148) refer to work by Wilson *et al.* (1994) showing that for a sample of 32 locations worldwide, limited access is the second most common traditional fishery regulation (after fishing area restrictions). However, the feasibility of a limited entry rights approach will depend on the particular small-scale fishery, and on how the approach is implemented. For example, if such rights were given out to community members, but not to outsiders, it could be a helpful means to protect local livelihoods – indeed, perhaps a mechanism to institute fishery rights that also reflect human rights. On the other hand, if it were seen as a means to give fishing rights only to some in a community but not to others, serious social and/or political conflict could result, unless there is broad acceptance of who constitutes the valid fishers.

Furthermore, it should be noted that even if licensing of this form is feasible, limited entry cannot be expected by itself to 'solve' all management problems. In particular, while limited entry specifies access rights, it does not limit the fishing of those with such rights. Over-harvesting could still occur. To deal with this, limited entry, if implemented at all, should be seen not as a sole measure by itself but rather as part of a 'management portfolio' that also includes approaches by which current fishers limit their own fishing activity.

Summary

Access rights have the advantage, from a fisher and fishing community perspective, that those with such rights – whether an individual fisher, fishers' organizations or a fishing community – are provided with some security over access to fishing areas. If access rights are managed well, they can reflect a desired balance of social, cultural, economic and environmental goals; they can assist in reducing rather than causing conflict; they can enhance food security and livelihoods for small-scale fishers and fishing communities; and they can protect local ecosystems (e.g. by preventing over-harvesting and potentially by favouring more conservationist gear types or fishing practices). However, there are significant issues to be addressed in restricting fishery access, notably relating to equity considerations, and to impacts on poverty and vulnerability of households and communities (see, e.g. Béné *et al.*, 2010).

Effort and harvest rights

Within the spectrum of possible fishery use rights, access rights may be extended through quantitative (numerical) use rights – rights to use a specific amount of fishing effort (*effort rights*, e.g. to fish for a certain amount of time or with a certain amount of gear) or to take a specific catch (*harvest rights* allocated to individual fishers, companies, cooperatives or communities, to catch a specified amount of fish). Such forms of fishery rights have relatively high information and management requirements, and thus are less common and indeed often inappropriate in small-scale fisheries. Nevertheless, as they are widely discussed in the fisheries literature, and may be suitable in certain circumstances, they will be briefly reviewed here.

Both effort rights and catch rights have parallels in fishery management regulations, namely in terms of fishing effort limits (e.g. 'How much gear can be used?') and catch quotas (e.g. 'How much fish can be caught?'), respectively – see, e.g. Pope (2002). Clearly, quantitative use rights like these incorporate or must be accompanied by access rights, but the converse need not be the case – many fisheries operate through access rights without there being any quantitative use rights specified.

Effort rights

As noted above, effort rights are related to fishing effort controls, i.e. restrictions on the activity of the fishing fleet (through limits on time fished, amount of gear, gear attributes, etc.) to keep that activity at levels compatible with resource sustainability. Effort rights typically designate a specific amount of fishing time and/or gear for each fisher, or vessel (Charles, 2001). This can serve conservation needs as well as spreading the effort across more vessels than would otherwise be the case, for equity reasons. A common example of such an effort rights approach arises in trap fisheries, notably those for lobster, crab and other invertebrates, where each fisher has the right to set a specified number of traps. It may be that all fishers have equal rights (i.e. to the same number of traps) or that the rights vary from one individual to another, perhaps based on location, boat size or some other criteria.

A key challenge for an effort rights programme arises if the rights relate to only one or two of the factors influencing fishing effort. In the above example, if rights relate only to the number of traps a fisher uses, that leaves the amount of time to use the traps unlimited. To overcome this, a multidimensional approach is needed, by implementing effort rights over not one but a range of inputs. Another challenge is the need to deal with the natural process of technological improvement that gradually increases the effectiveness of any given set of inputs over time. An effort rights programme must adjust for improvements in fishing efficiency by reducing the total number of allowable input units over time. Thus effort rights, while more costly than simple access rights, can be a viable approach if care is taken in defining the rights, if the rights cover a range of effort inputs and if a plan is put in place to deal with fishing efficiency improvements.

Harvest rights

The second main form of quantitative use rights is the harvest right (or 'catch quota'). If a fishery is managed through a total allowable catch (TAC), and that TAC is then subdivided into quotas held by sectors of the fishery, individual fishers, companies or communities, these shares of the TAC are the harvest (or catch) rights. They may be held *collectively*, whether by a sector of the fishery or by fishing communities (see the discussion of 'community quotas' later in this section). Alternatively, the rights may be allocated to *individual* fishers as trip limits (providing the right to take a certain catch on each fishing trip) or as individual quotas, rights to harvest annually a certain fraction of the TAC. In the latter case of individual quotas, these harvest rights may be non-transferable, or (mainly in industrial fisheries) there may be buying and selling of these quotas in a 'quota market' (i.e. for 'individual transferable quotas', or ITQs).

Harvest rights are widely promoted at present as a means of better matching catches to available markets, and avoiding the 'race for the fish' (so that catches can be taken at a lower cost and with less incentive for overcapacity e.g. Shotton, 2000). This is meant to increase profitability by reducing fishery inputs such as fleet size and the number of fishers, and by increasing product value. However, harvest rights raise economic and conservation concerns in small-scale fisheries (Copes and Charles, 2004). Perhaps most fundamentally, the costs of running a quota system can be prohibitive – in determining the suitable TAC, in monitoring catches and in enforcing catch allocations. There are also risks to conservation, including those arising with catch controls in general (notably the potential to overestimate biomass and thus TACs), and those arising if the catch rights are allocated to individuals. The latter risks are due to: (i) inherent incentives to cheat by under-reporting catches, since every caught fish that is unreported is one less that must be deducted from the quota; (ii) similar incentives to dump, discard and high-grade fish, since this allows the fishers with the quota directly to increase the value of what they actually land; and (iii) pressure on decision-makers to increase the TAC beyond sustainable levels, to help fishers who have gone into debt to purchase rights (quota) from others. The high costs and various negative impacts of harvest rights explain why individual quota systems (in particular) are rarely found

Management rights

The various use rights described above serve to specify and constrain who is to be involved in resource use, and this has the potential to improve the effectiveness of management and make conservation more likely. There is a parallel need to specify who is to be involved in fishery management – i.e. through what are called *management rights*. Management rights reflect the fifth 'right' noted by Kearney above – the right to participate in decisions affecting fishing. Such rights can be seen in parallel with use rights: the former specify the right to participate in fishery management just as the latter specify the right to participate in the fishery itself. Indeed, management rights are among the *collective choice* rights defined by Ostrom and Schlager (1996); these contrast with *operational-level* rights (including use rights) and in fact include the 'authority to devise future operational-level rights' (Ostrom and Schlager, 1996, p.131).

There is a widespread understanding that effective management requires a broader approach than conventional top-down methods – through new *co-management* arrangements that involve some degree of joint management by fishers, government and possibly local fishing communities (Pinkerton, 1989; Wilson *et al.*, 2003). In the language of fishery rights, this co-management requires allocation of management rights, the right to be involved in managing the fishery.

Who should hold management rights? Typically, the relevant government will have the responsibility to conserve the resource, to produce benefits from that resource and to suitably distribute those benefits, so it will certainly be among those holding management rights. Furthermore, successful management requires the support (or at least the acceptance) of fishers (who already hold use rights), and thus they should be among the holders of management rights. Finally, it may be that communities, non-governmental organisations (NGOs) and the general public could all be involved in management, but this is much more likely in the case of *strategic* management (dealing with the fishery's overall objectives and policy directions) than for *operational* matters (measures such as closed areas and seasons, or allowable hook or mesh sizes, that affect the fishing process directly). This is because strategic issues are typically ones of broad public interest, about which a wide spectrum of interested parties – and fishing communities in particular – should hold management rights. On the other hand, for operational matters, it is particularly important for fishers to hold management rights, but dealing with such operational aspects may attract little interest among communities, NGOs and the general public.

Parallel to the question of who should hold management rights is that of what situations are actually conducive to co-management arrangements. For example, Brown and Pomeroy (1999, pp. 567–568) suggest that for countries in the Caribbean:

> ... the near shore fisheries targeted by small-scale fishers for benthic species such as lobster and conch, coral reef fish, and coastal pelagics will have the best chances for successful comanagement. These fisheries usually have easily identified users and boundaries, similar gear and fishing operation patterns, and a small number of target species. Co-management can be either resource-specific or site-specific depending on the situation.

Similar conclusions may hold for a range of other small-scale fisheries.

Communities and fishery rights

Use rights and management rights can be allocated to individual fishers or they can be held in a collective manner by a community or a fishers' association. There is a long history in small-scale fisheries of fishing rights being held collectively within a particular community, but unfortunately, there has been relatively little attention in current debates over fishery rights to community-held rights (cf. Charles, 2006). Furthermore, such rights have not always been properly understood and incorporated into 'modern' management, leading to social and conservation problems. It is thus worth paying extra attention to such rights here, particularly since, as Panayotou (1982, p. 44) has suggested: 'The revival and rejuvenation of traditional community rights over coastal resources offer, perhaps, the best possible management option for scattered, remote and fluid, small-scale fisheries'.

The choice between individual and community rights should depend on both the historical context and the fishery objectives being pursued. For example, in the case of a fishery that has developed relatively recently and that has an industrial focus, there may be a natural inclination to an individual rights system, which may be viewed as compatible with the entrepreneurial independence of fishers. On the other hand, while community rights cannot be expected to work in every fishery, the approach seems more likely to be effective given: (i) cohesiveness of the community involved; (ii) experience in and capacity for local management; (iii) geographical clarity of the community; (iv) a modest overall size and extent; and (v) an institutional framework in which rights are specified through a combination of legislation, government decisions and traditional/informal arrangements.

Where community rights are feasible, they have the potential to: (i) utilize management institutions and moral pressure locally to create incentives for resource stewardship (conservation); (ii) increase management efficiency; and (iii) improve the implementation of local enforcement tools. In addition, with community rights, local 'fine-tuning' can help to achieve equity and fairness goals – e.g. by taking into account a broader range of fishery participants in a community, including not only current boat or licence owners but also crew members, shore workers and those (present and future) with an interest in participating in the fishery (Graham *et al.*, 2006).

Pursuing community rights may involve understanding and reviving former management systems. As Panayotou (1982, p.45) notes: 'Such revival would necessitate a removal of the factors responsible for the breakdown of these traditional management systems by: (a) explicitly allocating the coastal resources to artisanal fisheries; (b) dividing these coastal resources among fishing communities...'.

This allocation can take place with any desired combination of spatial access rights (such as TURFs), limited-entry licensing approaches and other use rights.

As but one example, while harvest rights in the form of catch quotas are most often inappropriate for small-scale fisheries, if they are to be implemented, then a promising approach is through 'community quotas', i.e. community-defined harvest rights in the form of portions of a TAC allocated to coastal communities. Defined on a geographical basis, they have the potential to bring people in a community together in a common purpose since, typically, the community as a whole (or the group of fishers in the community) manages the quota in such a way as to suit their specific local situation, to maximize overall benefits and to reflect community values and objectives (Charles, 2001). By having each community decide for itself how to utilize its quota, this can support community empowerment and enhance community sustainability. Examples of this approach in small-scale fisheries within industrialized countries are found in Alaska (specifically community development quotas (CDQs) and Atlantic Canada (Charles *et al.*, 2007).

Community rights contrast with market-based rights (such as individual transferable quotas) – see Copes and Charles (2004). Berkes (1986, p. 228) proposes that a community-based approach '... provides a relevant and feasible set of institutional arrangements for managing some coastal fisheries', particularly '... small-scale fisheries in which the community of users is relatively homogeneous and the group size relatively small'. On the other hand, he suggests that individual market-based rights may be appropriate '... for offshore fish resources and larger-scale, more mobile fishing fleets'. This indicates that a useful differentiation can be made between small-scale fisheries (with fishers closely connected to communities, and with history and tradition playing a major role) and those that are predominantly industrial and capital-intensive (in which profitability dominates over other societal goals). However, there are bound to be exceptions to any general direction, and a wide range of intermediate options can be contemplated as well, so allocation decisions must be made with great care.

Implementing Fishery Rights in Small-scale Fisheries

The previous section reviewed the various fishery rights, notably access, effort, harvest and management rights. In this section, we explore some major considerations in implementing these rights in small-scale fisheries, specifically: (i) the recognition of pre-existing rights, if they exist, or the choice among new rights systems, if needed; (ii) the approaches available for allocating rights; and (iii) choices relating to the duration of rights and whether transferability of those rights should be allowed.

Recognizing rights

In many existing small-scale fisheries, particularly those with a long history, rights have already developed naturally over time, perhaps put in place by fishers themselves or by their communities (see, for example, Dyer and McGoodwin, 1994; Hanna *et al.*, 1996). Indeed, Béné *et al.* (2010, p. 338) suggest that this situation of existing rights is a general reality: 'Anyone who has worked closely with small-scale fisheries in developing countries knows that the access to fisheries (in particular, small-scale coastal or inland fisheries) is always conditioned by some form of formal or informal, symbolic or substantial, control systems generally established at the local/community level'.

It is not surprising that access rights would have emerged, since there are clear benefits to defining the group of fishers entitled to fish in certain locations, both for the fishers themselves and for the well-being of the fishing community. If rights already exist, and holders of the rights are already specified, it will be important to assess the nature of those rights, how effective they are in meeting current objectives (as well as criteria of equity and sustainability) and whether there are available mechanisms to reinforce them. Certainly, it is likely to be less costly and easier politically to accept and reinforce traditional rights than to attempt the development of an entirely new regime.

Choosing among rights

If for some reason no use rights system is already in place (or alternatively, if use rights do exist but the current system is not functioning in a manner widely considered as effective or acceptable), then those involved in fishery management are faced with a choice among the various use rights options described above. However, given the biological, economic and social diversity of fisheries, no single-use rights approach will be applicable everywhere. The choice of use rights must fit into the culture, the historical reality and the policy directions of the specific fishery and overall jurisdiction. As the head of FAO's Fisheries and Aquaculture Department has noted (Nomura, 2006, p. 25): '... fisheries policies, management approaches – and fishing rights – need to be tailored to the specific context of countries and localities with respect to the fisheries in question, the social setting, culture, etc'. This reinforces the broad point of Kuperan and Raja Abdullah (1994, p. 306): 'Planning and setting objectives for management of small-scale coastal fisheries requires a good understanding of what is meant by small-scale coastal fisheries, the resource attributes, the traditional values of fishing communities, the institutional arrangements and the overall environment in which small-scale fisheries operate'.

This implies the need for a collaborative process to determine a framework of use rights that will meet objectives and be feasible in practice. The collaboration must be designed and implemented in an equitable manner that is widely recognized as legitimate, and involve fishery managers and planners working together with a suitable range of interested parties. It must also be recognized that each use rights option has its inherent advantages and limitations, so that what is 'best' will depend on the fishery in question. Thus it is important to understand how the particular fishery circumstances influence the desirability of certain options over others. Factors to take into account include: (i) the societal objectives; (ii) the relevant history and traditions; (iii) the relevant social, cultural and economic environment; (iv) the key features of the fish stocks and the ecosystem; and (v) the financial and personnel capacities of the particular fishery (Charles, 2002). It should not be surprising, given this reality, that there is no consensus about which use rights options are most compatible with which fishery features, only some trends (e.g. that sedentary fishery resources may be especially amenable to the use of TURFs).

Allocation of rights

In small-scale fisheries, as has been noted, rights may well already be allocated. However, if a new use rights system is being implemented for some reason, or if there is seen to be a need for adjustments to the existing system, how should the rights be allocated? There is no universally correct way to accomplish this and difficult choices are faced. Some approaches, such as one-time auctions or ongoing markets for rights, are not generally suitable for small-scale fisheries, since community and social values, while crucial in such fisheries, are typically ignored in these approaches. For example, as Panayotou (1982, p. 43) notes: 'Auctioning or market sale of a limited number of licences is certain to exclude many small-scale fishermen who have poor access to funds to bid for or purchase a licence'. The sale of fishing rights also tends to limit (especially financially) the capability of governments to undertake new policy directions, such as shifts in the fishery toward small-scale rather than industrial fisheries, or toward conservationist over destructive fishing gear.

Another allocation option is to assign rights on the basis of 'catch history'. This is common in industrial fisheries, where it is often done in proportion to each individual's past catches, or some other measure of participation in the fishery, possibly with adjustments to increase equity among the fishers. However, it is problematic to properly define historical participation, especially in small-scale fisheries where catches are rarely fully monitored.

A third option is for use rights to be allocated on a group/collective basis directly to participating communities, fishing sectors or other identifiable groups. Typically, the community or group holding the rights in common makes subsequent allocations (whether permanently or periodically) to participating individuals through methods that can be tailored locally. This approach has desirable features, in terms of empowering communities and allowing for local values to be reflected, but must ensure that possible imbalances in power within the community do not lead to inequitable results in the allocation of rights.

Duration of rights

In small-scale fisheries, the fishers and fishing communities involved typically have a long-term dependence on the fishery for their livelihood. The link between fishery rights and social, economic and human rights is therefore one in which access to the fishery is guaranteed to local fishers and communities. In return, the security of tenure and access can lead to local stewardship of coastal resources and an incentive to better 'plan for the future' in husbanding the resource. Thus in many small-scale or artisanal fisheries, access rights – which may well be available to all those in the local community – tend to be of indefinite duration, considered essentially permanent.

On the other hand, long-term rights can be problematic if a fishery was initially developed or exploited by industrial fishing companies or foreign fleets, but government now seeks to improve the situation of small-scale fishers by shifting rights to them. If the initial larger-scale operators had been given long-duration use rights, that might prevent the subsequent entry of small-scale fishers. In such situations, clearly there could be a benefit in shorter-duration rights, to provide greater management flexibility.

Transferability of rights

The transferability of use rights refers to the capability of rights holders to shift ownership of the right to someone else – whether permanently (e.g. by selling those rights, or handing them down in a family from one generation to the next) or temporarily (e.g. by transferring the rights to another fisher within a fishing season). The choices in this regard can have large impacts on small-scale fisheries and fishing communities.

If those holding use rights transfer these to their children, this may well be positive from the perspective of community stability. On the other hand, if the rights are able to be bought and sold, as advocated by some fishery commentators, this tends to lead to a concentration of those rights, as those with greater financial resources buy out others (Copes and Charles, 2004). Since small-scale fisheries are often the economic foundation of their communities, this concentration of rights is likely to produce negative impacts on community stability, because the rights typically shift out of small communities and into larger centres, together with a loss of rural livelihoods (employment) and detrimental effects on equity in the coastal economy. Given all these impacts, it will typically be important to place limits on (if not fully prohibit) the permanent transfer of use rights. This would be particularly important for market-based use rights, but even for the widely acceptable within-family process of handing down the rights from fishers to their children, there could be benefits in greater stability within the fishing community or region if transferability is restricted to *within* the particular sector or community in which the use rights reside.

On the other hand, there may be relatively few problems with temporary transferability, in which use rights can be transferred from one fisher to another within a fishing season, but then revert back to the original fisher at the end of the season. This provides occasional short-term flexibility (e.g. for fishers who happen to become sick or injured in a given year) while maintaining long-term stability in the distribution of the rights.

Conclusions

This chapter has focused in two main directions: (i) describing fishery rights from the specific perspective of small-scale fisheries; and (ii) linking fishery rights with human rights. Both of these areas of emphasis are very much in the spirit of a major meeting organized by the Food and Agriculture Organisation (FAO) of the United Nations in 2008, the Global Conference on Small-scale Fisheries. That meeting, which brought together a wide range of fishers, fishworkers, NGOs, governments and international organizations, reinforced a major shift in fisheries management, and the end of an era of simplistic thinking about rights in fisheries.

The simplistic view of rights revolved around an imbalance between fishery rights and human rights, with the focus on the first while ignoring the second. This led to an illusory view of the world in which it was assumed that, to achieve success in fisheries, one merely needs to assign the right to fish, regardless of whom gets those rights. In such a view, it really does not matter whether the rights holders are fishers, corporations or communities, only that rights are assigned.

This simplistic approach had an element of truth at its roots – that regardless of who holds rights, having secure access to the fishery does provide them with more security and makes it more worthwhile to take care of the resource into the future. However, other key realities in small-scale fisheries and fishing communities were neglected:

1. That rights may well already be in place in many small-scale fisheries, and these should be reinforced and supported, rather than ignored and replaced.
2. That who holds fishing rights, and how those rights are handled, makes a critical difference to the broader issues of community well-being, poverty alleviation, socioeconomic success and system resilience.
3. That fishing rights need to be closely linked with, and supportive of, social, economic and human rights.
4. That rights held by communities ('community rights') may be particularly effective in some small-scale fisheries.

Figuring out the right form of rights requires an understanding of all these realities. Indeed, moving to a more realistic vision of rights requires reinterpreting a term commonly used in the literature on fishery economics and management – 'rights-based management' (Neher et al., 1989). What is needed is an understanding that, for fisheries management to be 'rights-based', it must take place in the context of all the various forms of rights. Given their mandate, fisheries agencies may have been inclined to focus only on use rights (over fishery access) and management rights (as in co-management). A broader vision of rights involves adding social, economic and human rights to the picture – rights that are fundamental and cannot be given out or taken away by government.

Furthermore, along with rights go responsibilities. The FAO Code of Conduct for Responsible Fisheries (1995, para. 6.1) states: 'The right to fish carries with it the obligation to do so in a responsible manner …'. A key aspect in moving toward responsible fisheries thus lies in developing effective and accepted sets of both rights and responsibilities among fishers. As Jentoft et al. (1998, p. 434) note: 'When rights of management and property go together, property is not only a right but also a responsibility for the collective as well as the individual. Without that responsibility there is no guarantee that property rights may institute sustainable resource use'.

Understanding, assessing and dealing with the impact of fishery rights on livelihoods, poverty, community well-being and human rights are clearly critical topics. In a context of developing countries, attention is needed to the relationship of fishery rights to the overall objectives of fishery and development policy. For example, a more complete rights-based approach, one combining fishery and human rights, can contribute in a practical way to achieving a balance in the debate over 'wealth-based' and 'welfare function' perspectives on the priorities for small-scale fisheries (Cunningham et al., 2009; Béné et al., 2010).

Drawing on insights in an oft-quoted paper of Béné (2003) on poverty and fisheries, Hersoug (2006, p. 7) concludes that: 'The point is simple: rights-based fisheries management may secure some type of ownership, be it individual or collective. But we need to secure rights for the right people. That can only be done through institutional reforms …'.

A similar conclusion is reached by Jentoft (2007, p. 93): 'Property rights can lead to more inequity but they can also be employed for correcting inequities, as they can be used as a mechanism to protect those in need of protection, that is, the marginalized and impoverished among fishers'.

Herein rests a major challenge in linking human rights and fishery rights within a context of small-scale fisheries.

To this end, we need to move toward the 'bigger picture' that connects the fisheries 'silo' to broader policy and legal frameworks, and to the well-being of coastal communities, in order to address, in a holistic way, the many issues facing small-scale fisheries (Berkes et al., 2001; Charles, 2001). For example, ensuring access rights to subsistence fishing in coastal communities may well be closely related to enhancing local food security, and incorporating post-harvest aspects into rights discussions can be important to ensure consideration of the rights of women involved in marketing fish. Moving to a 'bigger picture' perspective will involve better understanding linkages among the various forms of rights, both within the fishery system itself and in a multi-sectoral context, so as to produce more comprehensive approaches to managing small-scale fisheries, ones that are better able to improve well-being and safeguard livelihoods.

Acknowledgements

This chapter has grown out of the author's previous research on fishery rights, referenced herein, as well as from helpful discussions with many colleagues, including Melanie Wiber, John Kearney, Merle Sowman, Chris Béné, Evelyn Pinkerton, Fikret Berkes, Patrick McConney, Bob Pomeroy, Serge Garcia, Kevern Cochrane, Chris Milley, Arthur Bull and Sherry Pictou. As usual, any remaining errors are the responsibility of the author. Financial support from the Sciences and Humanities Research Council of Canada is gratefully acknowledged.

References

Allison, E.H. and Ellis, F. (2001) The livelihoods approach and management of small-scale fisheries. *Marine Policy* 25, 377–388.

Allison, E.H. and Horemans, B. (2006) Putting the principles of the Sustainable Livelihoods Approach into fisheries development policy and practice. *Marine Policy* 30, 757–766.

Béné, C. (2003) When fishery rhymes with poverty: a first step beyond the old paradigm on poverty in small-scale fisheries. *World Development* 31, 949–975.

Béné, C., Hersoug, B. and Allison, E.H. (2010) Not by rent alone: analysing the pro-poor functions of small-scale fisheries in developing countries. *Development Policy Review* 28, 325–358.

Berkes, F. (1986) Local-level management and the commons problem: a comparative study of Turkish coastal fisheries. *Marine Policy* 10, 215–229.

Berkes, F., Mahon, R., McConney, P., Pollnac, R. and Pomeroy, R. (2001) *Managing Small-scale Fisheries: Alternative Directions and Methods*. International Development Research Centre, Ottawa, Canada.

Brown, D.N. and Pomeroy, R.S. (1999) Co-management of Caribbean Community (CARICOM) fisheries. *Marine Policy* 23, 549–570.

Charles, A. (2001) *Sustainable Fishery Systems*. Blackwell Science, Oxford, UK.

Charles, A. (2002) Use rights and responsible fisheries: limiting access and harvesting through rights-based management. In: Cochrane, K.L. (ed.) A fishery manager's guidebook. Management measures and their application. FAO Fisheries Technical Paper No. 424, FAO, Rome, Italy.

Charles, A. (2004) Rights-based fishery management: a focus on use rights. In: Petruny-Parker, M.E., Castro, K.M., Schwartz, M.L., Skrobe, L.G. and Somers, B. (eds) *Who Gets the Fish? Proceedings of the New England Workshop on Rights-Based Fisheries Management Approaches*. Rhode Island Sea Grant, Narragansett, Rhode Island, pp. 3–5.

Charles, A. (2006) Community fishery rights: issues, approaches and Atlantic Canadian case studies. In: Shriver, A.L. (ed.) *Proceedings of the Thirteenth Biennial Conference of the International Institute of Fisheries Economics & Trade*, July 11–14 2006, Portsmouth, UK. International Institute of Fisheries Economics and Trade, Corvallis, Oregon.

Charles, A. (2009) Rights-based fisheries management: the role of use rights in managing access and harvesting. In: Cochrane, K.L. and Garcia, S.M. (eds) *A Fishery Manager's Guidebook*. Wiley-Blackwell, Oxford, UK.

Charles, A., Bull, A., Kearney, J. and Milley, C. (2007) Community-based fisheries in the Canadian Maritimes. In: McClanahan, T. and Castilla, J.C. (eds) *Fisheries Management: Progress Toward Sustainability*. Blackwell Publishing, Oxford, UK, pp. 274–301.

Christy, F.T. (1982) Territorial use rights in marine fisheries: definitions and conditions. FAO Fisheries Technical Paper No. 227, FAO, Rome, Italy.

Civil Society Preparatory Workshop (2008) Statement from the Civil Society Preparatory Workshop to the Global Conference on Small-Scale Fisheries (4SSF), Bangkok, Thailand. SAMUDRA Report No. 51, pp. 7–9. Available at: http://sites.google.com/site/smallscalefisheries/statement (accessed 30 March 2010).

Copes, P. and Charles, A. (2004) Socioeconomics of individual transferable quotas and community-based fishery management. *Agricultural and Resource Economics Review* 33, 171–181.

Cunningham, S., Neiland, A.E., Arbuckle, M.A. and Bostock, T. (2009) Wealth-based fisheries management: using fisheries wealth to orchestrate sound fisheries policy in practice. *Marine Resource Economics* 24, 271–287.

Dolsak, N. and Ostrom, E. (2003) The challenges of the commons. In: Dolsak, N. and Ostrom, E. (eds) *The Commons in the New Millennium: Challenges and Adaptation*. The MIT Press, Cambridge, Massachusetts, pp. 3–34.

Dyer, C.L. and McGoodwin, J.R. (eds) (1994) *Folk Management in the World's Fisheries: Lessons for Modern Fisheries Management*. University Press of Colorado, Niwot, Colorado.

FAO (1995) *Code of Conduct for Responsible Fisheries*. FAO, Rome, Italy.

FAO (2007) Social issues in small-scale fisheries. Report prepared for the *27th session of the Committee on Fisheries (COFI)*, 5–9 March 2007. COFI/2007/6, FAO, Rome, Italy.

FAO (2008) *Global Conference on Small-Scale Fisheries. Securing Sustainable Small-Scale Fisheries: Bringing Together Responsible Fisheries and Social Development*, 13–17 October 2008, Bangkok, Thailand. Available at: http://www.4ssf.org/ (accessed 30 March 2010).

Graham, J., Charles, A. and Bull, A. (2006) *Community Fisheries Management Handbook*. Gorsebrook Research Institute, Saint Mary's University, Halifax, Canada. Available at: http://www.coastalcura.ca/resources.html (accessed 30 March 2010).

Hanna, S., Folke, C. and Maler, K.G. (1996) *Rights to Nature*. Island Press, Washington, DC.

Hersoug, B. (2006) Opening the tragedy? *SAMUDRA* 45, 3–7.

ICSF-WFFP (International Collective in Support of Fishworkers and World Forum of Fisher People) (2009) Briefing note for delegates to the *28th session of FAO's Committee on Fisheries (COFI)*, 2–6 March 2009, Rome, Italy. Available at: http://sites.google.com/site/smallscalefisheries/ (accessed 30 March 2010).

Jaffer, N. (2006) Fishing rights vs. human rights? *SAMUDRA* 44, 20–24.

Jentoft, S. (2007) The litmus test. In: ICSF (ed.) Sizing up: property rights and fisheries management. A collection of articles from SAMUDRA Reports. ICSF, pp. 92–93.

Jentoft, S., McCay, B.J. and Wilson, D.C. (1998) Social theory and fisheries co-management. *Marine Policy* 22, 423–436.

Johannes, R.E. (2002) The renaissance of community-based marine resource management in Oceania. *Annual Review of Ecology and Systematics* 33, 317–340.

Kearney, J. (2007) Fulfilled, healthy, secure? *SAMUDRA* 46, 18–21.

Kuperan, K. and Raja Abdullah, N.M. (1994) Small-scale coastal fisheries and co-management. *Marine Policy* 18, 306–313.

Neher, P.A., Arnason, R. and Mollett, N. (eds) (1989) *Rights Based Fishing*. Kluwer Academic Publishers, Dordrecht, Netherlands.

Nomura, I. (2006) No one-size-fits-all approach. *SAMUDRA* 44, 25.

Ostrom, E. and Schlager, E. (1996) The formation of property rights. In: Hanna, S., Folke, C. and Mäler, K.G. (eds) *Rights to Nature: Ecological, Economic, Cultural and Political Principles of Institutions for the Environment*. Island Press, Washington, DC, pp. 127–156.

Panayotou, T. (1982) Management concepts for small-scale fisheries: economic and social aspects. FAO Fisheries Technical Paper No. 228, FAO, Rome, Italy.

Pinkerton, E.W. (1989) *Cooperative Management of Local Fisheries*. University of British Columbia Press, Vancouver, Canada.

Pomeroy, R.S. (2001) Devolution and fisheries co-management. In: Meinzen-Dick, R., Knox, A. and Di Gregorio, M. (eds) *Collective Action, Property Rights and Devolution of Natural Resource Management: Exchange of Knowledge and Implications for Policy*. DSE/GTZ, Feldafing, Germany, pp. 108–145.

Pope, J. (2002) Input and output controls: the practice of fishing effort and catch management in responsible fisheries. In: Cochrane, K.L. (ed.) A fishery manager's guidebook. Management measures and their application. FAO Fisheries Technical Paper No. 424, FAO, Rome, Italy, pp. 75–93.

Ruddle, K. (1989) The organization of traditional inshore fishery management systems in the Pacific. In: Neher, P., Arnason, R. and Mollett, N. (eds) *Rights Based Fishing*. Kluwer Academic Publishers, Dordrecht, Netherlands, pp. 73–85.

Schumann, S. and Macinko, S. (2007) Subsistence in coastal fisheries policy: what's in a word? *Marine Policy* 31, 706–718.

Shotton, R. (ed.) (2000) Use of property rights in fisheries management. In: *Proceedings of the FishRights99 Conference*, Fremantle, Western Australia, 11–19 November 1999 (mini-course lectures and core conference presentations). FAO Fisheries Technical Paper No. 404/1, FAO, Rome, Italy.

Sowman, M. (2006) Subsistence and small-scale fisheries in South Africa: a ten-year review. *Marine Policy* 30, 60–73.

United Nations (1948) Universal Declaration of Human Rights. Adopted by UN General Assembly Resolution 217A (III) of 10 December 1948. United Nations, New York. Available at: http://www.un.org/Overview/rights.html (accessed 30 March 2010).

Veitayaki, J. (1998) Traditional and community-based marine resources management system in Fiji: an evolving integrated process. *Coastal Management* 26, 47–60.

Wilson, D.C., Nielsen, J.R. and Degnbol, P. (2003) *The Fisheries Co-management Experience: Accomplishments, Challenges and Prospects*. Kluwer Academic Publishers, Dordrecht, Netherlands.

Wilson, J.A., Acheson, J.M., Metcalfe, M. and Kleban, P. (1994) Chaos, complexity and community management of fisheries. *Marine Policy* 184, 291–305.

5 Managing Overcapacity in Small-scale Fisheries

Robert S. Pomeroy

Introduction

There is growing concern worldwide about the impacts of overfishing and overcapacity on the sustainability of fisheries and on the social and economic conditions of fishers and fishing communities. In fisheries

> ... that are unmanaged or managed as de facto open access, the race for fish soon tends to create a fishing capacity that is larger than that needed to catch the sustainable yield. If this is uncontrolled, this capacity generally leads to overfishing.
> (Cunningham and Gréboval, 2001, p. 5).

Garcia and Newton (1997) estimated that in 1989, the world's fishing fleet reflected an overcapacity of 25–53% with respect to maximum economic yield. Over the period 1970–1990, the capacity of the world's industrial fisheries grew eight times faster than the rate of growth of landings from world capture fisheries (Gréboval and Munro, 1999). A report published by the World Wildlife Fund (Porter, 1998) stated that the world fleet was 2.5 times (150%) greater than world fish stocks could sustain. FAO research shows that tuna fisheries worldwide have an average harvesting overcapacity of about 20%, although this varies from region to region. Similarly, a recent government study in the USA found that overcapacity exists in 55% of 73 important fisheries (FAO, 2004).

Fishing capacity is defined by FAO (2000, p. 3) as 'the amount of fish (or fishing effort) that can be produced over a period of time by a vessel or a fleet if fully utilized and for a given resource condition.' Two terms are important to the discussion of capacity – excess capacity and overcapacity (Ward et al., 2004). Excess capacity exists when potential catch or effort is greater than actual catch or effort in a given period. Excess capacity is a short-term issue that can arise when boats operate on average for fewer days than expected under more average conditions, as a result of, for example, lower prices, higher costs or management decision. The term 'overcapacity' conveys the fact that fishing capacity is greater than some optimal or desired level (in terms of catch and corresponding fleet size). Overcapacity is a long-term concept, in that there are excessive levels of capacity over the longer term in relation to some target level of yield or capital (boats, gear, fishers) used in the fishery. In simple terms, overcapacity refers to the fact that there are 'too many fishers chasing too few fish' in the long term. The objective of capacity management is to identify the desired level of capacity and bring existing capacity into line with this target level (Ward et al., 2004).

The problems of overcapacity have become a key issue in fisheries management. In 1999, the FAO Committee on Fisheries adopted 'The International Plan of Action for the Management of Fishing Capacity', which calls for states to prepare and implement national plans to effectively manage fishing capacity, with priority to be given to managing capacity on fisheries where overfishing is known to exist (Gréboval, 2000).

An FAO (2004) survey on capacity management found that 60 of 80 countries that responded to the survey had conducted, or planned to conduct, an initial assessment of national fishing capacity – but mostly for large-scale commercial fisheries. Half of the countries reported having national programmes in place for monitoring fishing capacity, but fewer (26) had established target capacity levels for their commercial fishery fleets. Eighty per cent reported having directly incorporated capacity considerations into their day-to-day fisheries management regimes. In the Asian region in 2007, six of ten countries (Australia, Bangladesh, Cambodia, Indonesia, Malaysia, Thailand) had developed national plans to address fishing capacity, although specific copies had not been provided to FAO and there was some doubt as to the accuracy of the reporting (Morgan et al., 2007).

International policy discussions of fishing overcapacity have largely ignored the small-scale fisheries subsector. The problem of reducing overcapacity in small-scale fisheries in developing countries is much more complex than that of reducing overcapacity in industrial fleets. Due to the complexities inherent in small-scale fisheries, countries are unlikely to prepare effective plans to address overcapacity in this subsector without support to analyse the problem and generate new policy options. Actions taken to date by resource managers to deal with overcapacity in small-scale fisheries, such as command and control regulation and vessel and gear buyback, have not been effective at dealing with the issue. Unless the core issues of fishing capacity, i.e. of overcapacity, are addressed, that is, access control and defined property or user rights and facilitating the exit of labour and capital from the fishery, any regulatory measures or other management strategy, such as marine protected areas, will simply be a stop-gap measure since more people will continue to enter the fishery.

The purpose of this chapter is to present and discuss the concept and assessment of overcapacity in small-scale marine fisheries, and the appropriate and integrated approaches to facilitating the removal of overcapacity. The chapter should assist governments and fisheries managers to prepare national and fishery-specific plans of action for the management of capacity in small-scale fisheries.

The Problem of Overcapacity in Small-scale Fisheries

Overcapacity in a fishery tends to develop as a result of some market imperfection, such as the absence of clear property rights. Overcapacity in fisheries leads to several problems, including the following (Metzner and Ward, 2002; Ward et al., 2004):

- over-investment in fishing boats and gear;
- too many fishers (captains and crew);
- reduced profit and decline in quality of life of fishers and their families;
- increasing conflict in the fishery; and
- political strife in the management process.

In addition, overcapacity is one of the leading causes of overfishing. Many near-shore fisheries around the world are overfished (Ward et al., 2004), resulting in severe depletion of coastal fish stocks and degradation of marine habitats (Morgan et al., 2007). Such declines have increased poverty among coastal fishers in developing countries. Overfishing has also reduced the contribution of coastal fisheries to employment, export revenue, food security and rural social stability in these nations (Ward et al., 2004). Unless remedial action is taken, declining resources, increasing poverty and impaired contribution to national development are expected to worsen as coastal populations increase.

Although the problem of overcapacity is well recognized and relatively easy to analyse, it remains one of the most intractable issues in fisheries. Integral elements of this situation that add to the complexity of finding solutions include:

- growing populations;
- poverty;
- open-access nature of the resource;
- a high dependence of fishers on the resource for food and livelihood;
- sluggish economies;
- high level of unemployment or underemployment;
- lack of ready alternative and supplemental employment and livelihood opportunities within the fishing community;
- pressure to find additional fisheries resources;
- increasing numbers of part-time and seasonal fishers;
- lack of management and enforcement;
- limited transferability of and rigidities in the movement of use-specific capital and labour;
- lack of credit and markets;
- government policies that encourage capital investment and exploitation of fish stocks;
- lack of institutional mechanisms for a coordinated and integrated approach to horizontal economic and community development blending fishery and non-fishery sectors; and
- lack of research and information.

Because the capital and labour employed in small-scale fisheries are generally use-specific, their exit is often difficult and painfully slow. Many small-scale fishers exist at the subsistence level and have a short-term survival strategy of taking care of themselves and their family that day (Pomeroy, 1991). Such fishers, due to limited mobility and lack of alternative employment, utilize whatever resources are available (technology, skill, capital) to harvest as much fish as possible before others do so. The fisher, living at the subsistence level, has a high discount rate concerning use of the resource; that is, profits and food are preferred today rather than a continual flow into perpetuity. Under such conditions, as long as small-scale fishers can obtain a positive return, they will continue fishing and try to circumvent any command and control regulatory measures, such as gear limitations, closure of fishing areas and other means (see also Hauck, Chapter 11, this volume).

Issues of overcapacity and sustainable resource use cannot be isolated from poverty, unemployment and declining quality of life in fishing communities. The main brunt of such economic and social distress is borne by women, children and unskilled fishers, as well as by those unskilled people who are directly and indirectly dependent on fishing.

Fisheries managers have become increasingly aware of the need to develop appropriate policies to facilitate the exit of capital and labour from overexploited fisheries. The focus on capacity reduction efforts at both international and national levels has been on industrial and commercial fisheries. There is a lack of a policy focus on the specific approaches that could be pursued in small-scale fisheries. Countries with small-scale fisheries with severe overcapacity are unlikely to implement effective plans to address fishing overcapacity without methods and approaches to analyse the problem and generate new policy options. Methods are needed to facilitate easily measurable fishing capacity in small-scale fisheries, and approaches are needed to address the issue.

Overcapacity in Small-scale Fisheries in South-east Asia

The countries of South-east Asia (Brunei, Cambodia, Indonesia, Laos, Malaysia, Philippines, Taiwan, Thailand and Vietnam) have a population of over 510 million, of whom approximately 35% live below the poverty line. The population of these nine countries is expected to reach 650 million by the year 2020. The average fish consumption for the region is relatively high, at 22 kg per capita per year, and is higher in coastal communities. In some countries and fishing communities, fish provides the main source of animal protein. In addition, fishing and the extraction of aquatic resources provides the main livelihood for millions of families (ADB, 2006).

It is estimated that the demand for food fish in the year 2010, calculated at a constant per capita consumption rate of 22 kg/year, would be 18–19 million t (Delgado et al., 2003). Production from marine capture fisheries is not expected to keep up with demand, causing concerns for food security in the region. The increasing demand for fish from the expanding population will create more stress on the already depleted coastal and inshore fishery resources targeted by small-scale fishers in the region.

It is now almost universally accepted that most of the nearshore fisheries in South-east Asia are overfished. It is also accepted that overcapacity is one of the leading causes of this overfishing. The results of a regional expert consultation on management of fishing capacity (SEAFDEC, 2006) concluded that overcapacity of fisheries was a primary cause of the major problems within the fishery sector, including declining catches. Silvestre et al. (2003, p.13) state that:

> The results of overfishing in South and Southeast Asia are that coastal fish stocks have been severely depleted. Resources have been fished down to 5–30 percent of their unexploited levels. Such declines have increased poverty among coastal fishers who are already among the poorest of the poor in developing Asian countries. Overfishing has also reduced the contribution of coastal fisheries to employment, export revenue, food security and rural social stability in these nations. The trends (resource decline, increasing poverty and impaired contribution to national development) are expected to worsen as coastal populations increase, unless remedial action is undertaken.

Other authors have echoed these concerns: Stobutzki et al. (2006, p. 116) state that '… there is an urgent need to reduce fishing capacity in the region', and Sugiyama et al. (2004, p. 21) predict that 'Based on current trends, production from capture fisheries in the Asia-Pacific region will decline over the next 10–20 years unless overcapacity and fishing effort is greatly reduced'.

As an example, in the Philippines, the total number of vessels in the municipal fishery sector was estimated at 20,000 units in the whole country in 1948, of which 83% were non-motorized. This grew to an estimated 500,000 units after 40 years, with a higher percentage of motorized boats (Dalzell and Corpuz, 1990). The total number and tonnage of commercial fishing boats rose from 3265 and 150,260 t, respectively, in 1988, to 4014 and 216,090 t (increases of 23% and 44%) in 1994 (Courtney et al., 1998). Catch per unit effort, as measured in tonnes/horsepower (t/hp) for the total small pelagic fish catch from municipal (small-scale) fisheries in the Philippines has declined from 2.9 t/hp in 1948 to an estimated 0.20 t/hp in 2000 (Green et al., 2003). The Lingayen Gulf, a major fishing ground in northern Luzon, Philippines, reached its maximum sustainable yield more than 20 years ago (Green et al., 2003). It is estimated that the fishery now has 400% too much effort for the available fish stocks. Catch rates in Lingayen Gulf are five times smaller than they were in 1990 (Green et al., 2003). In 1983, Pomeroy (1989) estimated that there were 767 full-time fishers using 25 different fishing gear types in the ten coastal *barangays* (villages) of the Municipality of Matalom, Leyte. Subsequent visits in 1993 and 2001 found an increase in the number of full-time fishers to 923 and 1087, respectively. Daily fish catch for line-fishers had declined from 2.1 kg in 1983 to 0.5 kg in 1993; for fish-trap fishers from 13.5 kg to 5.4 kg; and for gill-net fishers from 23.7 kg to 8.3 kg (Pomeroy, personal observation). Research conducted by the WorldFish Center (2002, p. 25) on coastal fish stocks in the Philippines found that, overall '… the level of fishing in the grossly modified stock is 30% higher than it should be (i.e. fish are being harvested at a level 30% more than they are capable of producing)'. The same general pattern of overfishing and overcapacity most probably holds true for other small-scale fisheries in South-east Asia.

An expert consultation on the management of fishing capacity in the South-east Asian region (SEAFDEC, 2006, p. 10) concluded that 'Improvement of fishery management could not be done without addressing the issue management of fishing capacity'. In addition, it was found (p. 10) that:

> No aggregated data on fishing capacity at national/regional level, however, the information is available in more site-specific and projects related forms than statistical information and national policy and plans exist for management of fishing capacity. However, among them there is a different degree of readiness, in most countries there is not any proper management system in place.

Morgan *et al.* (2007, p. 26) state that 'There is an increase in awareness of, and actions to address, fishing capacity issues by member countries'. They further state that (p. 26):

> While there has been an increase in the use of capacity reduction programmes in small-scale fisheries in countries of the region within the past few years, this has not occurred to the same extent in industrial fisheries. This parallels the increase in activity related to measurement of fishing capacity in small-scale fisheries, but not industrial fisheries.

While there may be an increase in awareness of the issue in the region, there is no quantitative confirmation that capacity reduction measures in small-scale fisheries in the region has brought about any substantive improvements in either the resource or the lives of people.

Fisheries Management

Ward *et al.* (2004, p. 7) state that 'Overcapacity is a relative measure, basically indicating that capacity is greater than some desired level'. For example, if having a large number of boats operating in the fishery and reducing the size of the fish stock is not compatible with the overall objectives of fisheries management, then overcapacity exists and something needs to be done to deal with this problem. Ward *et al.* (2004, p. 7) further state that:

> The fundamental objective of capacity management is to identify the desired level of capacity and bring the existing capacity into line with this target level. Further, this target level of capacity – either input or output based – also relates to some desired stock size and level of exploitation of the stock, so there is also an implicit (or, in some cases, explicit) target fishing mortality and stock level.

Thus, in order to manage capacity, managers must have a clear set of management objectives for the fishery in order to guide the identification of an appropriate target capacity. Identification of the appropriate target capacity level for the fishery is often not easy, given multiple and often conflicting objectives, such as conserving fish stocks, providing employment and maintaining healthy ecosystems.

The selection and adoption of a target level of capacity is often a critical stumbling block in many fishery management schemes. The emphasis on target levels is clearly appropriate when there is enough information to identify the level. When there is not enough information, management reference directions (Berkes *et al.*, 2001) may be an adequate basis for management action. This will often be the case in small-scale fisheries, particularly those showing signs of overexploitation. In fisheries where overcapacity exists and can be measured to some extent, the specific target capacity level may not be known but the need to move in that direction may be clear, and it may be possible to do so without knowing the target end point. This shifts the focus of management action from 'where do we want to be?' to 'how do we move in the desired direction?' Incorporation of the concept of reference directions into management planning will be a sufficient basis for action in many small-scale fisheries where problems have been qualitatively identified, but quantification is not possible (Berkes *et al.*, 2001).

In order to manage capacity, managers need to measure and understand how much capacity currently exists in the fishery and what is the desirable level of capacity (i.e. the target level of capacity) that best meets management objectives. Regular assessments of capacity are needed to determine how capacity is changing over time and what the impacts of these changes are.

Measuring Fishing Capacity at the Small-scale Fishery Level

Small-scale fisheries can generally be characterized by multiple fishing gears targeting multiple fish species using a variety of motorized and non-motorized vessels in coastal waters located near the fishing community (Berkes *et al.*, 2001). In any individual fishery, there may be 5 to 25 or more different fishing gears being used, ranging from spears to traps to handlines to nets, and operating 24 hours a day.

Fishers may range from women gleaners, to children spearing for octopus, to non-motorized handliners, to upland farmers undertaking seasonal fishing, to motorized gillnetters, to beach seiners, to migratory fishers chasing coastal pelagics, to commercial trawlers. The level of fishing effort from all this diversity of fishing gears and fishers is usually very high throughout the year.

To manage fishing capacity in these complex fisheries, it will be necessary to understand the current capacity. If excess capacity exists and if it is a problem, the level of capacity that best meets the objectives of management, what this means for the structure of the fishery, and how to move to the desired level of capacity (Ward et al., 2004) begs answers to several questions:

- How is fishing capacity defined?
- How is fishing capacity to be measured?
- What is the level of overcapacity in the fishery?

There are a number of different understandings of capacity depending upon disciplinary background. A recent FAO report (Ward et al., 2004, p. 2) states that:

> Fishing technologists often consider fishing capacity as the technological and practical feasibility of a vessel achieving a certain level of activity – be it days fishing, catch or processed products. Fisheries scientists often think of fishing capacity in terms of fishing effort, and the resultant rate of fishing mortality (the proportion of the fish stock killed through fishing). Fisheries managers generally have a similar view of fishing capacity, but often link the concept directly with the number of vessels operating in the fishery. Many managers express fishing capacity in measures such as gross tonnage or as total effort (e.g. standard fishing days available). Most of these ideas reflect an understanding of capacity primarily in terms of inputs (an input perspective). Economists tend to consider capacity as the potential catch that could be produced if the boat were to be operating at maximum profit or benefit (an output perspective).

While a definition of fishing capacity may be agreed, there is still no generally accepted and standardized definition of *how* capacity should be measured, particularly for small-scale fisheries. However, a number of methods for measuring fishing capacity, both quantitative and qualitative, are available.

Techniques for Measuring and Assessing Capacity in Fisheries

Measuring capacity or the degree of capacity utilization in a fishery is relatively easy, because it does not require any information on the status of the fishery resource per se. One can compare actual levels of fishing inputs use (such as number of vessels, gear or fishing effort) or output (using catch as an indicator) with potential ones, assuming unrestricted but normal full use of the available inputs (actual levels of capacity) (Pascoe and Gréboval, 2003; Ward et al., 2004).

To quantitatively assess overcapacity, existing capacity has to be compared to an optimal or target level of capacity (the target fishing capacity, e.g. in terms of catch, corresponding effort level and corresponding fleet size assuming 'normal' use of fishing units). Because of the mobile nature of fleet capacity, this comparison may be made at stock, fishery, area or exclusive economic zone level. Establishing a target level of exploitation is required to set a target level of capacity. Reference points may be used to indicate the 'optimum' level of capacity, depending upon management objectives.

Both input- and output-based measures of capacity have been developed (Pascoe and Gréboval, 2003). Pascoe (2007) reports that most countries have adopted an input-based measure of capacity (such as vessel numbers, engine power, gross tonnage, vessel length), and relatively few countries have adopted an output-based measure for capacity management. Pascoe (2007) further states that the reason for this is that input-based measures are preferred by policy makers since capacity management is viewed as fleet (or effort) management, rather than addressing the property rights issue.

Both quantitative and qualitative methods exist to determine the level of capacity in a fishery. While seemingly easy, and possibly being so for some single-gear and single-species fisheries, quantitative estimation of capacity can be complex and difficult. Any analytical method will require, and be limited by, adequate data and analytical skills. Analysis of overcapacity has to address the dynamic nature of the fishery, including, for example, seasonality of fisheries and mobility of stocks and fishers. Adequate data must be available, such as number of vessels, vessel characteristics, fish catch and landings, economic data, etc. to have sufficient information for analysis.

The vast majority of small-scale fisheries have limited or no formal, quantified information on catch or effort. Therefore, any method for use in small-scale fisheries must require limited data and be simple and cost-effective. The characteristics of small-scale fisheries (i.e. multiple gears, multiple species, open access, seasonal fluctuations in capacity and effort, interactions between small-scale and large-scale fleets) make use of available approaches to measurement of fishing capacity limited, and resulting estimates subject to some uncertainty.

It should be noted that all of the indicators discussed below have limitations. Most provide only information about overcapacity at one point in time. No indicator should be used alone, but a combination of indicators can at a minimum provide a determination of the existence of overcapacity and the need to take action.

Quantitative estimates

Several quantitative approaches have been developed to estimate overcapacity (Asche, 2007; Pascoe, 2007). These include: (i) peak-to-peak; (ii) factor requirements function when there are total allowable catch limits or a revenue function when outputs are unconstrained and freely chosen; (iii) frontier production function and output; (iv) dual-economic based; (v) data envelopment analysis and frontier; and (vi) maximum potential effort based on ideal, empirical and practical, and fishing power or fixed effects. Which method is used depends in large part upon the nature of the fishery, availability of data – especially cost data for variable inputs and a capital rental or service price – and the intended use of the capacity measure (Kirkley and Squires, 1999).

It is acknowledged that existing methods will be too complex for use in most small-scale fisheries and will need to be modified and/or new methods developed in order to reflect the unique characteristics of small-scale fisheries. Of the six quantitative methods identified above, the one with the most potential applicability for small-scale fisheries is the data envelopment analysis (DEA). DEA is a mathematical programming technique for estimating technical efficiency and capacity utilization. The methodology for measuring the technological-economic concept of capacity and capacity utilization using DEA was developed by Färe et al. (1989, 1994). The method derives a frontier output that corresponds to the output that could be produced given full and efficient utilization of variable inputs constrained by the fixed factors, the state of technology and the resource stock (Kirkley and Squires, 1999; Squires et al., 2003).

Overcapacity measures that utilize DEA estimate overcapacity levels relative to a biological target level of yield or to an economic target level of yield such as maximum economic yield. The DEA approach allows capacity estimates for a range of situations, including multi-species fisheries, multiple outputs and inputs, almost all data possibilities, and individual boat level or relative to fleet performance. Pascoe (2007) reports that DEA has been the preferred method for estimating capacity due to its ease in incorporating multiple outputs, lack of assumptions regarding functional forms and single technologies, and its ability to incorporate other information, such as fisher behavioural assumptions, prices and costs. Because it was developed for industrial fisheries and can be complex to use, the DEA method would need to be modified for analysis of small-scale fisheries and to meet data and analytical limitations. It cannot be used to rank different fisheries on their level of capacity, nor to accommodate the stochastic (random) nature of fisheries, and is based on observed outputs under prevailing conditions (short term in nature).

Bioeconomic models have also been used to estimate input-based measures of overcapacity. Using such models, the fleet size and composition that best conforms to the management objectives can be estimated and compared with current fleet sizes and composition to develop an estimate of the level of overcapacity. Bioeconomic models can be complex and time consuming, unreliable due to uncertainty in the biological relationships and the lack of good economic data, have a limited life and are slow to provide information (Pascoe, 2007).

Qualitative estimates

Because quantitative methods are inappropriate for estimating fishing capacity in most small-scale fisheries, non-quantitative estimates of capacity may be an alternative approach. Qualitative indicators show whether overcapacity exists at a point in time, but do not indicate the magnitude of the problem or the direction of change (Ward et al., 2004). A finding of overcapacity in the fishery may result if several qualitative indicators suggest that overcapacity exists.

Subjective estimates

Subjective measures, based on informed judgement about the fishery from knowledgeable individuals, can be used. Several commonly used socio-economic assessment and rapid-appraisal techniques can be utilized to derive subjective estimates. This might include a combination of socio-economic field data collection methods such as observation, focus group interviews and key informant surveys; Delphi technique; visualization techniques such as maps, transects and timelines; and use of indigenous and traditional knowledge of fishers. Such techniques can provide a wealth of information in a relatively simple and cost-effective manner and provide information on historical changes and trends in the fishery (Pido et al., 1996; Bunce et al., 2000; Berkes et al., 2001). It should be noted that the information provided by these techniques may be biased and should be used with caution.

Qualitative indicators of overcapacity

Qualitative assessments of overcapacity can be based on verifiable indicators, which themselves are based on scientific methods (Ward et al., 2004; Garcia et al., 2009, see also Evans and Andrew, Chapter 3, this volume). The indicator approach is used to apply a common measure to fisheries and to reduce subjective judgment. It has the advantage of making maximum use of existing biological, management and fleet-specific information. A combination of indicators, each indicating change over time, will be needed to determine qualitative capacity levels. These indicators may include:

- Biological status of the fishery. If signs of overfishing are observed for the target species in a directed fishery, it is probable that overcapacity exists, especially against a background of increasing capacity.
- Harvest/target catch ratio. Overcapacity is likely to exist when harvest levels regularly exceed the target catch, with a harvest-to-target catch ratio significantly exceeding 1.0. However, this indicator must be considered in the context of the management of the fishery. If a fishery is closed before the target catch is exceeded, the harvest level will not exceed the target, and no apparent overcapacity will be observed. Also, this indicator is not sensitive to any discarding that may take place in a fishery managed through quotas, and is therefore not a good indicator of overcapacity in fisheries that are managed through total allowable catch (TACs) or quotas. In addition, if the fishery has been overfished, and the harvest level is below the target level, the measure may be less than 1.0 in spite of the presence of overcapacity.
- TAC/season length. Using the ratio of the TAC level to the season length, an increase over time of this ratio indicates overcapacity.
- Conflict. Controversies surrounding the setting of the TAC and the sub-allocation

of the TAC among user groups may also indicate overcapacity.
- Catch per unit of effort. A decline over time in catch per unit of effort (CPUE) against a background of stagnating catches generally implies overfishing and, most likely, overcapacity. However, fluctuating total catches under a constant fishing mortality management strategy could mask this effect, and CPUE trends may remain constant or increase for schooling species even though overall stock abundance is declining.
- Value per unit of effort. The value of catches per unit of effort (VPUE) may be a potential indicator of overcapacity in multi-species fisheries, especially if the VPUE decreases as overall CPUE stagnates or decreases. VPUE is a useful capacity indicator in fisheries where it is impractical to record the catch of each species separately, but recording the total value of sale is feasible.
- Age of fleet. An increased age of the fleet is an indicator of lack of investment in the fishery and declining profitability.

Other considerations in assessing capacity in fisheries

Assessments of capacity in SSF may not provide enough information to formulate policies to address the problem. It is critical to understand the underlying context that has brought about overcapacity, and this will require a more detailed examination of linkages among people, the economy, the resource and government policies. For example, an analysis of demographic and economic development trends that influence patterns of coastal and marine use can be conducted. Using population census data, current occupational and migration patterns in the whole country and in coastal areas can be compared with historical occupational and migration patterns. The analysis should evaluate linkages between population, demographic changes and patterns, economic development, poverty and natural resource use. Labour-adjustment processes and occupational rigidities and their impacts may be examined. Policies for social and economic development and the fishery sector should be examined to assess the types of subsidies and economic incentives being provided to the fishing industry and to understand interactions and impact on current fleet capacity and resource sustainability. Because overcapacity is a consequence of absent or poorly defined property or user rights, it will be useful to undertake an analysis of prevailing access conditions and possible alternative arrangements.

Approaches to Reducing Overcapacity

If overcapacity is considered to exist in a fishery, a target or desired level of capacity will need to be identified and agreed upon, and appropriate measures for reducing capacity will need to be selected. A variety of management measures are available to reduce overcapacity, although the practical reality of usage of these management measures is constrained by political, economic, social and technical considerations. As with the use of any management measure in fisheries, there will be winners and losers as a result of capacity management programmes, and issues of equity will need to be addressed. It is important to keep in mind the social implications of management measures to address overcapacity given the lack of social safety nets, the limited number of employment opportunities in rural areas and the lack of skills needed to find other employment. The social and economic impacts of reducing jobs in the fisheries sector may include increased conflict and stress on individuals and families, the collapse of rural communities and economies and loss of cultural identity.

Ward *et al.* (2004), based on the use of capacity management measures to change the set of incentives facing fishers, describe the measures as either 'incentive blocking instruments' or 'incentive adjusting instruments'. Incentive blocking instruments attempt to restrict the level of activity in some form and include limited entry, buy-back programmes, gear and vessel restrictions, aggregate quotas, non-transferable vessel catch limits and individual effort quotas. Incentive adjusting instruments attempt to address the property rights issue (mentioned above that the existence of overcapacity is symptomatic of the absence of well-defined property or user rights) and allow the market to assist in reducing overcapacity and include individual transferable quotas, taxes and royalties, group fishing rights and territorial use rights.

A recent study on fishers' perceptions of the use of a variety of fishing capacity management measures in three South-east Asian countries (Cambodia, the Philippines and Thailand) found that some management measures are acceptable to the fishers, while others are not (Table 5.1; Salayo et al., 2008). Overall, measures concerning effort reduction are not acceptable. Although most measures within the cluster 'gear/area/temporal restrictions' are acceptable, there is certain ambivalence towards a closed season. Both effort reduction and gear/area/temporal restrictions are regulatory in nature. Alternative and supplemental livelihoods were well accepted in all three countries.

Incentive blocking schemes, such as limited entry, buy-back programmes and gear and vessel restrictions, have seldom been effective in small-scale fisheries in limiting overcapacity (Jensen, 2002). While they may be effective in certain situations in reducing capacity in the short term, they are seldom effective in limiting overcapacity in the longer term (Jensen, 2002). The primary reason they are ineffective in small-scale fisheries is that overall management of these fisheries is poor due to a lack of management plans and limited capacity, resources and political will to implement and enforce management measures. In addition, these schemes do not address the basic incentives that create overcapacity – the absence of well-defined property and user rights.

Limiting entry through vessel licensing limitations reduces the number of vessels and overall access to the resource stocks. Fishers often respond to limited-entry programmes by 'capital stuffing' (increasing a vessel's horsepower or length) or technological innovations in fishing gear or in fishing activity. Unless the fleet size is kept small, there will be competition among gear types and fishing in different areas. In small-scale fisheries, limiting entry to the fishery through licence limitation, for example, has not been used or effective due to a lack of political will to limit the number of fishers, and the social and economic implications, especially on poor fishers' access to the resource for food, income and livelihood. In the study by Salayo et al. (2008), having a steady year-round income and livelihood was the primary reason why the respondents were against the idea of limiting the number of fishers. Small-scale fishers in the Philippines particularly felt that as long as they used legally permitted gears, they should be allowed to fish within municipal waters. Many perceived that they did not have other employment opportunities. Most either lacked the skills for non-fishing-related

Table 5.1. Perceptions of respondents to strategies for exit from fisheries in Cambodia, the Philippines and Thailand. Data from Salayo et al. (2008).

Strategies for exit from fisheries	Cambodia	Philippines	Thailand
Effort reduction			
Catch limitation	Disagreed	Disagreed	n/a
Limiting the number of fishers	Disagreed	Disagreed	n/a
Gear/area/temporal restrictions			
Banning the use of some gears	Agreed	Agreed	Rec.
Closed season/non-fishing seasons	Disagreed	Ambivalent	n/a
Establishment of protected areas	n/a	Agreed	Rec.
Sustainable alternative livelihoods	Agreed	Agreed	Rec.

Disagreed, > 50% of all respondents not in favour of the strategy; agreed, > 50% of all respondents in favour of the strategy; ambivalent, percentages of respondents who agreed and disagreed similar; n/a, the question was not specifically asked to respondents; rec. (recommended), the strategy was identified, but the specific percentage of respondents was unknown.

jobs or did not possess the required educational qualifications. A few argued that fishing was a way of life, and hence those who would like to stay must be allowed to do so. The respondents reported that limiting their existing numbers was unacceptable, although they acknowledged overcapacity and the need to reduce the fishing effort through other means. It was stated that limiting new entrants may be more feasible, rather than reducing existing numbers of fishers. The Cambodian fishers also responded that limiting the number of fishers is quite impractical given that they have no other livelihood opportunities (Salato et al., 2008).

Buy-back programmes buy and remove vessels or licences from a fleet to decrease fishing capacity (Jensen, 2002; Curtis and Squires, 2007). Although removal of vessels, licences and/or gear may reduce capacity in the short term, Holland et al. (1999) concluded that buy-back programmes are generally not an effective way to address capacity reduction as long as open-access fishery incentives remain. Buy-back programmes do not remove the economic incentives for creation of overcapacity. Any improvements in stock abundance will attract new fishers and increase capacity in the longer term. There will also need to be proper monitoring and regulation in place to manage the fishery after the fleet reduction. For example, buy-back schemes for small-scale fishers have been tried in several locations in Vietnam. In one case in the Tam Giang Lagoon in TT Hue province in the central part of the country, boats of 'floating fishers', fishing families that live and work on their boats, operating in the lagoon were purchased by the government and the fishers were provided with land in the northern part of the lagoon for resettlement. In less than a year, the fishers had sold the land and re-purchased boats and were back fishing. While the family was provided with land, they were not provided with any training in new livelihoods and therefore returned to the sea (Ha Xuan Thong, Hanoi, Vietnam, 2008, personal communication).

While gear and vessel restrictions can provide an initial reduction in harvests, they are generally not effective in small-scale fisheries due to lack of resources and incentive to enforce the restrictions or fishers substituting new gear for those that have been restricted. In the longer term, capacity will increase and there will be more motive to engage in illegal fishing. Restrictions should consider what gear and where it is being used. For example, bottom trawls may be more destructive than upper or mid-water trawls. In addition to an overall ban on the use of certain fishing gears, there may be restrictions on use of the gear in space (certain fishing grounds) and time (season). However, Salayo et al. (2008) found that respondents reported that banning the use of some gears is a measure that is largely accepted as a way of restricting fishing effort, and for rehabilitating fishery habitats and increasing fish population. The fishers themselves were particularly against the use of destructive gears, such as dynamite and noxious substances. Some fishers in the Philippines suggested banning highly efficient gears such as Danish seines, because the fine-mesh nets catch juveniles. In Cambodia, many fishers agreed that banning certain gears would sustain the fish stocks.

Aggregate quotas, non-transferable vessel catch limits (a form of quota without transferability) and individual effort quotas will also not be effective in small-scale fisheries as they require the establishment of a total allowable catch (TAC) and strong enforcement, both of which do not normally exist in these fisheries (Berkes et al., 2001). If effort and entry into the fishery are unrestricted, the use of these management measures will bring increases in capacity in the longer term. Additional regulations may be required to control input substitution. Salayo et al. (2008) found that fishers felt that limiting catch would mean reduced income. Many small-scale fishers are already classified as poor, and limiting fish catch would only aggravate their poor living conditions. Cambodian fishers argue that limiting the amount of catch would mean not having enough food for fishing households. The fishers were apprehensive about the level of reduction to be instituted. To most of the fishers, catch limitation is an alien management measure. Many respondents felt that commercial fishers should be the ones to reduce their effort as they catch more fish. The majority of municipal fisher respondents in the Philippines claim that the volume of their fish catch has declined compared with 5 years ago. Hence, they could not comprehend the logic of limiting effort when in fact they have already been experiencing lower catch.

Incentive adjustment measures have also met with mixed success in small-scale fisheries. Individual transferable quotas (ITQs) will allow market forces to drive out overcapacity. However, ITQs require setting a TAC before the harvest shares are allocated. The vast majority of small-scale fisheries have limited or no information to set the TAC and are multi-species fisheries with a high degree of variability. ITQs seem better suited to some types of fisheries (discrete, single-species fisheries) over other types, such as the multi-species fisheries normally targeted by small-scale fishers (Copes, 1986). ITQs do not seem to be applicable to highly variable fish stocks, such as shrimp fisheries, due to the problem of determining an appropriate TAC each year (Ward et al., 2004). Managers have limited ability to control access of fishers. Further, funds and monitoring, control and surveillance systems to implement the ITQ are limited. The WHAT Commission (2000, p. 11) concluded that:

> For complex, small-scale, multi-species fisheries with limited scientific and enforcement capability, rights-based fisheries management with shares specified as catch quotas or ITQs are not a realistic option. Furthermore, where there are few alternative livelihoods for large coastal populations and weak or nonexistent social welfare systems, a rapid reduction in participation in fisheries would be disastrous.

Participants at a meeting of Asian fish workers came out strongly against privatization of resources, particularly individual property rights (ICSF, 2007). Participants rejected the notion that the sea could be owned or privatized. Rather, the participants stated a demand for more equitable and responsible sharing of fisheries resources and for guaranteeing preferential access rights to small-scale fishing communities.

The use of taxes on fish landings or royalty fees paid per unit weight of fish landed or on quota holdings would encourage less profitable vessels and vessels of low capacity utilization to leave fishing (Jensen, 2002). Taxes and royalties carry the problem of deciding the appropriate rate of tax or the fee to be paid. This approach is administratively intensive as it requires constant adjustment to maintain capacity at desired levels. In a number of Asian countries, taxes on landings caused widespread protests among small-scale fishers and consumers who expected the taxes to result in higher prices (FAO, 1998; Ward et al., 2004).

Group fishing rights and territorial use rights (TURFs) address the issue of market imperfections by transforming open-access property rights structure to a regulated common property (see also Charles, Chapter 4, this volume). This approach provides a defined group of users with a clearly defined area, and the group of users regulate themselves to promote cooperative behaviour. There are now well-known and documented examples of these approaches from all over the world. Group fishing rights and TURFs require the group's understanding of the value of the rights, the capability to co-manage the resource, the need to restrict group membership, and the ability to limit access.

An important consideration in the application of any of these capacity-reducing measures is the issue of equity. Capacity reduction will result in a reallocation of benefits and wealth in the fishery and some fishers having to leave the fishery. Those who will have to leave the fishery will, in most cases, be negatively affected.

Management measures to reduce overcapacity in small-scale fisheries tend to be used in isolation. In the Asia-Pacific region, for example, boat and gear restrictions and space/time restrictions are the most common measures used to reduce overcapacity in small-scale fisheries (Morgan et al., 2007). Social support programmes to get fishers to leave fishing are also reported to be used in many countries in the regions. Morgan et al. (2007, p. 19) report, however, that 'It is therefore of some concern that these programmes are being reported as being implemented without parallel programmes to achieve real fishing capacity reduction'. Morgan et al. (2007, p. 17) also conclude that '… capacity reduction programmes in the region to date have not been successful in limiting or reducing fishing capacity'.

An Integrated Approach to Reducing Overcapacity

There is no single, simple solution to the overcapacity problem in small-scale fisheries. The complexities of small-scale fisheries make the use of any single approach to reducing overcapacity in isolation ineffective. Given these realities, the only feasible solution may be one based on a coordinated and integrated approach involving resource management, resource restoration and conservation, livelihoods and economic and community development, and restructured governance arrangements. The reduction of overcapacity implies an increased focus on people-related solutions and on communities.

This approach recognizes that solutions involve targeting not just the individual fisher but the whole household and its broader economic livelihood strategies. To be effective, solutions must address not only resource and technical issues of overcapacity but the underlying non-resource-related issues of poverty, vulnerability and marginalization in coastal households and communities. The strategy needs to address multiple challenges including food security, employment, income generation, livelihoods, health, improved quality of life, social development and community services and infrastructure. This approach finds solutions to the problem of overcapacity in both the fishery sector and non-fishery economic sectors. This calls for a broader vision of the fisheries system as a whole, going beyond fisheries sector-specific policies to the vast array of seemingly unrelated policies that may have beneficial side effects for the fisheries sector. The broader policy context is justified by the understanding and development of linkages between fisheries resource management, social and community development, coastal community economies and regional and national economies. Departments or agencies of fisheries cannot undertake this approach alone. There will be a need to reach out and coordinate with other government ministries or departments with expertise in economic and social development, for example, and across different levels of government from national to local.

In small-scale fisheries it may take a long period of time to reduce overcapacity, since the mobility of labour and capital is limited. Timing and sequencing of interventions and actions are critical. For example, rather than trying to remove fishers all at once, it may make more sense to phase in reduction in order to reduce social and economic disruption. Supplemental and alternative livelihoods need to be in place before access control measures are implemented. Gear restrictions, for example, will still allow fishers to have a livelihood while reducing overall fishing effort.

At the core of this approach is a plan that has been developed and agreed upon through a participatory process, and which identifies goals and objectives, management and development strategies and actions, and roles and responsibilities of all partners. The plan is structured around the key components of resource management, capability development, community and economic development, livelihood development, resource restoration and conservation, and institutional development.

Resource management

Resource management must be innovative and utilize a mix of management measures. Difficult decisions will need to be made about the use and impacts of fishing rights and access control measures, as there will be positive and negative social and economic implications. Preferential access rights can be assigned to coastal areas for small-scale fishers through, for example, fish zones. Due to the characteristics of small-scale fisheries, they are well suited to community property rights systems. Group fishing rights and TURFs hold the promise of restructuring the resource into a regulated common property. An individual group of fishers can determine who has access to the area and how to harvest fish from the area. For implementation to be successful in small-scale fisheries, any of these measures must be simple and cost-effective due to limited resources for administration and enforcement. For example, boats allowed access to a particular fishery may all be painted the same colour, with the licence number prominently displayed.

In addition, resource management may involve the use of more conventional fisheries management measures, such as limits on gear, fishing time and season. Gear restrictions, for example, can be used to limit the types of fishing gear or fishers may be allowed alternate days or areas to fish. Fishers may still be allowed to fish, but certain fishing practices or gears that contribute to overfishing or overcapacity may be forbidden. This should be undertaken through a gradual process over time to reduce negative impacts. In all cases, there will be a need for effective monitoring, control and surveillance measures.

While access control may seem intuitively simple at first, the complexity of small-scale fisheries makes implementation difficult. One of the biggest issues is that of entitlements – 'Who is entitled to have access to the fishery?' This question will need to be addressed initially, and is best accomplished through participation from and negotiation with individuals and groups to ensure equity. For any small-scale fishery, there are a multitude of users from various backgrounds and needs. There are full-time fishers using a variety of fishing gears, there are part-time fishers, there are seasonal fishers (such as upland farmers and migratory fishers) and there are subsistence fishers (such as widowed women). For example, restricting access to the fishery of an upland farmer who has based his family's livelihood strategy on having access to fish to dry and be available during lean periods will impact upon their food security. These entitlements are often informal and based on tradition and indigenous rights. These individuals may not be able to argue their rights to the resources with a legal framework. However, a structure should be established to allow all who believe themselves stakeholders the right to argue their case for entitlement.

The impact of environmental degradation from both fishery and non-fishery activities on the ecosystem that supports fisheries is increasingly recognized as a major fishery management problem (Silvestre et al., 2003). Separating these impacts on exploited resources from the direct effects of fishing mortality may be one of the major challenges of fishery management. Since most small-scale fisheries are near-shore, non-fishery human impact is usually a more important issue in their management than in large-scale fisheries. Consequently, different types of management measures are likely to be useful, depending on distance from shore (Caddy, 1999).

Resource restoration and conservation

Marine protected areas (MPAs) can serve to protect target species from exploitation in order to allow their populations to recover through closing an area or a population(s) of species in an area from exploitation. Perhaps more importantly, MPAs can protect entire ecosystems by conserving multiple species and critical habitats such as spawning areas and nursery beds. Stocks inside these areas can serve as a 'bank account' or insurance against fluctuations in and the depletions of populations outside the protected area caused by mismanagement or natural variability. MPAs can also reduce conflicts between fishers and other users by providing areas where non-fishery users can pursue non-consumptive uses of the resources.

In addition to closing areas through MPAs, there will be a need for restoration of marine habitats (coral reefs, mangroves, seagrass, wetlands) that are susceptible to man-made pollution and physical destruction. The restoration of these habitats, particularly those that limit the abundance of a resource at some life-history stage, may be the most important step to increasing fish stock productivity.

Livelihoods and community and economic development

This approach recognizes that any policies that reduce the number of fishers in small-scale fisheries without creating non-fishery employment opportunities will inevitably fail. This is because fishers will merely fish illegally, obtain a new boat and gear or do whatever else is necessary to continue to make a living in order to feed their family.

While heavily advocated as a solution to the many problems facing small-scale fisheries, the provision of supplemental and alternative livelihoods has had only limited success in most cases (Pollnac et al., 2001). The reason is that most rural economies have only a limited number of employment opportunities available. In most cases, excess labour already exists in these rural economies. A resource such as land is not readily available or is too costly to purchase; credit is difficult to obtain and skills training in order to find other employment is not readily available, if at all. The rural economy may have weak links to the regional and national economy and is not growing enough to absorb the growing rural labour force.

It is necessary to give fishers and their families a broad range of livelihood options, both supplemental and alternative, to choose from in order both to support exit from the fishery and reduce the household's economic dependence on the fishery. Families tend to have a certain household income need. If a household livelihood strategy is taken, rather than just focusing on the fisher, it is possible to provide this broader range of livelihood options. A focus on all members of the family allows them to be trained in new livelihoods and better address the income and other needs of the household. This will allow, for example, for management measures that reduce overall effort or restrict access to the fishery to be put in place with more limited economic disruption to the household. It will be necessary to go beyond the commonly used solution of giving fishers 'pigs and chickens' as a supplemental livelihood to more innovative livelihood approaches involving micro-enterprise development, skills development and training and the use of information technology.

In addition to livelihoods, there is a need to improve the basic public services provided to coastal households and communities. Social and community development efforts can help to ensure the expansion of opportunities in communities by integrating population, health, education, welfare and infrastructure (roads, communication, water) programmes into the approach. Education, extension and skills training can support supplemental and alternative livelihood programmes. A formal social security mechanism can help to make fishers and their families feel more secure about change and more willing to transition to a new fishing management strategy or livelihood.

It is unlikely that the rural economy in most countries will grow fast enough to absorb excess labour and capital from the fishery sector. It will be necessary to understand regional and national economic development trends, projections and policies to determine future employment and investment opportunities and constraints. Working with economic development experts, an analysis of trends and projections in both the regional and national economies and in future occupational demands can provide direction for skills training and micro-enterprise development. Economic base studies can provide information useful for identifying economic linkages between the community economy and the regional and national economies.

Governance

The active participation of people in this approach, through a strategy of co-management, is mandatory in planning, formulating and implementing development and management activities. Building and strengthening fisher organizations allows for consultation, cooperation and seeking consensus on strategies to address overcapacity. Community-based co-management can provide a framework for such a coordinated and integrated approach (Berkes et al., 2001; McConney et al., 2003; Pomeroy and Rivera-Guieb, 2006). Empowered and organized people are more able to plan and engage in the often complex discussions and planning needed to realize this approach. Community-based co-management can serve as a mechanism for not only resource management, but also social, community and economic development by encouraging people actively to learn, solve problems, address needs in their community and adapt to change. Organized people are better able to network and advocate for their needs and the resources necessary for implementation.

Conclusions

The problem of addressing overcapacity in small-scale fisheries is much more complex than that of reducing overcapacity in industrial fleets. The complexity in small-scale fisheries is compounded by: growing populations, sluggish economies, fishers' high dependence on the resource for food and livelihood, a paucity of non-fishery employment, increasing numbers of part-time and seasonal fishers, limited transferability of and rigidities in the movement of use-specific capital and labour, conflicting policies and data availability, among others.

In order to manage capacity, managers need to measure and understand how much capacity currently exists in the fishery and what is the desirable level of capacity (i.e. the target level of capacity) that best meets the set of management objectives. Regular assessments of capacity are needed to determine how capacity is changing over time and what the impacts of these changes are.

The only feasible solution to overcapacity may be based on a coordinated and integrated approach involving a mixed strategy of resource management, resource restoration and conservation, livelihoods and economic and community development, and restructured governance arrangements. The reduction of overcapacity implies an increased focus on people-related solutions and on communities.

Acknowledgements

Thank you to Rebecca Metzner of FAO for comments on an early draft of this paper. Thank you also to Kristian Parker and the Oak Foundation for support.

References

ADB (Asian Development Bank) (2006) *ADB Fisheries Policy*. Operations Evaluation Department, Asian Development Bank, Manila, Philippines.

Asche, F. (2007) Capacity measurement in fisheries: what can we learn? *Marine Resource Economics* 22, 105–108.

Berkes, F., Mahon, R., McConney, P., Pollnac, R. and Pomeroy, R. (2001) *Managing Small-scale Fisheries: Alternative Directions and Methods*. International Development Research Center, Ottawa, Canada.

Bunce, L., Townsley, P., Pomeroy, R. and Pollnac, R. (2000) *Socioeconomic Manual for Coral Reef Management*. Australian Institute of Marine Sciences, Townsville, Australia.

Caddy, J. (1999) Fisheries management after 2000: will new paradigms apply? *Reviews in Fish Biology and Fisheries* 9, 1–43.

Copes, P. (1986) A critical review of the individual quota as a device in fisheries management. *Land Economics* 62, 278–291.

Courtney, C., Atchuee, J.A., Carreon, M., White, A.T., Pestano-Smith, R., Deguit, E., Sievert, R. and Navarro, R. (1998) *Coastal Resource Management for Food Security*. Coastal Resources Management Project, Cebu, Philippines. Available at: http://oneocean.org/download/db_files/food_security.pdf (accessed 25 March 2010).

Cunningham, S. and Gréboval, D. (2001) *Managing fishing capacity: a review of policy and technical issues*. FAO Fisheries Technical Paper No. 409. FAO, Rome, Italy.

Curtis, R. and Squires, D. (2007) *Fisheries Buybacks*. Blackwell Publishing, Ames, Iowa.

Dalzell, P. and Corpuz, P. (1990) The present status of small pelagic fisheries in the Philippines. In: PCAMRD, *Philippine Tuna and Small Pelagic Fisheries: Status and Prospects for Development*. PCAMRD Book Series No. 07, Philippines, pp. 25–51.

Delgado, C., Wada, N., Rosegrant, M., Meijer, S. and Ahmed. M. (2003) *Fish to 2020: Supply and Demand in Changing Global Markets*. International Food Policy Research Institute, Washington, DC and WorldFish Center, Penang, Malaysia.

FAO (1998) *Report of the Technical Working Group on the Management of Fishing Capacity*. La Jolla, USA, 15–18 April 1998. FAO Fisheries Report No. 586, FAO, Rome, Italy.

FAO (2000) *Report of the Technical Consultation on the Measurement of Fishing Capacity*. FAO Fisheries Report No. 615, FAO, Rome, Italy.

FAO (2004) *Progress on the Implementation of the International Plan of Action for the Management of Fishing Capacity*. Technical consultation to review progress and promote the full implementation of the IPOA to prevent, deter and eliminate IUU fishing and the IPOA for the management of fishing capacity, Rome, Italy, 24–29 June 2004. FAO, Rome, Italy.

Färe, R., Grosskopf, S. and Kokkenlenberg, E. (1989) Measuring plant capacity utilization and technical change: a non-parametric approach. *International Economic Review* 30, 655–666.

Färe, R., Grosskopf, S. and Lovell, C.A.K. (1994) *Production Frontiers*. Cambridge University Press, Cambridge, UK.

Garcia, S. and Newton, C. (1997) Current situation, trends and prospects in world capture fisheries. In: Pikitch, E.K., Huppert, D.D. and Sissenwine, M.P. (eds) *Global Trends: Fisheries Management. American Fisheries Society Symposium 20*, Bethesda, Maryland, pp. 3–27.

Garcia, S., Rey-Valette, H. and Bodiguel, C. (2009) Which indicators for what management? The challenge of connecting offer and demand of indicators. In: Cochrane, K.L. and Garcia, S. (eds) *A Fishery Manager's Guidebook,* 2nd edn. FAO and Wiley-Blackwell, Rome, Italy and Oxford, UK, pp. 303–335.

Gréboval, D. (2000) *The FAO International Plan of Action for the Management of Fishing Capacity*. FAO, Rome, Italy.

Gréboval, D. and Munro, G. (1999) Overcapitalization and excess capacity in world fisheries: underlying economics and methods of control. In: Gréboval, D. (ed.) Managing Fishing Capacity: Selected Papers on Underlying Concepts and Issues. FAO Fisheries Technical Paper No. 386, FAO, Rome, Italy, pp. 1–48.

Green, S.J., White, A.T., Flores, J.O., Carreon III, M.F. and Sia, A.E. (2003) *Philippine Fisheries in Crisis: a Framework for Management*. Coastal Resource Management Project of the Department of Environment and Natural Resources, Cebu City, Philippines. Available at: http://oneocean.org/download/db_files/philippine_fisheries_in_crisis.pdf (accessed 11 February 2010).

Holland, D., Gudmundsson, E. and Gates, J. (1999) Do fishing vessel buy-back programs work? A survey of the evidence. *Marine Policy* 23, 47–69.

ICSF (International Collective in Support of Fishworkers) (2007) *Asserting Rights … Defining Responsibilities … Perspectives from Small-scale Fishing Communities on Coastal and Fisheries Management in Asia*. Workshop and Symposium, 3–8 May, Siem Reap, Cambodia.

Jensen, C.L. (2002) Reduction of the fishing capacity in "common pool" fisheries. *Marine Policy* 26, 155–158.

Kirkley, J.E. and Squires, D.E. (1999) Measuring capacity and capacity utilization in fisheries. In: Gréboval, D. (ed.) Managing Fishing Capacity: Selected papers on Underlying Concepts and Issues. FAO Fisheries Technical Paper No. 386, FAO, Rome, Italy.

McConney, P., Pomeroy, R. and Mahon, R. (2003) *Guidelines for Coastal Resource Co-management in the Caribbean: Communicating the Concepts and Conditions that Favour Success*. Caribbean Conservation Association and Centre for Resource Management and Environmental Studies, University of the West Indies, Barbados.

Metzner, R. and Ward, J.M. (2002) Report of the Expert Consultation on Catalysing the Transition away from Overcapacity in Marine Capture Fisheries, Rome, 15–18 October, 2002. FAO Fisheries Report No. 691, FAO, Rome, Italy.

Morgan, G., Staples, D. and Funge-Smith, S. (2007) *Fishing capacity management and IUU fishing in Asia*. RAP Publication 2007/16. Asia-Pacific Fishery Organization, UN Food and Agriculture Organization, Regional Office for Asia and the Pacific, Bangkok, Thailand.

Pascoe, S. (2007) Capacity analysis and fisheries policy: theory versus practice. *Marine Resource Economics* 22, 83–87.

Pascoe, S. and Gréboval, D. (eds) (2003) Measuring capacity in fisheries. FAO Fisheries Technical Paper No. 445, FAO, Rome, Italy.

Pido, M.D., Pomeroy, R.S., Carlos, M.B. and Garces, L.R. (1996) *A Handbook for Rapid Appraisal of Fisheries Management Systems (Version 1)*. ICLARM Education Series 16, ICLARM, Manila, Philippines.

Pollnac, R.B., Pomeroy, R.S. and Harkes, I. (2001) Fishery policy and job satisfaction in three Southeast Asian fisheries. *Ocean and Coastal Management* 44, 531–544.

Pomeroy, R.S. (1989) Monitoring and evaluation of fishery and agriculture projects: a case study and discussion of issues. In: Pollnac, R.B. (ed.) *Monitoring and Evaluating the Impacts of Small-Scale Fishery Projects*. International Center for Marine Resource Development, the University of Rhode Island, Kingston, Rhode Island, pp. 41–55.

Pomeroy, R.S. (1991) Small-scale fisheries management and development: towards a community-based approach. *Marine Policy* 15, 39–48.

Pomeroy R.S. and Rivera-Guieb, R. (2006) *Fishery Co-management: a Practical Handbook*. International Development Centre and CABI Publishing, Wallingford, UK.

Porter, G. (1998) *Estimating Overcapacity in the Global Fishing Fleet*. World Wildlife Fund, Washington, DC.

Salayo, N., Garces, L., Pido, M., Viswanathan, K., Pomeroy, R., Ahmed, M., Siason, I., Seng, K. and Masae, A. (2008) Managing excess capacity in small-scale fisheries: perspectives from stakeholders in three Southeast Asian countries. *Marine Policy* 32, 692–700.

SEAFDEC (2006) *Regional Technical Consultation on Management of Fishing Capacity and Human Resource Development in Support of Fisheries Management in Southeast Asia*, 19–22 September 2006, Phuket, Thailand. Southeast Asian Fisheries Development Center, Bangkok, Thailand.

Silvestre, G., Garces, L., Stobutzki, I., Luna, C., Ahmed, M., Valmonte-Santos, R.A., Lachica-Alino, L., Munro, P., Christensen, V. and Pauly, D. (eds) (2003) *Assessment, Management, and Future Directions for Coastal Fisheries in Asian Countries. WorldFish Center Conference Proceedings 67*. WorldFish Center, Penang, Malaysia.

Squires, D., Omar, I.H., Jeon, Y., Kirkley, J., Kuperan, K. and Susilowati, I. (2003) Excess capacity and sustainable development in Java seas fisheries. *Environment and Development Economics* 8, 105–127.

Stobutzki, I., Silvestre, G. and Garces, L. (2006) Key issues in coastal fisheries in South and Southeast Asia, outcomes of a regional initiative. *Fisheries Research* 78, 109–118.

Sugiyama, S., Staples, D. and Funge-Smith, S.J. (2004) *Status and potential of fisheries and aquaculture in Asia and the Pacific*. RAP Publication 2004/25, FAO Regional Office for Asia and the Pacific, Bangkok, Thailand.

Ward, J.M., Kirkley, J.E., Metzner, R. and Pascoe, S. (2004) Measuring and assessing capacity in fisheries: basic concepts and management options. FAO Fisheries Technical Paper No. 433/1, FAO, Rome, Italy.

WHAT (2000) *Governance for a sustainable future: fishing for the future*. Report of the Commission on fisheries, World Humanity Action Trust, London.

WorldFish Center (2002) *Strategies and options for increasing and sustaining fisheries and aquaculture production to benefit poorer households in Asia, ADB-RETA 5945: project completion report*. WorldFish Center, Penang, Malaysia.

6 Adaptive Management in Small-scale Fisheries: a Practical Approach

John Parks

Introduction

Managers of small-scale fisheries (SSF) face many challenges. Some challenges may be controlled but others may not be so easily predicted or resolved. Of all the challenges that managers face, perhaps *uncertainty* is among those that are most difficult to address. Uncertainty is the commonly found condition in which managers have incomplete and/or incorrect information or knowledge, making it impossible for them to predict or describe future outcomes, and thereby plan adequately for effective action in addressing them. Examples of sources of uncertainty in natural resource management are numerous, and well documented (e.g. Shannon and Antypas, 1997).

Even when uncertainty can be reasonably addressed by fishery managers, how are they expected to be able practically to describe and understand uncertainty as it relates to, for example, the particular fishery they are managing? Science-based professions such as engineering, meteorology and natural resource management describe uncertainty as the margin of *error*. These are measurements that deviate outside the range of values that are likely to enclose observed values, or standard deviation (Drosg, 2009). This margin of error may be very important to understand when managing SSF. At the same time, fishery managers are often operating in a state of great uncertainty, even if they do not know it. Given the challenge of uncertainty, fishery managers may understandably attempt to be proactive in minimizing both the sources and degrees of error within which they are working.

Because of uncertainty, even competent managers with all the required technical, human and financial resources to implement and enforce their fisheries management efforts may not succeed. More often, fishery managers do not have all the technical and financial resources that they need, and have incomplete and/or inaccurate data with which to manage fishery populations and predict population trends. Given all of the natural variability present within marine ecosystems, coupled with human-induced global changes, how is it that fishery managers can realistically be expected by management authorities and society to perform effectively when working conditions can be characterized by high levels of uncertainty and sudden, unpredictable changes? In such a situation, what is a manager to do?

Fortunately, fishery managers have a powerful tool at their disposal to mitigate the effects of uncertainty. If used effectively, this tool can help offset – and sometimes even limit – the effects of uncertainty. This tool is *adaptive management*. In the first part of this

© CAB International 2011. *Small-scale Fisheries Management*
(eds R.S. Pomeroy and N.L. Andrew)

chapter I discuss the origins and theory of adaptive management. The bulk of the chapter presents a step by step summary of actions needed to design and implement an adaptive management programme. The text is written in a practical style that is designed for practitioners rather than researchers.

Defining adaptive management

Adaptive management (AM) is a relatively simple concept that is easy to understand and practice: it is the formal process of systematically testing management assumptions through time, learning periodically from the evaluation of such testing and using this learning to revise and improve management practices (Holling, 1978; Walters, 1986; Walters and Holling, 1990; Salafsky et al., 2001; CMP, 2007). In other words, AM is the process of testing assumptions in order to learn and adapt future action.

The intention of using this test-learn-adapt, or 'learning by doing' (Holling, 1978; Walters and Holling, 1990), approach is that results of testing and learning allow decision-makers and managers to adapt and make decisions regarding future management in a timely and informed manner (CMP, 2007). Lee (1993) and Bormann et al. (1996) further explain AM as an experimentally designed process that rigorously tests hypotheses about management interventions and/or policies.

Conceptual development

Adaptive management and deliberate, focused 'learning by doing' is not a new concept. In fact, similar approaches have been used for centuries. Some traditional knowledge and customary management practices of indigenous groups have relied on periodic *observations* to make judgments about resource state and respond to actual or perceived changes in resource availability or scarcity by altering local harvest behaviour (Alcorn, 1993; Berkes, 1999; Berkes et al., 2000). For example, traditional ecological knowledge and customary practices using periodic prohibitions on resource harvesting (*tabus*) in many indigenous Pacific Island societies depended upon many of the same principles used within a modern AM approach (Johannes, 1978a,b; Ruddle et al., 1992; Roberts et al., 1995; Hviding, 1996).

The use of AM as a guiding management concept in contemporary, Western science-based natural resources management efforts developed over a period of more than three decades: from the late 1960s and early 1970s through to the 1990s. During the 1960s and early 1970s, distinct scholarly discussions across several academic fields converged, focusing on the concept of uncertainty, including chaos theory and the inherent complexity of natural and human systems. These discussions began to cast doubt on man's ability reliably to use science to predict behaviour within systems, even if such systems were not complex and followed deterministic rules (Prigogine and Stengers, 1984). Key principles recognized during this time as necessary in overcoming the challenges of uncertainty and complexity included *resilience*, flexibility, and adaptive potential (Holling, 1973; Gunderson, 1999). Ecologists recognized that maintenance or strengthening of resilience – that is, the capacity to withstand and recover from the effects of a disturbance – within natural systems was a desired outcome of natural resource management efforts (Holling, 1973; Walters and Hilborn, 1978; Berkes et al., 2003; Walker et al., 2004). Traditional AM practices of some indigenous societies had emphasized the concepts of natural resource complexity, renewability and resilience long before they entered into mainstream Western practices (Alcorn, 1993; Folke et al., 1998; Berkes et al., 2000). Building and maintaining resilience is now recognized as a long-term social process that requires learning, innovation and adaptation (Berkes and Folke, 1998; Berkes et al., 2007).

During the late 1960s and early 1970s, C.S. Holling, Carl J. Walters and others at the University of British Columbia, Canada initiated the first field investigations into applying an AM approach in real-world natural resource management efforts. Based on these experiences, Holling was the first to define

AM in the scientific literature, calling it 'adaptive environmental assessment and management' in his book *Adaptive Environmental Assessment and Management* (1978). In this book, Holling describes how AM can be applied within natural resource management, and emphasizes how management decisions and policies should be viewed as hypotheses, and that uncertainty can serve as an opportunity for learning. Based on such learning, Holling recommended that research findings from applied scientific study of natural systems should be explicitly incorporated into decision-making and future management actions and policies. In this way, Holling and others viewed AM as a continuous method for management to be revised and strengthened, despite uncertainty and even if opposed to public opinion.

Holling's seminal work of the 1970s was subsequently built upon during the 1980s and 1990s. Walters (1986) wrote a book entitled *Adaptive Management of Renewable Resources*, describing (p. 9) 'adaptive management' as beginning '… with the central tenet that management involves a continual learning process that cannot conveniently be separated into functions like research and ongoing regulatory activities, and probably never converges to a state of blissful equilibrium involving full knowledge and optimum productivity'.

This challenge to conventional natural resource management brought into question many of the basic assumptions that underwrote accepted practice at that time. The concept of adaptive management was subsequently adopted by a few natural resource management agencies in the USA (e.g. Lee and Lawrence, 1986).

Based on this early experience, in 1993 Kai N. Lee brought the concept of AM into mainstream discussions between natural resource managers and conservation professionals through his book *Compass and Gyroscope* (Lee, 1993). In it (p. 6), he uses the compass and gyroscope as metaphors for the approach, in which science, when '… linked to human purpose is a compass, a way to gauge directions when sailing beyond the maps', and where democratic policy making is a gyroscope used '… to maintain our bearing through turbulent seas'.

In this regard, policies should be seen as experiments to be evaluated and learned from (Lee, 1993). The approach also promoted greater public participation and thoughtful discourse in policy issues and decision-making. As a critical part of this approach, and to ensure public trust in the process and outcomes, managers would need to embrace *empiricism* as the guiding principle of their work, thereby requiring management decisions to be based on observation and experimental evidence, rather than a priori reasoning[1]. Bormann et al. (1994) emphasize the need for identifying and measuring appropriate indicators that are accurate and sensitive enough to detect changes occurring in natural systems as a result of management interventions, suggesting the potential for adaptive management efforts also to serve predictive purposes.

By the late 1990s, it was becoming more widely recognized and accepted that such a transformational shift in how natural resource management actions were conceived and approached was necessary in order to curb the tide of rising environmental change and destruction across the global, even if doing so would be likely take decades to become established as common practice (Gunderson and Holling, 2002).

Adaptive management theory

The AM approach is built upon a number of shared premises: (i) natural systems are characterized by persistent complexity and uncertainty, which must be planned for; (ii) many questions we have about the natural world can only be addressed through observation and

---

[1] Knowledge gained through a priori reasoning does not require experience, and can be derived through reason alone. In contrast, under an empirical approach, knowledge is dependent on and gained through experience, where theories and hypotheses must be independently tested against observations made in the real world or validated through experimentation.

experimental evidence; (iii) our understanding and knowledge of the natural world is inherently limited, and we will never have complete information from which to take action; (iv) experience and knowledge that is gained is not necessarily documented or shared, and so can be lost; (v) some of what we think that we know may in fact be incorrect, and while we are unaware of this we will proceed to use our inaccurate or simplified understanding as a basis upon which to take action (Holling, 1978; Walters, 1986; Salafsky et al., 2001).

The deliberate, experimentally designed 'learning by doing' process used under an AM approach is thought to be in direct contrast to a random or opportunistic 'trial-and-error' approach to learning characterized by non-AM approaches in natural resources management (Holling, 1978; Walters and Holling, 1990; Lee, 1993, 1999; Berkes et al., 2001; Salafsky et al., 2001). Similarly, distinctions are made between active versus passive approaches to AM, whereby active AM approaches are focused on regular, systematic learning and improvements in management effectiveness made through real-world experimentation and observational evidence, as opposed to passive AM approaches that consistently incorporate new learning into management efforts when such learning is made available, such as through secondary sources of information or predictive modeling (Walters and Holling, 1990; McCarthy and Possingham, 2007).

Berkes et al. (2007, p. 328) note that a prevalent characteristic defining the evolving practice of AM is '… complex systems thinking, with an attention to scale, self-organization, uncertainty, resilience, and other characteristics of complex systems'. The theoretical intention of using an AM framework to guide natural resource management efforts within complex systems generates three sets of results: (i) improved effectiveness of management actions taken (e.g. Hockings et al., 2006); (ii) reduced uncertainty and enhanced resilience (e.g. Walters, 1986; Folke et al., 2002; Berkes et al., 2007); and (iii) empirical learning leading to knowledge that improves our understanding of how our management and policy interventions affect natural systems (e.g. Lee, 1993; Salafsky et al., 2001; Folke et al., 2002; ICSU, 2002). With these results, there should be a better chance of achieving management objectives. By employing an AM approach, in theory uncertainty actually becomes a tool for focused learning, rather than merely being a challenge.

Given its continuous, iterative nature AM is often described and visualized as a *cyclical feedback loop* that narrows and focuses management as knowledge builds and uncertainty shrinks with each 'loop' in the cycle. Over the past decade, several AM guides and other materials aimed at natural resources managers and conservation professionals (including fishery managers) have been published (e.g. Salafsky et al., 2001; Stankey et al., 2005; CMP, 2007; Williams et al., 2009). There are several examples of the generally similar, cyclical illustrations of the iterative AM process steps (USDA, 1994; Murray and Marmorek, 2003; CMP, 2007; TNC, 2007; WWF, 2007b).

Adaptive management applied in current practice

During the last two decades, AM has been used in many countries and is today widely embraced as a fundamental concept in natural resources management. Successful applications of AM are well documented in, for example, fisheries (e.g. Berkes, 2006; Wilson et al., 2006; Raakjær Nielsen et al., 2007), forestry (e.g. Bormann et al., 1996; Murray and Marmorek, 2003; Marmorek et al., 2006; Wilson et al., 2006) and wildlife management (Williams et al., 1996, 2009; Johnson and Williams, 1999; Varley and Boyce, 2006; Nichols et al., 2007).

While AM is no longer a new concept, the systematic application of AM in fisheries management is relatively recent. Because of this, there is both increased interest and learning regarding how to best apply AM in fisheries. Wilson et al. (2006) illustrate how, when fisheries management collaboratively works across local and national administrative scales, such cross-scale linkages can not only empower an AM approach to managing fisheries, but that government agencies can serve a valuable role in balancing and moderating agendas and interactions across these scales in ways that would otherwise not be possible. Recent experience highlights the benefits of taking an AM approach to fisheries management in that the emerging cross-scale institutional linkages under a collaborative framework promote management transparency, encourage passage of appropriate fisheries legislation and empower civil society engagement (Wilson *et al.*, 2006).

A Practical Approach in Using Adaptive Management

With a thorough conceptual development completed and active implementation now under way, the process of using AM has become clear and practical. This is good news for small-scale fishery managers, who not only must manage highly complex ecological and social systems, but are also practising within a specialized field of natural resource management that is still in relative infancy and often operating without sufficient scientific information on the near-shore marine ecosystems and coastal habitats typically utilized by small-scale fishers. Such complexity, coupled with the high uncertainty associated with fisheries management, makes the adoption of a practical approach to AM appealing and useful.

While the specific language used and number of steps cited varies by expert and publication, generally speaking most guidance provided on the steps in taking an AM approach follows a similar process and recognizes four broad steps (Table 6.1). These four steps are: (i) develop a plan; (ii) take action; (iii) evaluate progress; and (iv) adjust future action.

Each of these four steps has a recommended set of tasks that should be completed before moving on to the next step. Upon completion of these four steps, one cycle of AM process has been achieved, returning to step one with the cycle beginning again. This iterative process continues through time, with the aim of improving outcomes. A summary description of each of the four AM steps and associated tasks follows.

Developing a plan (Step 1)

The first step for managers in applying the AM process to SSF is developing an adaptive management plan that is realistic. Existing fisheries management plan, legislated regulations or traditional management practices can all serve as a starting point for this first step. In some

Table 6.1. Similarity between steps outlined in this chapter for AM of SSF (first column) and equivalent steps outlined within four examples from the established AM literature.

This chapter	USDA (1994)	CMP (2007)	Margolius and Salafsky (1998)	TNC (2007)
1: Develop a plan	1: Plan	1: Conceptualize. 2: Plan actions and monitoring	A: Develop conceptual model B: Develop management plan C: Develop monitoring plan	1: Define your project 2: Develop strategies and measures
2: Take action	2: Act	3: Implement actions and monitoring	D: Implement management and monitoring plans	3: Implement strategies and measures
3: Evaluate progress	3: Monitor	4: Analyse, use and adapt	E: Analyse data and communicate results	3: Implement strategies and measures
4: Adjust future action	4: Evaluate	5: Capture and share learning	Iterate: Use results to adapt and learn	4: Use results to adapt and improve

cases, existing laws, plans or practices may actually contain most or all of the information and guidance necessary to complete the associated tasks under this step. In such cases, managers may simply need to collate, organize and review such existing sources under the auspices of completing the necessary tasks under this step. To develop an adaptive fisheries management plan and complete Step 1 of the process, six tasks must be completed in sequential order, as described below.

Task 1(a): Engaging fishers and other stakeholders

It is generally accepted that the collaborative process, through which *stakeholder groups* are engaged and invited to be actively involved in adaptive management planning, is critical to long-term success (CMP, 2007; TNC, 2007; Williams *et al.*, 2009). Stakeholder groups are sets or collections of individuals having distinct interests and/or influences over the resource(s) in question. Stakeholders may also have something to gain or lose through a specific management action. Stakeholder groups are important to understand because such interests or influences may positively or negatively impact the health of the resource(s) in question. Fishery-related stakeholders may hold the most in-depth knowledge and expertise regarding the status and trends of certain fish stocks, particularly when scientific information and/or existing fisheries data are poorly documented or not available. In such instances, a manager must learn how effectively to access the experience and expertise of knowledgeable fishers and fishery stakeholders (Berkes *et al.*, 2001). Engaging stakeholders both early on and regularly throughout the adaptive management process can not only provide a wealth of information and useful ideas and perspectives to managers on how best to manage a particular fishery, but may also result in encouraging their support and buy-in on the implementation of your adaptive management plan later (see Step 2). Doing so may also help to avoid or reduce conflicts with stakeholder groups later on during the adaptive management planning process.

Managers should complete a *stakeholder analysis* to identify the range of stakeholders with an interest in or influence on the small-scale fishery in question. Stakeholder analysis is a participatory process used to identify, characterize and distinguish the type and level of influence and/or interest each group has on the fishery. The results of stakeholder analyses can be useful for managers, not only in identifying and understanding the influence and/or interest of each group, but also in terms of prioritizing which stakeholder groups must be consulted carefully and closely (primary stakeholders) versus those groups that would be useful to engage but are less critical, or are intermediaries and so themselves do not directly influence (secondary stakeholders). Practical guidance on how to conduct a stakeholder analysis is found in WWF (2005b) and Evans and Andrew (see Chapter 3, this volume, for references to the primary literature).

Managers can initially engage and gather information from stakeholders through *key informant interviews* with knowledgeable individuals or during *focus group interviews* with selected representatives from stakeholder groups, and/or from meetings of multiple stakeholder groups. Feedback can be solicited and documented on the issues or problems perceived relating to the small-scale fishery of concern, as well as suggested possible management solutions. Practical guidance for managers on the design and specific social survey methods used in such stakeholder engagement is found in Bunce *et al.* (2000).

Task 1(b): Describing the current situation in the fishery

As Holling (1978) and Walters (1986) argue, because some of what a manager thinks that they know about a fishery may in fact be incorrect, before proceeding to taking action using an inaccurate or simplified understanding of what is happening within the fishery it is important first, explicitly and openly, to describe what the believed current situation in the fishery is.

To begin this process, the manager should research, review and summarize all relevant *secondary data*, or existing information, relating to the small-scale fishery in question. In some cases, secondary data may be incomplete, inaccurate or out of date. However, a documented summary of all existing and relevant information can serve as an important starting point for better understanding and description of the fishery in question. Once the manager has reviewed all relevant secondary data that could be found, s/he should summarize the results of this information into a background document that serves as a 'review and synthesis' of what is known about the fishery based on secondary data.

Once secondary data have been reviewed and summarized, the fishery manager should now invite knowledgeable fishers and other priority stakeholders, along with representatives from fishery management partners (e.g. government agencies, concerned community groups), to come together and complete a *situation analysis* (see WWF, 2006a). The purpose of this activity is to discuss and agree upon what the status and trends are within the fisheries of concern, including priority issues that need to be addressed. As a starting point in this process, key findings from the 'review and synthesis' document of existing secondary data should be shared with stakeholders and partners. Similar scoping or diagnostic phases are described in the fisheries literature (e.g. Andrew *et al.*, 2007; see Evans and Andrew, Chapter 3, this volume for further discussion).

When initiating a situation analysis, two critical elements must first be identified and defined: management targets and threats. Management *targets* are the specific ecosystems under management (e.g. a coral reef ecosystem, including associated mangroves, or seagrass beds), communities (e.g. the reef fish community living on a reef crest and fore-reef slope) and/or species (for example, all parrotfish species or certain surgeonfish species on a fore-reef slope) that will serve as the focus for management efforts under the SSF management plan (CMP, 2007; TNC, 2007; WWF, 2009). Identifying a reasonably limited number (between 3 and 6) of the highest priority target species, communities and/or ecosystems to focus on is important, because these targets will not only serve as the basis for setting fishery management goals and carrying out certain management activities, but will also be the ultimate measure of how effective the fisheries management plan is. Based on secondary data and knowledgeable stakeholder and partner feedback, the viability of each target (i.e. the ability of the target to persist through time and remain resilient in the face of stressors, sometimes thought of as 'health') should be assessed by the manager, including whether or not the target's current state of health is in decline, stable or improving. Guidance on defining targets and methods for conducting in-depth assessments of target viability can be found in TNC (2007) and WWF (2009).

Once priority targets have been identified and assessed, the manager should develop a list of *threats*, which are factors that negatively affect the priority management targets. Threat factors can be divided into either those that are direct (that immediately degrade or destroy a target; for example, dynamite fishing and overharvesting) or indirect (that cause, underlie or exacerbate direct threats; for example, the need for dietary protein and human population growth). In an effort to identify, define and list all known direct threats operating globally, the International Union for the Conservation of Nature and the Conservation Measures Partnership released the *Unified Classification of Direct Threats* (IUCN, 2006a; Salafsky *et al.*, 2009). The manager should begin by identifying and defining a list of direct threats that are affecting priority targets. Next, direct threats should be prioritized through a threat rating process (e.g. each rated by scope, severity and irreversibility; WWF, 2006a; FOS, 2009). In this way, a limited number of the most critical direct threats can be identified and focused upon within the fisheries management plan. Once a limited set of the most critical direct threats has been developed, a list of the indirect threats (root causes) should be identified that drive or influence the direct threats. Guidance on how to identify and prioritize direct and indirect threats can be found in CMP (2007), TNC (2007) and WWF (2007a), all available free online.

With the priority targets and critical threats now identified, the next step of the situation analysis is to develop a *conceptual model*. A model is a simplified but plausible representation of the dynamic natural world (Williams *et al.*, 2009); a conceptual model may be a visual representation of a set of relationships between different factors that are believed directly or indirectly to affect the management targets (Margoluis and Salafsky, 1998; WWF, 2005a; CMP, 2007; TNC, 2007; FOS, 2009; Margoluis *et al.*, 2009). A useful conceptual model is one that illustrates the commonly shared understanding of how various factors are assumed to relate, or 'link', to one another. Practical guidance for managers on how to lead a group through the development of a conceptual model may be found in WWF (2005a) and FOS (2009).

There are a number of reasons why conceptual models are a useful starting point for fishery managers (FOS, 2009), including the fact that they clearly and visually reflect the group's assumed cause-and-effect relationships that are to be systematically tested under an AM approach (Margoluis *et al.*, 2009). Root causes analysis (WWF, 2000) and brainstorming (WWF, 2006a) are two additional approaches in conducting a situation analysis that may complement or support conceptual modeling.

Task 1(c): Defining management goals and strategies

Once a manager has completed a situation analysis, s/he should review the priority targets and related direct threats within their conceptual model. For each priority target, the manager should define a *goal*. A goal is a formal statement of the ultimate, desired state that is to be achieved for a priority target under the management plan (CMP, 2007). Goal statements should be simple enough to be easily understood and ambitious enough to be easily recalled, but realistic and specifically linked to a priority target (or targets) and measurable within a defined period of time (WWF, 2006b).

Once a limited number of goals have been defined encompassing all priority targets, the manager should identify management *strategies* to achieve these goals. Strategies are broad courses of action taken to achieve stated goals by abating or reducing critical threats (TNC, 2007; see also IUCN, 2006b; Salafsky *et al.*, 2009 for discussion of strategies). In this sense, strategies are designed to intervene on direct or indirect threats in order to end or limit their impact on priority targets. In some cases, more than one strategy may be necessary to intervene sufficiently on a threat or set of threats.

Strategies are composed of multiple, associated management activities that work together to reduce threats. In order to be practical, each strategy must be: (i) focused around a specific set of actions linked to a specific set of threats; (ii) feasible given human and financial resources available and other operating constraints; and (iii) appropriate given the cultural, social and political norms and biological setting (WWF, 2006b; CMP, 2007). Salafsky *et al.* (2002, 2009) and IUCN (2006b) identify seven categories of conservation strategies through which management activities (fisheries or otherwise) occur: (i) land and water protection; (ii) land and water management; (iii) species management; (iv) education and awareness; (v) law and policy; (vi) livelihood, economic and other incentives; and (vii) external capacity building.

Once a set of management strategies has been identified to achieve the stated goals, it may be useful to share and discuss these with partner organizations and selected stakeholders to obtain their review and feedback. Guidance on identifying and defining goals and strategies can be found in WWF (2006b, 2007b), CMP (2007) and TNC (2007).

Task 1(d): Outlining the shared assumptions to be tested

Stem *et al.* (2005) argue that taking an AM approach primarily differs from other management frameworks in that it requires managers to adhere rigorously to systematically testing whether or not assumptions are valid regarding how specific management actions will reduce or eliminate threats facing targets. To test the *underlying assumptions* regarding why certain strategies are believed to abate certain threats, managers must be explicit during the planning process about the causal results that they assume will follow from implementation of the strategy in question. Following a strategy's full and successful implementation, a logical 'chain' of cause-and-effect results is assumed by the manager (either explicitly or implicitly) to occur. Sometimes, managers proceed with implementing a management strategy without explicitly defining their underlying assumptions of how the results of the implemented strategy will result in abating a specific threat. Under an adaptive management plan, not only must the assumed chain of cause-and-effect results be clearly stated, but these must also be measured in order to test whether or not the assumed result actually occurs and, if so, to what degree. Managers are encouraged not only to state, explicitly and openly, their assumptions, but to do so as *testable hypotheses* that can serve as experiments through which managers can learn (Holling, 1978; Walters, 1986, 1997; Gunderson *et al.*, 1995) in order to inform and empower future decision-making (Lee, 1993).

One practical tool to help managers think carefully through and document this thought process is the results chain. A *results chain* is a tool that visually represents *how* a manager believes actions taken under a particular strategy will lead to a series of desired results that incrementally reduce a critical threat affecting a priority target (FOS, 2007). Results chains are derived from conceptual models, but differ in that while conceptual models are a visual representation of the world prior to management action, results chains represent the world after strategies have been implemented and action taken. A results chain should be developed for each pairing of a specific strategy with a specific direct threat that the strategy will address. Results chains are *not* a visual representation of the activities or steps taken in implementing a strategy; in fact, results chains assume that all steps taken in implementing a strategy have been completed. Rather, a results chain is made up of a sequence of incremental, desired results that are assumed will occur and can be measured once a strategy is implemented. A results chain outlines the assumed 'if ... then' logic that causally links the implemented strategy to successive intermediate results, through to a desired *threat reduction result* or measurable statement of the degree to which the direct threat will be reduced or eliminated as a result of the strategy's successful implementation. For results chains to be useful, sound and successive logic from each intermediate result to the next must be clear and defensible, demonstrating how change will occur and be observable. Ideally, results chains should be relatively simple, while also being reasonably complete, from strategy through at least one or two intermediate results, to the threat reduction result (direct threat) and target (FOS, 2007).

As part of the results chain process, *indicators* are identified as proxy measures as to whether or not the assumed changes (e.g. incremental results) are actually occurring along the chain, particularly in terms of the predicted threat reduction. An indicator is a unit of information that points to, describes the state of or provides information about a specific environmental or social factor operating within the observed system (Hockings *et al.*, 2006). If consistently and reliably measured over time, indicators allow managers to observe and document changes occurring in factors within the system over time (Pomeroy *et al.*, 2004; Hockings *et al.*, 2006). Walters (1986) and Bormann *et al.* (1994) emphasize how indicators form an important learning cornerstone of the entire adaptive management process. Under the SSF plan developed, they will serve as primary metrics to test whether or not the strategy has effectively led to the assumed changes along the chain, and either: (i) if so, the degree to which incremental results have been achieved and thus the strategy's effectiveness; or (ii) if not, whether the strategy was simply ineffectively implemented or perhaps the underlying assumptions flawed or erroneous regarding how the strategy would theoretically lead to the desired change. In this regard, the causal assumptions between incremental results in the results chain become testable hypotheses regarding the manager's 'theory of change' that can be measured through time, much like an experiment (FOS, 2007; Salafsky *et al.*, 2009).

Task 1(e): Defining the fishery management objectives and strategic activities

In the same way that management goals are linked to the priority targets within the conceptual model, the fishery management objectives are linked to key results within result chains (FOS, 2007). An *objective* is a formal statement detailing a desired result that is to come from management action taken in support of achieving a specific goal (WWF, 2006b; CMP, 2007). A useful objective is one that is stated to be SMART: that is, Specific, Measurable, Achievable, Relevant and Time-limited (Pomeroy *et al.*, 2004; TNC, 2007; Williams *et al.*, 2009). Goals and objectives differ greatly. Goals are broad and ambitious, whereas objectives are narrow and specific. Goals speak to the ultimate outcome in terms of the health of priority targets, whereas objectives focus on intermediate results in terms of reducing or eliminating threats affecting the targets. Writing useful management goals and objectives requires practice, so peer review and feedback may be very useful in the development of a new fisheries management plan that largely does not build from previously specified efforts.

Results chains can assist managers in defining a set of specific objectives logically to accompany a given management goal. Typically, for any given results chain there is one goal (relating to the target), usually one strategy and at least two to four objectives (relating to the most important intermediate results along the chain), one of which links to the threat reduction result (relating to the direct threat). Each objective should therefore logically relate to the strategy being employed. Once the manager identifies at least two and no more than five objectives linked to the results chain and associated with the specific goal, these objectives and their results chain should be shared with management partner representatives for peer review, and ideally shared with fishers and priority stakeholders to solicit feedback and group discussion.

Once a final set of peer-reviewed objectives has been defined under each goal, a set of strategic management *activities* must be identified which, if completed, should lead to full implementation of the strategy and thus the assumed causal achievement of each objective. A range of possible activities should be evaluated and rated against the following three criteria: (i) the level of benefit derived by completing the activity, in terms of scale, degree of contribution, duration of outcome and leverage; (ii) the overall feasibility of completing the activity, in terms of time, institutional support, motivation of key stakeholders and ease of implementation; and (iii) cost, in terms of staff time, supplies and equipment required and number of years to complete (TNC, 2007). Once all activities have been rated, they can be sorted by priority. Prioritized activities should be reviewed to ensure that they logically relate back to the strategy in question and cumulatively are sufficient to achieve full completion of the stated objectives. Priority activities should be organized by the manager within an activity work plan, specifying *who* (responsible party) will ensure the activity is completed, *what* resources will be required to complete the activity, *where* the activity will occur and *when* it will be completed over what time frame. Guidance on developing management objectives and activities is detailed in WWF (2006b, 2007c), FOS (2007) and TNC (2007).

Task 1(f): Defining how the effectiveness of the plan will be measured

By this stage, the manager will have identified a proposed set of small-scale fishery management goals, strategies, objectives and activities, all logically tied to a completed situation analysis (conceptual model) and sets of results chains, and with stakeholder consultation and input throughout the process. The final task in developing the plan is to define how management will be evaluated (Pomeroy *et al.*, 2004; Hockings *et al.*, 2006). Regular monitoring of the plan's effectiveness is critical to learning whether or not implemented strategies are having their assumed affects and, if not, whether strategies need to modified. Management effectiveness assessment differs from status assessment. The measurement of the status of a particular target (e.g. the abundance of bivalves in a seagrass meadow) differs from the measurement of the relative effectiveness of management efforts to protect the target (e.g. the degree to which newly implemented size/class restrictions in the collection of these bivalves are being obeyed by community residents).

Periodic *monitoring* of the plan's effectiveness can be achieved through measuring key indicators as specified under completed results chains (see Task 1d). Monitoring progress and evaluating results (see Step 3) for the purpose of adaptation is a minimum requirement for fishery managers working under a high degree of uncertainty (Andrew *et al.*, 2007). These indicators should be tied to the measurement of the threat reduction result and those key intermediate results that are linked to management objectives. Methods of measurement should be accurate, reliable, cost-effective, feasible and appropriate (Margoluis and Salafsky, 1998; TNC, 2007). Each indicator should then be checked to ensure that it is measurable (able to be recorded in quantitative or discrete qualitative terms), consistent (methods of measurement do not change over time), precise (defined to ensure that its meaning is the same to all people) and sensitive (changes proportionately in response to actual changes in the item being measured) (WWF, 2005c; TNC, 2007; see Evans and Andrew, Chapter 3, this volume). Wilson *et al.* (2006) advise that fisheries management indicators should be simple, cost-effective, relevant and linked to management objectives, and reflect the understanding and interests of stakeholders. Walters (1986) suggests that the manager 'bound' the system being monitoring to the current shared understanding of the system (as illustrated through conceptual models and results chains), including known uncertainties, rather than focusing on investigating measures outside of the system's current understanding.

Priority indicators should be organized within a *monitoring work plan*, within the larger management activity plan developed from Task 1(e). For the monitoring work plan to be useful, it should specify *who* will ensure that indicators are measured in a timely and reliable manner, *what* resources (including skill sets, equipment and funding) will be required to complete periodic measurement efforts, *where* monitoring will occur and *when* each indicator will be measured, and over what time frame. Periodic monitoring of the plan's administrative progress should also be outlined under the monitoring work plan. To learn more about the three levels of administrative monitoring that should be reflected within the monitoring work plan, see Task 3(a). Guidance on developing monitoring plans is detailed in WWF (2005c) and TNC (2007).

Taking action (Step 2)

Once the plan is in place, the next step is to initiate positive action. The level of time and effort required effectively to complete this second step is often underestimated by managers, and sometimes completely overlooked. As a result of this, an ineffective or inefficient start to the management plan could result, or indeed even plan failure. Therefore, careful thought and sufficient time and effort must be invested in the process of implementing the new management plan. Key elements in how to do this are outlined under the following five tasks.

Task 2(a): Securing resources and permissions to implement the plan

Prior to taking any action, human and financial resources need to be gathered to implement the management and monitoring plans. Based on the outputs of Step 1, the manager should have an understanding of the skills and funds required. The manager should differentiate between which of these human resources require funding and which do not. Volunteers and no-cost (in-kind) partner contributions should be secured, as appropriate. Based on the technical difficulty and requirements, the manager should also determine the level of skills and training necessary. Where training and skills development is not an option, the manager may need to secure outside technical assistance. Because citizen involvement is essential during the adaptive management process (Lee, 1993; Murray and Marmorek, 2003), stakeholders should be engaged and encouraged to participate in the implementation of the adaptive management plan, as identified in Task 1(a).

Based on the list of human resource needs, coupled with the outputs of Tasks 1(e) and 1(f) regarding equipment, supply and other costs, the manager must next estimate the financial resources required to implement the plan. Activity costs (including monitoring) under the plan should be tallied by objective, so that the manager can understand which objectives will require the most financial resources. Based on the estimated total costs, the manager should now develop an estimated annual or monthly budget to implement the plan, reflecting both currently available and secured future funding (income) and the total cost of the plan's implementation (expenses) through time. This budget should differentiate between one-off or infrequent costs (such as equipment purchases) versus ongoing costs on an annual or monthly basis. If income levels are insufficient to cover total projected expenses, then before proceeding with plan implementation, additional financial resources will need to be secured or some of the proposed activities temporarily suspended until adequate resources can be found. Proceeding without sufficient funds to implement the plan should not be considered an option. Guidance on creating budgets is detailed in WWF (2007c).

In addition to addressing the required human and financial needs for the plan, managers should also consider whether or not there is a need to secure any necessary permits, local community and governmental approvals or other activity permissions required to implement activities under the management plan.

Task 2(b): Forming an implementation team

The next task is to create an implementation team, whose primary role is to ensure an effective completion of stated activities under the management plan. The recommended size for an implementation team is between three and six members. Team members should include the manager and ideally at least: (i) one representative from relevant management partners (such as a government agency with fisheries management authority or non-governmental organization with relevant natural resource management interests); (ii) one fisher who is representative of the small-scale fishery in question being targeted for management; and (iii) one community representative, community leader and/or representative from a non-fishery priority stakeholder group(s) as identified from the completed stakeholder analysis under Task 1(a).

The reasons for creating an implementation team rather than a manager implementing on their own include: (i) to ensure the burdens and responsibilities of implementation are shared among relevant parties and individuals, rather than being placed on a manager alone; (ii) to encourage multiple opinions, perspectives and sources of review on all aspects of implementation in order to avoid or minimize process mistakes, faulty logic or missed details during the implementation of the plan; (iii) to reduce the dependence on a single individual; and (iv) to provide the necessary level of energy required to make the management initiative sustainable in the long term.

Task 2(c): Implement monitoring of work plan and collecting baseline data

Focused biological and social monitoring should be done before the interventions in order that *baseline data* (information on the targets and threats prior to intervention) can be used to assess the performance of the management strategy following implementation. Collecting adequate baseline data may require months of time and effort. This task is one of the most crucial steps in the adaptive management process, because comparison of pre- and post-intervention data will inform decisions as to whether or not strategies are effectively intervening on threats as predicted based on assumed causality, and as such how management efforts should be adapted through time under the plan. As Williams *et al.* (2009, p. 51) emphasize, because '... success in adaptive management ultimately depends on effectively linking monitoring and assessment to objective-driven decision making' and effective implementation of monitoring efforts will allow managers to '... assess system models, update their confidence measures, and reduce system uncertainty'.

Baseline data should be collected for those indicators and methods identified from Tasks 1(d) and 1(f). In some cases, this may require collection of both biological and social information. For example, a baseline survey to determine the target viability of a stock of bivalves might require several months of both relative abundance and size/class distribution estimation within seagrass meadows where management activities are to be focused, as well as periodic interviews with fishers to characterize harvest levels and average sizes of clams being landed.

As part of an experimentally designed adaptive management plan, baseline and ongoing monitoring data should be collected both at *treatment and control sites* so that changes through time that are attributable to the influence of an intervention (treatment site) can be systematically compared with changes observed that are occurring independently of management efforts (control sites). Such paired data should be collected from control sites within the same geographic area and biological and/or social characteristics as treatment sites, and across a sufficient number of *replicates*, or repeat sample areas, within both treatment and control sites. While time consuming and challenging, commitment to this experimental design is a central requirement in taking an adaptive management approach. Success of the adaptive management planning process is in part dependent upon the manager's commitment to involving controls and replication within the management process (Murray and Marmorek, 2003; Stankey et al., 2005). The ultimate value of monitoring data collected through an adaptive management approach is in its ability to contribute to adaptive decision-making (Williams et al., 2009).

Task 2(d): Implementing activity work plans

Once baseline monitoring has been completed and the monitoring plan activated, work can begin on the activity work plans determined in Task 1(e). Some activities can occur at certain times or be constrained by natural events (e.g. seasonal cycles, tidal fluctuations and life history patterns of specified targets). Social drivers such as fishing behaviour, partner availability and stakeholder events or holidays will also constrain the time frame for when things can get done.

Contingency plans may be needed to maintain flexibility during implementation so that management activities can continue despite complications or unforeseen disturbances.

Task 2(e): Initiating a communication plan

Because of the importance of consistent stakeholder engagement and involvement in adaptive management planning, particularly with respect to adaptive decision-making, it is critical to implement a communications strategy that is focused on specific *target audiences* (intended recipients of communicated information) around key *messages* that are effectively internalized by target audiences and encourage dialogue or new behaviour (see also McConney and Haynes, Chapter 10, this volume).

Evaluating progress (Step 3)

By this stage, the management plan should be fully implemented. The implementation team should have all necessary baseline data collected, activities outlined in the work plan getting under way or fully implemented, and ongoing active communication with target audiences regarding the implementation of the plan. The next step in the AM process is to evaluate the performance of the plan. Progress evaluation must be done at two levels: (i) evaluation of administrative progress of the plan's implementation against the annual work plan and budget; and (ii) evaluation of the effectiveness of the plan in meeting its stated management goal(s) and objectives. Key elements in how to do both levels of evaluation are outlined under the following four tasks.

Task 3(a): Regular evaluation of the plan's administrative progress

The time and effort required effectively to check on the plan's administrative progress is often underestimated by managers. The plan's *administrative progress* is the degree to which the completion and timing of implemented activities and expenses agree with the stated work plan, budget and deliverables. If progress is sound and on schedule, there may be a tendency for less attention and energy to be allocated towards regular evaluation until problems arise that could have otherwise been identified early and avoided. If progress against the plan is not proceeding well, this may be in part due to a manager not having carefully evaluated the plan's progress through time and who thus may have missed early signs that the plan was off track or heading into trouble. Either way, insufficient attention to periodic evaluation of the plan's administrative progress could result in ineffective implementation or even failure.

The administrative progress of the plan's implementation can be evaluated at three levels, as outlined under the monitoring work plan generated from Task 1(f): (i) an end-of-month review of monthly progress made against each activity outlined under the monthly plan; (ii) a quarterly review of progress on the annual work plan to monitor the timing and thoroughness of completed activities during the quarter, production of required deliverables and outputs from completed activities, and expenditure rate and totals against the annual budget; and (iii) an end-of-year review of annual progress made against sets of activities outlined under each of the plan's objectives within the annual work plan, and review and assessment of the level of progress and completion against required outputs, stated activity milestones and annual budgets (see WWF, 2006b, 2007c).

Task 3(b): Periodic evaluation of the plan's management effectiveness

As stated previously, periodic monitoring and evaluation of the effectiveness of the implemented plan in achieving its stated goal(s) and objectives is one of the most critical tasks in taking an adaptive approach to SSF management (Andrew *et al.*, 2007). Building from the completion of Tasks 1(f) and 2(c), the periodic collection of biological and social data for selected indicators using the specified methods and within the designated time line under the monitoring work plan must be completed under this Task. Full discussion and guidance on the need for and step-wise process of evaluating *management effectiveness* may be found in Pomeroy *et al.* (2004) and Hockings *et al.* (2006). Pomeroy *et al.* (2004)

highlight six general steps in evaluating management effectiveness: (i) select relevant indicator; (ii) identify an evaluation team; (iii) develop an evaluation activity work plan; (iv) collect, manage and analyse data (see Task 3c); (v) share evaluation results with target audiences (see Task 3d); and (vi) adapt management practices as needed (discussed under Step 4).

The *evaluation team* should include the specified responsible parties under the monitoring work plan from Task 1(f). However, it may be useful to invite and include other partners on to the evaluation team, including scientific experts such as fishery biologists and social scientists. Consistent with the participatory approach outlined under Task 1(a), creation of an evaluation team should be viewed by managers as an opportunity to involve priority stakeholders who may not be represented within the implementation team but who are either primary beneficiaries of the plan or perceive themselves somehow to be adversely affected by the management plan (Pomeroy *et al.*, 2004; WWF, 2005c; TNC, 2007).

Task 3(c): Managing and analysing monitoring data

As monitoring data are periodically collected through time at control and treatment sites, these data must be carefully handled and protected for future retrieval and use in analysis. The process of doing so is called *data management*. Data management includes several discrete and important tasks, most notably data compilation, cleaning (review and correction when in error), coding, entry, organizing and storage within a database, and maintenance and safeguarding (e.g. data storage back-up) during future tasks that use the data such as retrieval, export and analysis (see below). Data management is a critical but frequently underestimated task. A member of the evaluation team should be designated to serve as the data manager, who will be responsible for receiving, handling and managing the data collected for each indicator (Pomeroy *et al.*, 2004). The data manager should clearly understand and be prepared to handle the full range of data as they are collected by the evaluation team, both in terms of their form (written, digital file, audio recording, image, etc.) and type: *quantitative* (numerical information in continuous form), *qualitative* (textual information or non-continuous numerical representation of textual information, such as rankings or ratings) or graphical (Pomeroy *et al.*, 2004).

Practical guidance regarding specific tasks of data management, including data cleaning, coding, entry and storage, is included in Pomeroy *et al.* (2004). Baseline data collected under Task 2(c) should be managed at this stage, even if the first round of data collection for management effectiveness evaluation purposes, as outlined under the monitoring work plan, is not called for until several months ahead.

Periodically, the monitoring work plan will call for an *analysis* to evaluate the management effectiveness of the plan's implementation. The period between evaluations will vary depending on the fisheries plan and specified objectives. However, as a general rule of thumb, comprehensive data analysis should not be necessary more than once every year, but no fewer than once every 3–5 years. The evaluation work plan should specify not only how frequently data are to be collected for each indicator, but also how frequently data analysis is to be conducted.

When the time comes to analyse the degree to which predicted changes assumed to occur as the result of the implementation of strategies and activities under the plan's objectives are being observed, comparison will be made between baseline conditions against changes or trends observed through time at both control and/or treatment sites. The implementation team should begin this process by reviewing the assumptions held and outlined as testable hypotheses under the management plan in Task 1(d), and then by *systematically evaluating* the degree to which each assumption is validated through data collected. Where continuous numerical data are available, quantitative statistical analyses may assist the implementation team in identifying and characterizing observed change. In other cases, non-parametric statistical analysis may assist in identifying correlations or patterns in qualitative information.

Task 3(d): Sharing evaluation results with target audiences

Once the plan's administrative progress and/or management effectiveness have been evaluated, the next task is to summarize and prepare analytical results for presentation and discussion with relevant target audiences, including government and non-government partners, donors and/or stakeholders. The appropriate form (e.g. digital slide show, printed document, summary poster, flyers and summary handouts) taken by exported analytical results should will depend upon the target audience and messages to be delivered (see also McConney and Haynes, Chapter 10, this volume).

In sharing evaluation results, target audiences should be encouraged to *interpret results* in such a way that they come to their own findings and conclusions, rather than being given the findings and conclusions as interpreted by the manager and implementation team, which will help the implementation team avoid projecting their biases or skewing interpretive opinion with target audiences. Given the participatory nature of adaptive management, evaluation results should be openly shared with target audiences to ensure transparency and accountability.

Adjusting future action (Step 4)

This final, critical step focuses on discovering whether predicted outcomes and causal assumptions were accurate or not, and in learning which activities lead towards achievement of desired objectives. Causal assumptions and predicted outcomes were previously identified from the conceptual model and results chains in Task 1(d). Where casual assumptions are not upheld and a case made for rejecting stated hypotheses, such findings may warrant review and modification of future management effort related to the associated strategy and activities, along with modification of original assumptions and creation of a revised hypothesis to be tested during the next evaluation cycle. In order to practise adaptive management, full and careful completion of the four tasks under this step must occur so that informed revision and refinement of management efforts through time are ensured based on iterative rounds of evaluation results.

Task 4(a): Systematic reviewing and checking of assumptions against evaluation findings

Using the evaluation results, the implementation team should invite partners and stakeholders, collectively, to revisit original assumptions and *systematically compare predicted versus observed changes* relating to each stated hypothesis. This process will require the manager first to organize evaluation findings by each stated hypothesis, and then present to the group the following:

- a list of the change(s) that was (were) originally predicted would occur as a result of implementing the management action relating to the stated hypothesis;
- what actual, observed change(s) was (were) measured and documented from the evaluation findings relating to the stated hypothesis; and
- a side-by-side summary of the predicted versus observed change(s) relating to each stated hypothesis.

As outlined by Murray and Marmorek (2003), once observed evaluation results have been compared with the predicted changes specified under each hypothesis, the next step is to identify the potential implications arising from these comparisons. The implementation team should carefully document the group's interpretation of these comparisons for future reference, through either written records or audio recording. The following questions may be useful in generating group discussion regarding interpretation of these comparisons:

- Has a sufficient time elapsed to conclude confidently that measurable change can be observed?
- Based on the evaluation results, does it appear that some hypotheses might be incorrect, and therefore should be rejected and/or reformulated?
- Are any hypotheses strongly supported by the evaluation results?
- Which hypotheses are neither strongly supported nor rejected by the evaluation results?

Measurable change occurring as a consequence of management action may take time to manifest and observe, even years. In such cases, the implementation team should not immediately dismiss their original assumptions or reject hypotheses simply because no measurable change was observed within a single evaluation time frame.

Task 4(b): Identifying and discussing implications for future management effort

Potential implications of changes to designing interventions and larger policy settings need to be discussed. The implementation team, management partners and primary stakeholders should now agree on what, if any, potential changes to future management efforts should be considered. The implementation team can elucidate such *management implications* by discussing the following questions (adapted from Williams et al., 2009).

- Is it clear how evaluation results are generally understood and interpreted by the group, particularly with respect to consensus regarding potential needs to modify assumptions and re-formulate hypotheses?
- Have thresholds of change (either predicted or not) in the operating conditions been set and reached?
- If the identified threshold for change has been reached, what implications are there for future management action?
- How should management actions be adapted to encourage improved effectiveness of the strategy and activities? Which management actions should be abandoned?
- What is the overall effectiveness of management efforts under the plan, by objective? How can progress be improved against certain objectives by adapting future management efforts?

Recommendations generated through group discussion regarding how to adapt future management practices should be carefully documented for future reference. In some cases, evaluation results may suggest no need for adaptation of current management practices. This may be because evaluation results are either inconclusive or too premature to provide clear guidance on how future management actions should be modified, and may require additional time and monitoring before clear evidence of the need to adapt management efforts can be recommended and warranted. Alternatively, this may be because no significant differences were observed between predicted and observed changes across stated hypotheses, in which case the evaluation results may agree with hypothesized changes, serving perhaps to validate and encourage the maintenance of current management efforts. In such instances, the implementation team should feel justified in continuing current management efforts, and in proceeding with iteration of the adaptive management cycle without completion of Tasks 4c and 4d.

Task 4(c): Capturing learning and adapting management efforts as necessary

Once the implementation team and management partners have discussed the implications and agreed what they mean for future management (Task 4b), the implementation team should formally capture this group learning by summarizing the key points of agreement into a concise and specific set of recommended management changes that require be made regarding future activities to be carried out under the plan. Such recommended changes should be clearly linked to specific management objectives and activities under the plan, and how these are to be adapted or modified. If a specific management activity or objective is to be rewritten or abandoned, clear and logical justification must be clearly evidenced between the monitoring data collected (all relevant time periods, including baseline), evaluation results generated (for a specific period of time following implementation of management activities), findings derived from the group comparison between predicted versus observed changes, and the implementation team and management partners' discussion regarding potential implications for future management.

Once such evidence has been outlined and a defensible case has clearly been made for modified future management effort, as discussed and agreed upon by the implementation team and management partners, the manager should now convene and inform the relevant management authorities and decision-makers of the findings generated from the evaluation results and subsequent recommended adaptation of future management efforts. In doing so, the manager should seek authority agreement and permission to proceed with recommended changes. In some cases, proceeding with recommended changes to management efforts may first require alteration of existing policies or practices of management authorities (Murray and Marmorek, 2003). In other cases, authority may have been devolved from an agency or decision-making body to the manager to make modifications to management efforts without formal review and approval by such authorities or decision-makers. Even if this is the case, an attempt should be made at least to share the implementation team's learning with authorities and/or decision-makers, and ideally to secure an opportunity to consult with them on the proposed changes that will be made to future management efforts, as recommended.

Once the necessary authorities and decision-makers have approved and provided the relevant permissions, it is time to *adapt management practice*. To get this process started, the manager and implementation team should agree upon the time line and steps necessary to implement the necessary changes and adapt management efforts. This will include modification of the existing activity work plan, but may also include modification of one or more management objectives under the plan, as well as the monitoring work plan. From here, it is now the manager's responsibility to ensure that such adapted management practices are not only implemented, but also maintained through time.

Task 4(d): Communicating adaptive response taken with target audiences

Once management practices have been adapted, agreed upon and approved by any necessary authorities, the final task in the adaptive management cycle is to communicate what changes and modifications to management efforts will be made in the future with relevant target audiences, including small-scale fishers, community groups and other local stakeholders. Key messages regarding why such changes are being made and how they will benefit local interests should be developed and effectively communicated with target audiences.

At this stage, the manager should also consider sharing the knowledge and adaptive learning generated from Steps 3 and 4 with other fishery managers and conservation professionals. This can be done not only within one's own country, but also regionally and internationally. Sharing results and adaptive management learning will not only benefit others, but may also help to put the manager and implementation team in touch with others who are engaged in similar small-scale fishery management efforts and learning from their own experiences. Such sharing of experience and knowledge can strengthen fishery management efforts in ways that could not be done alone.

Iteration

By this stage, the four steps of the adaptive management process should have been completed, marking the completion of one cycle in the AM process. Once Step 4 of each cycle is complete, the manager should return to Step 1 and initiate a new cycle. With each completion of a cycle, the management plan is being honed and strengthened, adaptively, and based on empirical evidence. Over time, this should increase the likelihood of effective management.

In the case where management practices must be adapted, the manager should lead the implementation team, management partners and primary stakeholders through each of the tasks under Step 1, so that the necessary and relevant revisions to the SSF management plan can be made, as recommended and approved. This will include developing a revised set of assumptions and/or creation of a new hypothesis reflecting the new learning that has emerged from the recently completed cycle. These new assumptions should then be monitored and tested systematically through time, along with the others that were upheld or remain from the completed cycle. Adaptation under Step 2 is likely to focus largely on Task 2e (work planning), although some modifications to the securing of human or financial resources and changes to the implementation team may also be warranted.

Conclusion

A successful AM approach to SSF management is one that is committed to evaluation and focused on learning. With successful adaptation comes increased knowledge and improved ability of how effectively to manage SSF. With improvements in fishery management practice comes an increased awareness of how to address uncertainty and strengthen resilience while also accepting complexity and anticipating disruption. While managers must acknowledge and recognize that natural systems are too complex fully to understand and control, they take comfort in the knowledge that, by taking an adaptive approach, they are encouraging management to follow an evolutionary path, through which interventions become more honed and effective with time. Despite the considerable investment of resources, time and effort that must be made in taking an adaptive approach to SSF management, the consequences of management failure are too great to attempt any alternative approach less proven and championed around the world.

Acknowledgements

My thanks to Bob Pomeroy and Neil Andrew for their comments, discussion and patience on draft versions of this chapter. Inspiration and coaching me on making complex adaptive management concepts both approachable and useful to marine resource managers working on the ground and in the water has been consistently provided by Richard Margoluis and Bob Pomeroy over the past decade, for which I am extremely grateful.

References

Alcorn, J.B. (1993) Indigenous peoples and conservation. *Conservation Biology* 7, 424–426.
Andrew, N.L, Béné, C., Hall, S.J., Allison, E.H., Heck, S. and Ratner, B.D. (2007) Diagnosis and management of small-scale fisheries in developing countries. *Fish and Fisheries* 8, 227–240.
Berkes, F. (1999) *Sacred Ecology: Traditional Ecological Knowledge and Management Systems*. Taylor & Francis, Philadelphia, Pennsylvania and London.
Berkes, F. (2006) From community-based resource management to complex systems. *Ecology and Society* 11. Available at: http://www.ecologyandsociety.org/vol11/iss1/art45/ (accessed 18 July 2010).
Berkes, F. and Folke, C. (eds) (1998) *Linking Social and Ecological Systems: Management Practices and Social Mechanisms for Building Resilience*. Cambridge University Press, Cambridge, UK.
Berkes, F., Colding, J. and Folke, C. (2000) Rediscovery of traditional ecological knowledge as adaptive management. *Ecological Applications* 10, 1251–1262.
Berkes, F., Mahon, R., McConney, P., Pollnac, R.C. and Pomeroy, R.S. (2001) *Managing Small-Scale Fisheries: Alternative Directions and Methods*. International Development Research Centre, Ottawa, Canada.
Berkes, F., Colding, J. and Folke, C. (2003) *Navigating Social-ecological Systems: Building Resilience for Complexity and Change*. Cambridge University Press, Cambridge, UK.
Berkes, F., Armitage, D. and Doubleday, N. (2007) Synthesis: adapting, innovating, evolving. In: Armitage, D., Berkes, F. and Doubleday, N. (eds) *Adaptive Co-management: Collaboration, Learning, and Multi-level Governance*. University of British Columbia Press, Vancouver, British Columbia, Canada.
Bormann, B.T., Cunningham, P.G., Brookes, M.H., Manning V.W. and Collopy, M.W. (1994) Adaptive ecosystem management in the Pacific Northwest. Gen. Tech. Rep. PNW-GTR-341. US Department of Agriculture, Forest Service, Pacific Northwest Research Station, Portland, Oregon.
Bormann, B.T., Cunningham, P.G. and Gordon, J.C. (1996) Best management practices, adaptive management, or both? In: *Proceedings of the National Society of American Foresters Convention*, Portland, Maine, 28 October–1 November 1995, p. 6.
Bunce, L., Townsley, P., Pomeroy, R. and Pollnac, R. (2000) *Socioeconomic Manual for Coral Reef Management*. Australian Institute of Marine Science and Global Coral Reef Monitoring Network, Townsville, Australia, p. 251. Available at: http://www.aims.gov.au/pages/reflib/smcrm/pdf/smcrm-2000.pdf (accessed 18 July 2010).
CMP (Conservation Measures Partnership) (2007) *Open Standards for the Practice of Conservation*, Version 2.0. Available at: http://www.conservationmeasures.org/wp-content/uploads/2010/04/CMP_Open_Standards_Version_2.0.pdf (accessed 18 July 2010).
Drosg, M. (2009) *Dealing with Uncertainties: A Guide to Error Analysis,* 2nd edn. Springer, New York.
Folke, C., Berkes, F. and Colding, J. (1998) Ecological practices and social mechanisms for building resilience and sustainability. In: Berkes, F. and Folke, C. (eds) *Linking Social and Ecological Systems*. Cambridge University Press, London, UK, pp. 414-436.
Folke, C., Carpenter, S., Elmqvist, T., Gunderson, L. and Holling, C. (2002) Resilience and sustainable development: building adaptive capacity in a world of transformations. *AMBIO* 31, 437–440.
FOS (Foundations of Success) (2007) *Using Results Chains to Improve Strategy Effectiveness.* A FOS How-To Guide. Foundations of Success, Bethesda, Maryland. Available at: http://fosonline.org/Site_Documents/Grouped/FOS_Results_Chain_Guide_2007-05.pdf (accessed 18 July 2010).
FOS (Foundations of Success) (2009) *Using Conceptual Models to Document a Situation Analysis*. A FOS How-To Guide. Foundations of Success, Bethesda, Maryland. Available at: http://fosonline.org/Site_Documents/Grouped/FOS_Conceptual_Model_Guide_April2009.pdf (accessed 18 July 2010).
Gunderson, L. (1999) Resilience, flexibility, and adaptive management – antidotes for spurious certitude? *Conservation Ecology* 3(1), art. 7. Available at: http://www.consecol.org/vol3/iss1/art7/ (accessed 18 July 2010).
Gunderson, L.H. and Holling, C.S. (2002) *Panarchy: Understanding Transformations in Human and Natural Systems*. Island Press, Washington, DC.

Gunderson, L.H., Holling, C.S. and Light, S.S. (eds) (1995) *Barriers and Bridges to the Renewal of Ecosystems and Institutions*. Columbia University Press, New York.

Hockings, M., Stolton, S., Leverington, F., Dudley, N. and Courrau, J. (2006) *Evaluating Effectiveness: a Framework for Assessing Management Effectiveness of Protected Areas*, 2nd edn. IUCN, Gland, Switzerland and Cambridge, UK, p. 105. Available at: http://data.iucn.org/dbtw-wpd/edocs/PAG-014.pdf (accessed 18 July 2010).

Holling, C.S. (1973) Resilience and stability of ecological systems. *Annual Review of Ecology and Systematics* 4, 1–23.

Holling, C.S. (1978) *Adaptive Environmental Assessment and Management*. John Wiley and Sons, London.

Hviding, E. (1996) *Guardians of Marovo Lagoon: Practice, Place, and Politics in Maritime Melanesia*. Pacific Islands Monograph Series 14, University of Hawai'i Press, Honolulu, Hawaii.

ICSU (International Council for Science) (2002) *Resilience and Sustainable Development: Science Background Paper Commissioned by the Environmental Advisory Council of the Swedish Government in Preparation for WSSD*. Series on Science for Sustainable Development Number 3, International Council for Science, Paris.

IUCN (International Union for Conservation of Nature) (2006a) *IUC–CMP Unified Classification of Direct Threats*, Version 1.0. IUCN, Washington, DC.

IUCN (International Union for Conservation of Nature) (2006b) *IUCN–CMP Unified Classification of Conservation Actions,* Version 1.0. IUCN, Washington, DC.

Johannes, R.E. (1978a) Reef and lagoon tenure systems in the Pacific islands. *South Pacific Bulletin* 4th quarter, 31–34.

Johannes, R.E. (1978b) Traditional marine conservation methods in Oceania and their demise. *Annual Review of Ecology and Systematics* 9, 349–364.

Johnson, F. and Williams, K. (1999) Protocol and practice in the adaptive management of waterfowl harvests. *Conservation Ecology* 3(1), art. 8. Available at: http://www.consecol.org/vol3/iss1/art8/ (accessed 18 July 2010).

Lee, K.N. (1993) *Compass and Gyroscope: Integrating Science and Politics for the Environment*. Island Press, Washington, DC.

Lee, K.N. (1999) Appraising adaptive management. *Conservation Ecology* 3(2), art. 3. Available at: http://www.ecologyandsociety.org/vol3/iss2/art3/ (accessed 18 July 2010).

Lee, K.N. and Lawrence, J. (1986) Adaptive management: learning from the Columbia River Basin fish and wildlife program. *Environmental Law* 16, 431-460.

Margoluis, R. and Salafsky, N. (1998) *Measures of Success: Designing, Managing, and Monitoring Conservation and Development Projects*. Island Press, Washington, DC.

Margoluis, R., Stem, C., Salafsky, N. and Brown, M. (2009) Using conceptual models as a planning and evaluation tool in conservation. *Evaluation and Program Planning* 32, 138–147.

Marmorek, D.R., Robinson, D.C.E., Murray, C. and Grieg, L. (2006) *Enabling Adaptive Forest Management – Final Report*. Prepared for the National Commission on Science for Sustainable Forestry. ESSA Technologies Ltd, Vancouver, British Columbia, Canada.

McCarthy, M.A. and Possingham, H.P. (2007) Active adaptive management for conservation. *Conservation Biology* 21, 956–963.

Murray, C. and Marmorek, D. (2003) Adaptive management and ecological restoration. In: Freiderici, P. (ed.) *Ecological Restoration of Southwestern Ponderosa Pine Forests*. Island Press, Washington, DC, pp. 417–428.

Nichols, J.D., Runge, M.C., Johnson, F.A. and Williams, B.K. (2007) Adaptive harvest management of North American waterfowl populations: a brief history and future prospects. *Journal of Ornithology* 148 (Suppl. 2), 343–349.

Pomeroy, R.S., Parks, J.E. and Watson, L.M. (2004) *How is Your MPA Doing? A Guidebook of Natural and Social Indicators for Evaluating Marine Protected Area Management Effectiveness*. International Union for the Conservation of Nature (IUCN), Gland, Switzerland and Cambridge, UK. Available at: http://data.iucn.org/dbtw-wpd/edocs/PAPS-012.pdf (accessed 18 July 2010).

Prigogine, I. and Stengers, I. (1984) *Order out of Chaos: Man's New Dialogue With Nature*. Bantam Books, New York.

Raakjær Nielsen, J., Son, D.M., Stæhr, K., Hovgård, H., Thuy, N.T.D., Ellegaard, K., Riget, F., Thi, D.V. and Hai, P.G. (2007) Adaptive fisheries management in Vietnam: the use of indicators and introducing a multi-disciplinary marine fisheries specialist team to support implementation. *Marine Policy* 31, 143–152.

Roberts, M., Norman, W., Minhinnick, N., Wihongi, D. and Kirkwood, C. (1995) Kaitiakitanga: Maori perspectives on conservation. *Pacific Conservation Biology* 2, 7–20.

Ruddle, K., Hviding, E. and Johannes, R.E. (1992) Marine resource management in the context of customary tenure. *Marine Resource Economics* 7, 249–273.

Salafsky, N., Margoluis, R. and Redford, K. (2001) *Adaptive Management: a Tool for Conservation Practitioners.* Biodiversity Support Program, Washington, DC.

Salafsky, N., Margoluis, R., Redford, K. and Robinson, J. (2002) Improving the practice of conservation: a conceptual framework and agenda for conservation science. *Conservation Biology* 16, 1469–1479.

Salafsky, N., Salzer, D., Stattersfield, A.J., Hilton-Taylor, C., Neugarten, R., Butchart, S.H.M., Collen, B., Cox, N., Master, L.L., O'Connor, S. and Wilkie, D. (2009) A standard lexicon for biodiversity conservation: unified classifications of threats and actions. *Conservation Biology* 22, 897–911.

Shannon, M.A. and Antypas, A.R. (1997) Open institutions: uncertainty and ambiguity in 21st-century forestry. In: Kohm, K.A. and Franklin, J.F. (eds) *Creating a Forestry for the 21st Century: the Science of Ecosystem Management.* Island Press, Washington, DC, pp. 437–445.

Stankey, G.H., Clark, R.N. and Bormann, B.T. (2005) Adaptive management of natural resources: theory, concepts, and management institutions. Gen. Tech. Rep. PNW-GTR-654, U.S. Department of Agriculture, Forest Service, Pacific Northwest Research Station, Portland, Oregon.

Stem, C., Margoluis, R., Salafsky, N. and Brown, M. (2005) Monitoring and evaluation in conservation: a review of trends and approaches. *Conservation Biology* 19, 295–309.

TNC (The Nature Conservancy) (2007) *Conservation Action Planning Handbook: Developing Strategies, Taking Action and Measuring Success at Any Scale.* The Nature Conservancy, Arlington, Virginia. Available at: http://conserveonline.org/workspaces/cbdgateway/cap/resources/2/1/handbook/download (accessed 18 July 2010).

USDA (United States Department of Agriculture) (1994) Record of decision for amendments to Forest Service and Bureau of Land Management planning documents within the range of the northern spotted owl. US Department of the Interior (USDI), USDA Forest Service, Bureau of Land Management, Washington, DC.

Varley, N. and Boyce, M.S. (2006) Adaptive management for reintroductions: updating a wolf recovery model for Yellowstone National Park. *Ecological Modelling* 193, 315–339.

Walker, B., Holling, C.S., Carpenter, S.R. and Kinzig, A. (2004) Resilience, adaptability and transformability in social–ecological systems. *Ecology and Society* 9(2), art. 5. Available at: http://www.ecologyandsociety.org/vol9/iss2/art5/ (accessed 18 July 2010).

Walters, C.J. (1986) *Adaptive Management of Renewable Resources.* Macmillan Publishing Company, New York.

Walters, C.J. (1997) Challenges in adaptive management of riparian and coastal ecosystems. *Conservation Ecology* 1(2), art. 1. Available at: http://www.ecologyandsociety.org/vol1/iss2/art1/ (accessed 18 July 2010).

Walters, C.J. and Hilborn, R. (1978) Ecological optimization and adaptive management. *Annual Review of Ecology and Systematics* 8, 157–188.

Walters, C.J. and Holling, C.S. (1990) Large-scale management experiments and learning by doing. *Ecology* 71, 2060–2068.

Williams, B.K., Johnson, F.A. and Wilkins, K. (1996) Uncertainty and the adaptive management of waterfowl harvests. *Journal of Wildlife Management* 60, 223–232.

Williams, B.K., Szaro, R.C. and Shapiro, C.D. (2009) *Adaptive Management: the U.S. Department of the Interior Technical Guide,* 2nd edn. US Department of the Interior, Washington, DC.

Wilson, D.C., Ahmed, M., Siar, S.V. and Kanagaratnam, U. (2006) Cross-scale linkages and adaptive management: fisheries co-management in Asia. *Marine Policy* 30, 523–533.

WWF (World Wildlife Fund) (2000) Users Guide to Assessing the Socio-Economic Root Causes of Biodiversity Loss. Unpublished manuscript. WWF Macroeconomics for Sustainable Development Program Office, World Wildlife Fund, Washington, DC.

WWF (World Wildlife Fund) (2005a) Basic Guidance for Cross-Cutting Tools: Conceptual Models. Resources for Implementing the WWF Standards of Conservation Project and Programme Management. Unpublished manuscript. World Wildlife Fund, Washington, DC. [online] Available at: http://assets.panda.org/downloads/1_4_conceptual_models_03_23_09.pdf (accessed 18 July 2010).

WWF (World Wildlife Fund) (2005b) Cross-Cutting Tool: Stakeholder Analysis. Resources for Implementing the WWF Standards of Conservation Project and Programme Management. Unpublished manuscript. World Wildlife Fund, Washington, DC. [online] Available at: http://assets.panda.org/downloads/1_1_stakeholder_analysis_11_01_05.pdf (accessed 18 July 2010).

WWF (World Wildlife Fund) (2005c) Monitoring Plan: Basic Guidance for Step 2.2. Resources for Implementing the WWF Standards of Conservation Project and Programme Management. Unpublished manuscript. World Wildlife Fund. Washington, DC. [online] http://assets.panda.org/downloads/2_2_monitoring_plan_01_11_05.pdf

WWF (World Wildlife Fund) (2006a) Step 1.4: Define Situation Analysis. Resources for Implementing the WWF Conservation Project and Programme Management Standards. Unpublished manuscript. World Wildlife Fund, Washington, DC. [online] Available at: http://assets.panda.org/downloads/1_4_situation_analysis_2007_02_19.pdf (accessed 18 July 2010).

WWF (World Wildlife Fund) (2006b) Step 2.1: Define Action Plan Goals, Objectives, and Activities. Resources for Implementing the WWF Conservation Project and Programme Management Standards. Unpublished manuscript. World Wildlife Fund, Washington, DC. [online] Available at: http://assets.panda.org/downloads/2_1_action_plan_02_26_07.pdf (accessed 18 July 2010).

WWF (World Wildlife Fund) (2007a) Step 1.4: Define Threat Ranking. Resources for Implementing the WWF Conservation Project and Programme Management Standards. Unpublished manuscript. World Wildlife Fund, Washington, DC. [online] Available at: http://assets.panda.org/downloads/1_4_threat_ranking___july_13__2007.pdf (accessed 18 July 2010).

WWF (World Wildlife Fund) (2007b) WWF Standards of Conservation Project and Programme Management. Unpublished manuscript. World Wildlife Fund, Washington, DC. [online] Available at: http://assets.panda.org/downloads/0_0_wwf_standards_overview_02_09_07.pdf (accessed 18 July 2010).

WWF (World Wildlife Fund) (2007c) Step 3.1: Work plans and Budgets. Resources for Implementing the WWF Conservation Project and Programme Management Standards. Unpublished manuscript. World Wildlife Fund, Washington, DC. [online] Available at: http://assets.panda.org/downloads/3_1_work plans_and_budgets_02_27_07_1.pdf (accessed 18 July 2010).

WWF (World Wildlife Fund) (2009) Step 1.3: Targets and Target Viability. Resources for Implementing the WWF Conservation Project and Programme Management Standards. Unpublished manuscript. World Wildlife Fund, Washington, DC. [online] Available at: http://assets.panda.org/downloads/1_3_targets_and_viability___january_5__2009.pdf (accessed 18 July 2010).

7 Conditions for Successful Co-management: Lessons Learned in Asia, Africa, the Pacific and the Wider Caribbean

Robert S. Pomeroy, Joshua E. Cinner and Jesper Raakjær Nielsen

Introduction

Fisheries and coastal resources offer a unique opportunity and challenge for the development of co-management due, in part, to the independent nature of the resource users and the dynamic nature of aquatic resources. Co-management should be viewed not as a single strategy to solve all problems of fisheries and coastal resources management, but rather as a process of resource management – maturing, adjusting and adapting to changing conditions over time. Thus, the co-management process is inherently adaptive, relying on systematic learning and the progressive accumulation of knowledge for improved resource management (Pomeroy and Rivera-Guieb, 2006).

Over the last two decades, research and case studies undertaken at different locations around the world have documented many cases, both successful and unsuccessful, of co-management in fisheries and other coastal resources (Jentoft and Kristoffersen, 1989; White et al., 1994; Berkes et al., 1996; Hoefnagel and Smit, 1996; DeCosse and Jayawickrama, 1998; Normann et al., 1998; Raakjær Nielsen et al., 2004, 2007). From the results of this research, key conditions are emerging that are central to developing and sustaining successful co-management arrangements (Pinkerton, 1989, 1993, 1994). The list is long and varied, and is continually growing as new insights emerge from both theoretical and empirical research. It should be noted that these conditions are not absolute or complete. There can be successful co-management without having met all of the conditions. However, consensus is growing that the more of these conditions that are satisfied in a particular situation, the greater the chances for successful implementation of co-management.

The purpose of this chapter is to present and discuss key conditions for the successful implementation of fisheries and coastal co-management identified in South-east Asia, Africa, the Pacific and the wider Caribbean. These four regions were selected as several recent research and development projects have produced outputs in which key conditions have been identified. The conditions are reported on a regional basis not for a specific country, as this is how the authors have presented their results. It is expected that specific conditions would differ by country. These conditions will embrace the wide range of aspects that can affect the implementation and performance of co-management and activities, from resources and fisheries to cultural and institutional dimensions. The chapter will conclude with a discussion of policy implications for fisheries and coastal co-management.

Definitions and concepts

The term 'key condition' is used in the sense of Ostrom (1990, p. 88) as:

> ... an essential element or condition that helps to account for the success of these institutions in sustaining common property resources and gaining the compliance of generation after generation of appropriators to the rule of use.

Berkes *et al.* (2001) regard key conditions as variables or attributes that emerge as being central to the chances that co-management can be developed and sustained. For the purposes of this chapter, the term 'successful' co-management is defined here as better overall institutional performance, in terms of efficiency (optimal rate of resource use; transaction costs), equity (equitable distribution of benefits; pattern of redistribution of benefits) and sustainability (stewardship towards the resource; resilience of the management system; rule compliance), as compared with other resource management arrangements, such as centralized management (ICLARM/IFM, 1996). The term 'co-management', as used in this paper, includes various partnership arrangements and degrees of power-sharing, ranging from instructive (where the community is informed about decisions that government has already made) to community control (where power is delegated to the community and they inform government of decisions) (Berkes, 1994; Sen and Raakjær Nielsen, 1996).

Conditions for Successful Community-governed Commons

Theoretical work on commons institutions (Pinkerton, 1989; Ostrom, 1990, 1992, 1994; Agrawal, 2002) and empirical work on fisheries co-management from Asia (White *et al.*, 1994; Pomeroy *et al.*, 2001), Africa (Sverdrup-Jensen and Raakjær Nielsen, 1998; Geheb and Sarch, 2002; Hauck and Sowman, 2003; Khan *et al.*, 2004, Cinner *et al.*, 2009a, b) and the wider Caribbean (CANARI, 1999; McConney *et al.*, 2003; Pomeroy *et al.*, 2004) have identified that aspects of both the institutional design in co-management arrangements and situational context within which the co-management system is embedded are relevant to the success of co-management arrangements. Here, we group the conditions for successful co-management according to four categories:

1. Institutional design principles: these include aspects of how co-management institutions are designed.
2. Supra-community level: supra-community conditions affecting the success of co-management include those that are external to the community, including enabling legislation and supportive government administrative structures at the national level, and markets. They can also include demographic factors and technological change.
3. Community level: community-level conditions affecting the success of co-management include those found within the community, and include both the physical and the social environment in terms of potential relationships with fisheries and coastal resource management.
4. Individual and household level: the individual is responsible for making the decision to participate in co-management. Individual and household decision-making and behaviour is thus central to the success of co-management.

Key Institutional Design Principles

To help ensure cooperation and avoid free-loading, common property institutions often require specific design principles to ensure a credible commitment that resource users will follow the rules, and also effective monitoring of these commitments (Ostrom, 1990). Research on adaptive management, common property resources and effective governance of complex socio-ecological systems has found a number of common design characteristics shared by successful and long-enduring community-based management systems. These include the following.

Appropriate scale and defined boundaries

Here, we integrate two concepts that some commons literature treats as separate conditions for co-management: clearly defined boundaries and whether the scale and scope of the management system is congruent to local conditions (Ostrom, 1990; Cinner *et al.*, 2009a). Pomeroy *et al.* (2001) and Raakjær Nielsen *et al.* (2004) found that scale for co-management may vary a great deal, but should be appropriate to the area's ecology, people and level of management. This includes the size of the physical area to be managed and how many members should be included in a management organization so that it is representative, but not too large, so as to be unworkable. An important scale issue with co-management is whether and to what extent institutions that affect behaviour at one level or social organization, such as small-scale or micro-level societies, also play key roles at other levels of social organization, including national (meso-level) societies and international (macro-level) society, and vice versa. Geheb and Sarch (2002) found that having international boundaries traversing a fishery significantly impedes its co-management. If the unit around which management occurs is the landing site, the community or an access area, then the size of the fishery becomes largely irrelevant. In the Caribbean, McConney *et al.* (2003) found that resources are generally more easily co-managed if: (i) they are sedentary; (ii) the resource distribution corresponds with human settlements; and (iii) they fall under one political jurisdiction. McConney *et al.* (2003) further state that boundaries and scale for co-management should match the abilities of the resource users to manage the area. Boundaries allow stakeholders to know where their responsibilities lie.

Membership is clearly defined

In Asia, Pomeroy *et al.* (2001) found that the individual fishers or households with rights to fish in a bounded fishing area, to participate in management and to be an organization member should be clearly defined. The numbers of fishers or households should not be so large as to restrict effective communication and decision-making. In the Caribbean, Pomeroy *et al.* (2004) found that membership should be clearly defined as to who really has a stake in the fishery.

Participation by those affected

Most individuals affected by co-management arrangements are included in the group that makes decisions about and can change the arrangements (Pomeroy *et al.*, 2001). In South Africa, Hauck and Sowman (2003) state that fundamental to the concept of co-management is the active participation and involvement of resource users and their commitment to the co-management process. Without the commitment and willingness of resource users to participate in the process, sharing of management responsibility cannot be achieved. Hara and Raakjær Nielsen (2003) and Khan *et al.* (2004) found that active participation by all resource users can bring about legitimization of laws and the harmonization of traditional and modern management and enforcement systems. In evaluating experiences with participatory planning and management in the Caribbean, CANARI (1999) found that initiatives that incorporate all relevant stakeholders from the outset are most likely to be the most enduring. Pomeroy *et al.* (2004) found that participation in co-management in the Caribbean is constrained because, in many cases, fishers expect government to do things for them and they are reluctant to get involved in management. White *et al.* (1994) state that all stakeholders need to participate in the co-management process in order to ensure a politically neutral process. They also state that there needs to be an ongoing feedback of information on the co-management process to sustain and increase community participation. However, this may not always be the case. In Zimbabwe, Nyikahadzoi and Raakjær Nielsen (2009) show how government completely controlled the flow of information used for management purposes, resulting in a far from neutral process. In South Africa, Isaacs *et al.* (2007), Hara and Raakjær Nielsen (2009) and Nyikahadzoi *et al.* (2010) argue that for more than a decade the state had underestimated both the need for capacity building and the time it takes to build capacity among the new actors in order

Conflict management mechanisms

Arbitration and resolution of disputes are imperative when conflicts arise over co-management. If resource users are to follow rules, a mechanism for discussing and resolving conflict and infractions is needed (Pomeroy et al., 2001). Sverdrup-Jensen and Raakjær Nielsen (1998) report that mechanisms for conflict resolution need to be given high priority in the design of co-management arrangements, and management approaches that minimize conflict should be adopted wherever feasible. Geheb and Sarch (2002) state that a co-management structure needs to be established based on forums within which negotiation and conflict management can occur. In Laos, Raakjær Nielsen et al. (2004) identified a co-management programme that found an innovative way to channel the motivation to exclude gears used by outsiders into both conflict resolution and effective resource management. The programme operated on the principle that any community can ban any gear within their zone as long as everyone, insiders and outsiders, is equally affected.

Graduated sanctions

Sanctions need to increase with the number or the severity of offences. Education and compliance should be the first option, and enforcement used last. Rules and regulations should be compatible with stakeholders' resource use practices. The implementation of regulations should remain flexible by maintaining an open dialogue between managers and stakeholders. In order to foster compliance and ease enforcement, regulations and penalties should be simple, clear, understandable and appropriate to the socio-cultural context of the area. Enforcement requires an integrated and coordinated approach among the various agencies responsible for it. Enforcement should seek to use innovative means, such as social influences and sanctions, to improve compliance and lessen costs to management agencies. Meaningful but graduated and context-dependent penalties are more legitimate than draconian, one-size-fits-all, penalties. When imposed, enforcement must be swift and public, consistent with the due process of law. In this context, 'swift' means that enforcement should be directed at a specific and identified incident. Enforcement should be public so that others will be aware of the consequences of offending.

Nested institutions

These involve a critical level of decentralization and delegation of authority, but also require coordination between government and community. Successful co-management require that government has established formal policy and laws for decentralization and delegation for management. For some fisheries co-management operational rules, such as rotational closures, this also requires that the legal system is fast and flexible to allow for rules to be developed to respond adaptively to social or environmental conditions.

Importantly, though, it is not only delegation of authority that creates conditions for successful co-management, but also there must be mechanisms to coordinate local management arrangements, resolve conflict and reinforce local rule enforcement. For example, in the Seri fishing community in the Gulf of California, Mexico, a community-based fisheries management system was effective at creating incentives for resource users in the community to restrain their harvesting, which led to a rapid increase in resource abundance (Cudney-Bueno and Basurto, 2009). However, this institution lacked cross-scale linkages with enforcement agencies and, when outsiders began poaching, the social norms used to constrain effort were no longer sufficient and the system soon because a free-for-all (Cudney-Bueno and Basurto, 2009).

In Sofala Bay, Mozambique, the management of the shallow-water shrimp fishery is a good example of a co-management arrangement that operates on a relatively large scale and within a nested system (Raakjær Nielsen, 2009). A number of local fisheries co-management committees (mainly composed of artisanal fishermen) have been established. They mediate conflicts among artisanal fishermen, and between artisanal and semi-industrial/industrial fishermen, and are engaged in the formulation and implementation of local rules. At the national level, co-management is undertaken by the Committee of Fisheries Management (CAP), which is a consultative forum composed of representatives of industrial, semi-industrial and artisanal fishermen. Artisanal fishermen are represented within a kind of nested arrangement, in which local committees are represented at the provincial co-management committee, which again has representation in the CAP.

Monitoring of resources

Jul-Larsen et al. (2002), in their study of southern African freshwater fisheries, emphasized the need for co-management to be tuned to voices and needs at the local level; and that co-management can certainly represent an option, a process of empowering communities and a means to integrate the regulation of fisheries in the general development of the communities; and can become involved in monitoring the fishery and in communicating/evaluating the results to the various users.

Accountability

Co-management means having a process in which business is conducted in an open and transparent manner (Pomeroy et al., 2001). All partners must be held equally accountable for upholding the co-management agreement. Common property literature emphasizes the need for accountability of enforcers (Ostrom, 1990). There need to be accepted standards for monitoring and evaluation of management objectives and outcomes. White et al. (1994) state that monitoring with community participation can provide information that helps the community understand what is happening in the co-management process, and also maintains openness of the process. Hauck and Sowman (2003) state that resource users must establish a local-level institution that provides a voice for their contribution to management and that is accountable to them.

Design Principles in Practice

To illustrate the incorporation of these design principles into co-management arrangements, we highlight a case study from two countries in the western Indian Ocean that have recently developed frameworks to co-manage marine resources: Kenya and Madagascar (Cinner et al., 2009b; Table 7.1). As a result of ineffective top-down management, the Kenya Ministry of Fisheries Development began developing legal frameworks to share management responsibility for fisheries in the 1990s (Ogwang et al., unpublished report). This shared responsibility (or co-management) of fisheries resources was undertaken through a structure that enabled resource users to manage their landing sites, in a forum known as a Beach Management Unit (BMU). BMUs were first established on Lake Victoria and practised by the three countries bordering the lake (Kenya, Uganda and Tanzania) as a way of improving fisheries management. Guidelines have since been developed to supplement the provisions of the fisheries' regulation to increase stakeholder understanding in setting up BMUs.

On the Kenyan Coast, the BMUs were promoted by the government to create partnership between itself and local communities in the management of coastal resources. In 1996, the first legal framework to introduce the sharing of responsibility over the natural resource management among users was created, known as *Gestion Locale Sécurisée* (GELOSE; Antona et al., 2004). This law allowed communities to define their own goals and develop regulations for resource use and management in the form of by-laws, as long as these rules were consistent with national policy (i.e. one could not develop a regulation that allowed the use of a gear

Table 7.1. Design principles in practice in Kenya and Madagascar. Table organized from top to bottom by principles fully present in both countries, those partially shared and those present in only one country. Adapted from Cinner et al. (2009b).

Design principle	Description	BMU	GELOSE	Notes on implementation
Clearly defined membership rights	Clear delineation of membership rights to co-managed area	Yes	Yes	Membership clear and registered. In both Kenya and Madagascar, individuals may join only one CBO or GELOSE organization
Congruence	Whether scale and scope of rules are appropriate for the local conditions	Yes	Yes	Rules are developed by the resource users themselves and can build on local social norms; scale of management is roughly matched to scale of resource: many BMU and GELOSE sites have coral reef-based fisheries, where reef fish have small home ranges on a scale similar to what is being managed
Rights to organize	Whether resource users have rights to make, enforce and change the rules	Yes	Yes	In both countries, by-laws must be approved by agencies and codified before enforced; enforcement can be conducted by members, but law enforcement officers are needed to make arrests
Conflict resolution mechanisms	Rapid access to low-cost resolution forum	Yes	Yes	In Kenya, management committee, provincial administration and Ministry of Fisheries Development; in Madagascar, management committee from local to regional administration and Ministry of Environment
Nested institutions	Nested within lead agencies or partner organizations at critical stages	Partially	Partially	In both countries organizations are nested, but connections at different scales are missing at some key stages of the co-management process
Accountability of monitors	Whether there are accountability mechanisms for those enforcing the rules	Partially	Partially	Partially available in Madagascar when wardens paid to enforce rules – not part of general framework; in Kenya, some BMUs employ multiple monitors to reduce the possibility of corruption
Clearly defined geographic boundaries	Clear delineation of co-management area	Partially	Partially	Both countries generally use landmarks for coastal boundaries and outward extent of shallow water ecosystems for offshore boundaries (e.g. a reef edge)
Collective choice arrangements	Whether individuals affected by the rules can participate in changing the rules	Partially	Partially	Only members can participate in changing rules in both systems; if by-laws limit access of non-members there is no forum for input
Graduated sanctions	Whether sanctions increase with numerous offences or the severity of the offence	Partially	Partially	In practice, a first offence generally generates a warning, a second offence is dealt with locally and a third offence is dealt with by law enforcement/legal system
Monitoring of resources	Quantitative or qualitative monitoring of resource conditions	Partially	No	BMUs monitor catch and prices, but not *in situ* resources; in both countries, some scientists from NGOs monitor resources in selected sites, but not restricted to BMU or GELOSE areas

CBO, community-based organization; GELOSE, Gestion Locale Sécurisée; BMU, Beach Management Unit.

banned by national legislation; Antona *et al.*, 2004). As with many other forms of environmental policy in Madagascar, GELOSE was developed for terrestrial ecosystems and then applied to marine resources. In 1999, the first GELOSE site was applied to a mangrove socio-ecological system in Tulear.

In both systems, geographic boundaries were only partially clear because boundaries were often submerged landmarks (e.g. the edge of a reef), and the seaward distance covered was undefined. This is more reflective of the system being governed, rather than the framework. Likewise, collective choice arrangements are in place in both systems, but participation and input is limited to group members. For example, if a BMU decided to prohibit access to its fishing grounds to non-BMU members, these non-members would be unable to change the rules. In both cases, graduated sanctions were not specifically mentioned in the regulations, but applied *de facto* by not generally sanctioning a violator through legal channels unless it is a repeat offence. First offences were often dealt with by warnings or within a community, even though there is no legal requirement to do so. It is interesting that communities filled some gaps in the design principles in a *de facto* manner outside of the legal framework. On the one hand this shows that the frameworks can be somewhat adaptive to local circumstances. However, issuing warnings when locally developed rules do not provide for this risks undermining confidence in the management system if violators are viewed as not facing the penalties.

Providing for graduated sanctions in the framework may improve transparency of the system. Monitoring of the monitors was not conspicuously present in either system, however, but it also happened *de facto* in some cases. For example, the Kiruwitu community protected area in Kenya attempted to provide some transparency within the monitors by employing four monitors at any given time, which they hoped through peer pressure would decrease tendencies for corruption. In other BMUs in Kenya monitors may be monitored, although in practice this is done loosely without formal rules or protocols. The main components missing from these systems related to monitoring of resources. In some instances, monitoring of the resources themselves were conducted by scientists and conservation groups, although key informants in both countries expressed concerns with how the data were returned to the communities.

Addressing these design issues early in the development of co-management arrangements in these countries may enhance the chances of building robust institutions (Yandle, 2003). Consequently, examining ways to fill these potential gaps should be considered a priority for institutional capacity building by donors and governments. However, this should be done in a way that acknowledges that heterogeneity of institutions is critical and that there is not a one-size-fits-all policy to model institutions (Low *et al.*, 2002). Some of these design principles may not be appropriate in a specific local context and should not be 'forced' on local institutions. Critics of the design principle approach suggest that the emphasis on internal characteristics of the institutions can lead to a blueprint approach and a lack of adequate consideration for important contextual factors that are often critical to the success or failure of commons institutions (Steins *et al.*, 2000; Blaikie, 2006). This analysis of the design principles in BMU and GELOSE frameworks is not intended to be used as a metric of their success, but rather is a means to identify possible institutional gaps that may be the focal point for communities, NGOs and other bridging organizations. Further research is needed to examine how these design principles and the contextual factors in which they are situated relate to aspects of successful co-management.

Supra-community-level Context Considerations

Enabling policies and legislation

Pomeroy *et al.* (2001), presenting results of a research project in Asia on co-management, stated that if co-management initiatives are to be successful, basic issues of government action to establish supportive legislation, policies, rights and authority structures must be addressed. Policies and legislation need to: (i) spell out jurisdiction and control; (ii) provide legitimacy to property rights and decision-making arrangements; (iii) define and clarify local responsibility and authority; (iv) clarify the rights and responsibilities of partners; (v) support local enforcement and accountability mechanisms; and (vi) provide fisher groups or organizations with the legal right to organize and make arrangements related to their needs. The legal process formalizes rights and rules and legitimizes local participation in co-management arrangements.

In South Africa, Hauck and Sowman (2003) found that government's reluctance to relinquish a large degree of power was one of the most difficult challenges facing co-management, especially as government was wary of the capacity of people to manage resources. In this case, while laws and policies existed to support co-management, there was a need for a fundamental shift in government attitude and behaviour. Geheb and Sarch (2002), summarizing management challenges facing Africa's inland fisheries, state that too many political agendas are being pursued, resulting in a poor performance for co-management.

For any given context, there needs to be a fixed definition of co-management that is accepted by all stakeholders and policies that support and direct implementation. Khan et al. (2004), in a review of co-management in nine African countries, state that an honest willingness on the part of governments to relinquish exclusive control of aquatic resources is needed in order to establish trust and confidence among the various partners. In the Caribbean, Pomeroy et al. (2004) found that management approaches of governments for coastal resource management were not flexible and responsive to changing circumstances. Pomeroy et al. (2004) also stated that limited trust between government and fishers restricts the development of co-management. Sverdrup-Jensen and Raakjær Nielsen (1998), summarizing findings from eight co-management case studies in Africa, reported that governments should not leave the local partners with management responsibilities they are not capable of shouldering. They also reported that a balance needs to be struck between the responsibilities given to communities and the means at their disposal.

External agents

External change agents, such as non-governmental organizations, academic or research institutions, religious organizations and others, can facilitate the co-management process (Pomeroy et al., 2001). Hauck and Sowman (2003) found that external agents provide impartiality, knowledge, training, logistical support and financial aid, and often act as intermediaries between the resource users and government. McConney et al. (2003), presenting guidelines for establishing coastal resource co-management in the Caribbean, found that it is useful to have a trained facilitator guide the co-management process. Pomeroy et al. (2004) found that external agents provide support for co-management, but must not encourage dependency upon them by the community.

Alliances and networks

White et al. (1994) found that alliances and networks can help to solve larger issues. Mutually beneficial alliances and networks can be formed to counteract conflicting and often powerful interests outside the community. Alliances and networks can also be used to further policy agendas supported by many organizations from different sectors of society.

Community Level

Group cohesion

The group permanently resides near the area to be managed. There is a high degree of homogeneity, in terms of kinship, ethnicity, religion or fishing gear type, among the group. The group members have a common understanding of the problem and of alternative solutions. Group size, in terms of the number of individuals involved in the management arrangements, is relatively small. There is an incentive and willingness among the group members to engage in collective action. This does not mean, however, that co-management projects cannot succeed in socio-economically and culturally heterogeneous communities. For example, in the oxbow lakes of Bangladesh, Muslim and Hindu fishers were able to work together on the lake fisheries teams.

Existing organizations

Attempts to integrate customary institutions into co-management initiatives have proved successful under certain conditions (Aswani and Hamilton, 2004; Cinner and Aswani, 2007), but in other situations they have created barriers and in some cases have even eroded customary institutions (e.g. Gelcich et al., 2006). In Kimbe Bay, Papua New Guinea, the community had difficulty understanding the rationale behind a supposedly co-managed marine protected area (MPA) because it did not fit their customary model of reef closures (Cinner et al., 2003). For generations, the community had closed their reefs to fishing when a person of stature in the community died. After 3–12 months the reefs were usually opened, to harvest fish for a feast to mark the end of the mourning period. Thus, the community believed the goal of a reef closure was to build up fish stocks that could then be exploited when needed. Their prior experience with customary management made for a difficult transition to a co-management situation with different operational rules. In Cap Masoala, Madagascar, attempts to integrate the customary closure with the MPA resulted in sentiments that the sacred area was being desecrated (Cinner, 2007; Cinner et al., 2009b).

Leadership

Leaders set an example for others to follow, set courses of action and provide energy and direction (Pomeroy et al., 2001). While a community may already have leaders, they may not be the correct or appropriate leaders for co-management. Local elites may not be the most appropriate leaders, and new leaders may need to be identified and developed. Hauck and Sowman (2003) state that one or two people often become involved in the co-management process as 'champions', facilitating communication and interaction among stakeholders. White et al. (1994) found that organization formation is strategic in identifying and developing leaders. Sverdrup-Jensen and Raakjær Nielsen (1998), in an analysis of case studies on co-management in eight African countries, found that traditional leadership systems, often having a high legitimacy with local people, should be reflected in the design of co-management arrangements. Despite the importance of local leaders for co-management, Pomeroy et al. (2004) found a lack of effective leadership among fishers in the Caribbean to guide change and the co-management process.

Empowerment, capacity building and social preparation

Individual and community empowerment is a central element of co-management (Pomeroy et al., 2001). Empowerment is concerned with capability building of individuals and the community in order for them to have greater social awareness, to gain greater autonomy over decision-making, to gain greater self-reliance and to establish a balance in community power relations. Empowerment is enhanced by capacity building through education and training that raises the knowledge and information of those involved in the co-management process. Hauck and Sowman (2003) found that in South Africa, empowerment and capacity building are important in order for resource users to understand the concepts and principles of sustainable resource use and co-management.

CANARI (1999) stated that true participation can only be achieved when participants are provided with the information required to make decisions. McConney et al. (2003) found that in the Caribbean, building stakeholder capacity is essential for engagement in co-management. White et al. (1994) cautioned that education and training alone are not sufficient to change major behaviour patterns that have consequences for people's livelihoods. Changes in behaviour are bounded by community values. Pomeroy et al. (2004) reported that effective communication among stakeholders, brought about through capacity building, can improve the success of co-management. Capacity must be built so that local management institutions remain flexible to changing needs and conditions as co-management matures over time (Pomeroy et al., 2004).

Community organizations

The existence of a legitimate (as recognized by the local people) community or people's organization is a vital means for representing resource users and other stakeholders and for influencing the direction of policies and decision-making (Pomeroy et al., 2001). These organizations must have the legal right to exist and make arrangements related to their needs. They must be autonomous from government. Geheb and Sarch (2002) found that for the inland fisheries of Africa, traditional pre-existing management organizations and institutions should be a part of any new management structure. CANARI (1999) stated that in the Caribbean, participation in fisheries and coastal resource management requires the existence and support of effective local organizations. McConney et al. (2003) stated that community organizing and the establishment of stakeholder organizations is a critical component in the process of co-management in the Caribbean. Authorities need to support community organizing rather than just steer it towards management roles. However, Pomeroy et al. (2004) found that in the Caribbean, organizational capacity to engage in co-management is weak. White et al. (1994) state that co-management is not possible in the absence of community organizations.

Long-term support of the local government unit and political elites

The cooperation of the local government unit and the local political elite is important to co-management (Pomeroy et al., 2001). Local government can provide a variety of technical and financial services and assistance to the co-management process. There must be local political will to share benefits, cost, responsibility and authority with the community members. In Africa, Geheb and Sarch (2002) found that there is a strong role for local government in co-management that includes enforcement, sanctions, extra-community issues, extension and information. McConney et al. (2003) found that in the Caribbean the inclusion of the government as a partner is essential for establishing and sustaining co-management. White et al. (1994) stated that local government can provide appropriate support for co-management that the community members cannot, such as local ordinances to support management measures and enforcement. CANARI (1999) reported that participation in co-management requires changes in the attitude of government staff and political elites towards co-management and towards other stakeholders in the process. There needs to be an awareness that powerful stakeholders may circumvent participatory processes when it serves their interest to do so. McConney et al. (2003) stated that co-management is likely to redistribute power and to be resisted by those who want to avoid losing, or sharing, power. Geheb and Sarch (2002) stated that new co-management initiatives may be used by one or more stakeholders to improve or consolidate their power or position. In designing new co-management systems, it must be assumed that such struggles will affect the outcome of any intervention, and the objective becomes one of trying to ensure that any resulting social or economic disequilibrium is minimized.

Property rights over the resource

Property rights, either individual or collective, should address the legal ownership of the resource and define mechanisms (economic, administrative and collective) and the structures required for allocating use rights to optimize use and ensure conservation of resources, and the means and procedures for enforcement (Pomeroy et al., 2001). Without legally supported property rights, resource users have no standing to enforce their claims over the resource against outsiders. Hauck and Sowman (2003) stated that, while co-management arrangements in South Africa have focused on increased user participation in management, a fundamental first step has been the need to clarify and secure property rights to resources. McConney et al. (2003) stated that in the Caribbean, partners in co-management are unlikely to contribute significantly to the effort over the long term if they do not expect to be able to maintain or increase the benefits of their investment in participation. A key to success is to reduce the open-access nature of marine resources through the establishment of property rights. Pomeroy et al. (2004) reported that in the Caribbean, legislation providing property rights over marine and coastal resources is absent.

Adequate financial resources/budget

Pomeroy et al. (2001) reported that co-management requires financial resources to support the process. Funds are needed to support various operations and facilities related to planning, implementation, coordination, monitoring and enforcement, among other activities. Funding, especially sufficient, timely and sustained funding, is critical to co-management. Hauck and Sowman (2003) reported that in South Africa, limited funding and unrealistic time frames are a constraint to co-management. It is critical to recognize the time and resources required to develop and implement co-management arrangements. Unreliable funding can create significant obstacles to collaborative working relationships between stakeholders. Khan et al. (2004) reported that in Africa, the provision of adequate financial and technical resources is key to any effort for sustainable co-management. In the Caribbean, CANARI (1999) stated that the implementation of participatory decisions and management actions requires not only political support but also adequate technical and financial resources.

Partnerships and ownership of the co-management process

In Asia, Pomeroy et al. (2001) reported that active participation of partners in the co-management planning and implementation process is directly related to their sense of ownership and commitment to co-management arrangements. Partners involved in co-management need to feel that the process not only benefits them, but that they have a strong sense of participation in, commitment to and ownership of the process. Partnerships must grow out of a mutual sense of trust and respect among the partners. McConney et al. (2003) reiterated that trust and respect among partners is necessary for successful co-management in the Caribbean context. White et al. (1994) also stated that trust and respect between community workers, outside organizations and community members must be established and maintained. White et al. (1994) further stated that communities respond to an intervention when they believe that it is needed, that it will be effective in meeting their needs and that they 'own' the intervention process.

Clear objectives from a well-defined set of issues

The clarity and simplicity of objectives help to steer the direction of co-management (Pomeroy et al., 2001). Clear and simple objectives based on an understanding of the issues by the stakeholders are essential for success of co-management. Fundamental to co-management is a common understanding of the situation, comprehension of the root causes of the problems and the issues, and an agreement on appropriate solutions to the identified problems. Hauck and Sowman (2003) stated that the objectives of co-management must be agreed upon by all parties. Co-management originates as a result of varying objectives. People who are affected by management decisions must be involved in developing the objectives and setting the parameters to be achieved. In the Caribbean, Pomeroy et al. (2004) found that clear objectives for co-management need to be defined by the stakeholders based on the problems and their interests. White *et al.* (1994) stated that clear, salient objectives and issues are crucial early on, because many people need to know, from the outset, where the co-management process is headed. Key identification of the issues and clear objectives are key to motivating individuals and organizations to engage in co-management.

Management rules enforced

In Asia, Pomeroy *et al.* (2001) found that the enforcement of management rules was of great importance for the success of co-management. Rules must be simple and enforceable. Vigorous, fair and sustained rule enforcement requires the participation of all partners. Hauck and Sowman (2003) found that resource users should be consulted and actively involved when rules are developed to bring about greater legitimacy. There is a need for mutual agreement on what constitutes legitimate rules as a means of fostering trust and increasing compliance. Monitoring for enforcement should be an integral part of the co-management process, should involve local resource users and be backed up with government support. McConney *et al.* (2003) reported that weak enforcement undermines co-management by increasing the uncertainty of resource sustainability and decreasing the returns on participation. Pomeroy *et al.* (2004) found that in the Caribbean, the success of co-management was enhanced when management rules are enforceable by both resource users and the management authority.

Knowledge of resource

McConney *et al.* (2003) reported that co-management is more likely to succeed if the resource is one of which stakeholders have a good knowledge. The integration of good traditional knowledge, practices and tenure systems must be brought out and made a part of the co-management process (White *et al.*, 1994). Geheb and Sarch (2002) stated that if communities of resource users are to assume or retain responsibilities for controlling access to the fisheries, then their knowledge about and perception of the resources needs to be understood.

Individual and Household Level

Individual incentive structure

Pomeroy *et al.* (2001) stated that the success of co-management hinges on an individual incentive structure (economic, social, political) that induces individuals to participate in the process. CANARI (1999) stated that co-management efforts that appeal to the motivations (most often economic) of the stakeholders are the most likely to secure their participation. White *et al.* (1994) stated that individuals who are not dependent upon a finite resource will not respond quickly to co-management. Sverdrup-Jensen and Raakjær Nielsen (1998), in a summary of findings from eight co-management case studies in Africa, reported that when expectations are high among stakeholders, unmet expectations can lead to an unwillingness to participate in co-management. McConney *et al.* (2003) reported that incentives may not always work in favour of co-management unless there is some level of personal gain from participation.

Benefits exceed costs

Hauck and Sowman (2003) stated that while it is difficult to measure benefits and costs to individuals engaged in co-management because they are measured in different ways, and some factors are intangible and unmeasureable, there must be clear benefits that outweigh costs or disadvantages. McConney *et al.* (2003) stated that co-managers need to be concerned about benefits or incentives for all of the participating stakeholders in order to ensure that motivation is sustained, especially in the early stages of co-management. Individual stakeholders have their own real costs and need real benefits for themselves, often to justify participation to a larger constituency that they represent or with which they interact. Hauck and Sowman (2003) further stated that only when benefits become tangible can people afford to adopt a long-term view and behaviour about using resources sustainably.

Adaptive Co-management

Adaptive management recognizes that management is necessary even when all desirable information is not available and when the effects of management cannot be predicted. It views management not only as a way to achieve objectives, but also as a process in which we learn more about the resource and the fisheries system being managed (see also Andrew and Evans, Chapter 2 this volume, and Parks, Chapter 6, this volume). Learning is thus an inherent objective of adaptive management. As we learn more, we can adapt our policies to improve management and be more responsive to future conditions. A key feature of adaptive co-management is the combination of the iterative learning dimension of adaptive management and the linkage dimension of collaborative management in which rights and responsibilities are jointly shared. Although with co-management much of the focus is on the local scale where issues of management performance are felt most directly, adaptive co-management is a flexible system for environment and resource management that operates across multiple levels and with a range of local and non-local organizations (see Armitage *et al.*, 2007 and Parks, Chapter 6, this volume). Key features of adaptive co-management include a focus on learning-by-doing, integration of different knowledge systems, collaboration and power-sharing among community, regional and national levels, and management flexibility (Olsson *et al.*, 2004).

Many of the conditions identified in this chapter are critical factors in adaptive co-management, such as the integration of different knowledge systems, collaboration across different scales and management flexibility.

This is not to say that the features of adaptive co-management are not mentioned in the various studies. For example, Hauck and Sowman (2003, p. 335) stated:

> Success is more likely to be achieved if stakeholders involved in these various co-management initiatives share experiences, learn from past mistakes and are willing to modify their management strategies and rules to suit changing circumstances and management capabilities.

Further, as McConney *et al.* (2003, p. 11) stated:

> One approach is to manage by trial and error, without paying much attention to accumulating knowledge about the systems. A better approach is to learn through adaptive management. ... It involves institutional learning where all of the co-management stakeholders share information and record conclusions or decisions about the human and natural resource systems. By careful analysis and documentation, the co-management institution, as a whole, learns together for improvement.

Discussion

A number of studies in Asia, Africa, the Pacific and the wider Caribbean published in recent years have identified key conditions that help to account for the success and sustainability of co-management. These conditions show both similarities and differences between regions. It should be noted that these are generalized conditions for the region and that the key conditions may differ for an individual country within the region, or even locality within a country. They must be viewed in the distinct political, biological, cultural, technological, social and economic context of that region and the individual countries within the region. We also need to bear in mind the role these unique characteristics play in shaping the process and implementation of co-management in the region. While in Asia, where the use of co-management is more mature, there is more delegated co-management (government lets formally organized users/stakeholders make decisions) (Pomeroy and Viswanathan, 2003), in Africa and the Caribbean, where co-management is still a relatively new concept, there is more consultative co-management (government interacts often with users/stakeholders but makes most decisions) (Sverdrup-Jensen and Raakjær Nielsen, 1998; Hara and Raakjær Nielsen, 2003).

It is important to note that adaptive management is not explicitly mentioned as a key condition for successful co-management in any of the studies, although several individual features of adaptive co-management (integration of knowledge systems, collaboration across scales and management flexibility) are identified as key conditions. The iterative learning dimension of adaptive co-management is, however, discussed in several papers as an important element in the co-management process. The lack of acknowledgement of adaptive co-management in these papers may be due to either a lack of formal recognition of adaptation and learning in the co-management process or how relatively new the approach is.

Several of the key conditions stand out as being more common across the regions than others, including participation by those affected by the co-management arrangements, empowerment and capacity building, community organizations and individual incentive structure.

Some of the conditions can be met by attributes internal to the community, while others are external to the community. The number and variety of conditions illustrate that the planning and implementation of co-management must be conducted at several levels. These levels include: (i) the individual (i.e. individual incentive structure; benefits exceed costs); (ii) the stakeholder (i.e. participation by those affected; empowerment and capacity building; community organizations); (iii) the local government (i.e. long-term support of the local government unit and political elites); (iv) the national government (i.e. enabling policies and legislation); (v) the external agent; (vi) the resource (i.e. appropriate scale and boundaries); and (vii) the overall co-management process (i.e. clear objectives from a well-defined set of issues; management rules enforced; adequate financial resources).

None of the conditions exist in isolation, but each supports and links to another to make the complex process and arrangements for co-management work. In addition, all the stakeholders (resource users, external agents, government) have different but mutually supportive roles to play in co-management. The fulfilment of these complementary roles is crucial to the operation and sustainability of co-management.

Implementation is often a balancing act to meet these conditions, as timing and linkages in the co-management process and arrangements are important. For example, empowerment and capacity building are needed to support community organization development. Developing trust between partners is associated with effective communication. The recognition of resource management problems is associated with the development of clear objectives from a set of well-defined issues.

There will be a need to transform to adaptive co-management. As mentioned, many of the key conditions for adaptive co-management are already identified from previous studies. The importance of adaptation and learning as key conditions for success will need to be more explicitly recognized and utilized by co-management practitioners and communities. A key element will be to further formally integrate adaptation and learning into the process of co-management and the evolution to adaptive co-management.

References

Agrawal, A. (2002) Common resources and Institutional sustainability. In: Ostrom, E., Dietz, T., and Dolsak, N. (eds) *The Drama of the Commons*. National Academy Press, Washington, DC, pp. 41–86.
Antona, M., Bienabe, E.M., Salles, J.M., P'Echard, G., Aubert, S. and Ratsimbarison, R. (2004) Rights transfers in Madagascar biodiversity policies: achievements and significance. *Environment and Development Economics* 9, 825–847.

Armitage, D., Berkes, F. and Doubleday, N. (2007) *Adaptive Co-management: Collaboration, Learning, and Multi-level Governance*. UBC Press, Vancouver, Canada.

Aswani, S. and Hamilton, R.J. (2004) Integrating indigenous ecological knowledge and customary sea tenure with marine science and social science for conservation of bumphead parrotfish (*Bolbometopon muricatum*) in the Roviana Lagoon, Solomon Islands. *Environmental Conservation* 31, 69–83.

Berkes, F. (1994) Co-management: bridging the two solitudes. *Northern Perspectives* 22, 18–20.

Berkes, F., Fast, H. and Berkes, M.K. (1996) *Co-management and Partnership Arrangements in Fisheries Resource Management and in Aboriginal Land Claims Agreements*. University of Manitoba, Winnipeg, Canada.

Berkes, F., Mahon, R., McConney, P., Pollnac, R. and Pomeroy, R. (2001) *Managing Small-scale Fisheries: Alternative Directions and Methods*. International Development Research Center, Ottawa, Canada.

Blaikie, P. (2006) Is small really beautiful? Community-based natural resource management in Malawi and Botswana. *World Development* 34, 1942–1957.

CANARI (Caribbean Natural Resource Institute) (1999) *Evaluation of Caribbean Experiences in Participatory Planning and Management of Marine and Coastal Resources*. Caribbean Natural Resource Institute, St. Lucia.

Cinner, J. (2007) The role of taboos in conserving coastal resources in Madagascar. *Traditional Marine Resource Management and Knowledge Information Bulletin* 22, 15–23.

Cinner, J. and Aswani, S. (2007) Integrating customary management into marine conservation. *Biological Conservation* 140, 201–216.

Cinner, J., Marnane, M. and Ben, J. (2003) How socioeconomic monitoring can assist marine reserve management: Kimbe Bay, Papua New Guinea. In: Wilkinson, C. and Green, A. (eds) *Monitoring Coral Reef Marine Protected Areas*. Australian Institute of Marine Science and IUCN Global Marine Program, Townsville, Australia, pp. 28–29.

Cinner, J., Wamukota, A., Randriamahazo, H. and Rabearisoa, A. (2009a) Toward institutions for community-based management of inshore marine resources in the Western Indian Ocean. *Marine Policy* 33, 489–496.

Cinner, J., Fuentes, M.M.P.B. and Randriamahazo, H. (2009b) Exploring social resilience in Madagascar's marine protected areas. *Ecology and Society* 14(1), art. 41. Available at: http://www.ecologyandsociety.org/vol14/iss1/art41 (accessed 11 March 2010).

Cudney-Bueno, R. and Basurto, X. (2009) *Lack of Cross-scale Linkages Reduces Robustness of Community-based Fisheries Management*. Available at: http://plosone.org/article/info:doi/10.1371/journal.pone.0006253 (accessed 11 March 2010).

DeCosse, P.J. and Jayawickrama, S.S. (1998) Issues and opportunities in co-management: lessons from Sri Lanka. In: Kothari, A., Pathak, N., Anuradha, R.V. and Taneja, B. (eds) *Communities and Conservation: Natural Resource Management in South and Central Asia*. Sage Publications, New Delhi, India, pp. 245–289.

Geheb, K. and Sarch, M. (2002) *Africa's Inland Fisheries: the Management Challenge*. Fountain Publishers, Kampala, Uganda.

Gelcich, S., Edwards-Jones, G., Kaiser, M.J. and Castilla, J.C. (2006) Co-management policy can reduce resilience in traditionally managed marine ecosystems. *Ecosystems* 9, 951–966.

Hara, M. and Raakjær Nielsen, J. (2003) Experiences with fisheries co-management in Africa. In: Wilson, D.C., Raakjær Nielsen, J. and Degnbol, P. (eds) *The Fisheries Co-management Experience: Accomplishments, Challenges and Prospects*. Kluwer Academic Publishers, Dordrecht, The Netherlands, pp. 81–98.

Hara, M. and Raakjær Nielsen, J. (2009) Policy evolution in South African fisheries: the governance of the sector for small pelagics. *Development Southern Africa* 26, 649–662.

Hauck, M. and Sowman, M. (eds) (2003) *Waves of Change: Coastal and Fisheries Co-management in Southern Africa*. University of Cape Town Press, Lansdowne, South Africa.

Hoefnagel, E. and Smit, W. (1996) *Co-management Experiences in the Netherlands*. Agricultural Economics Research Institute, Fisheries Department, The Hague, The Netherlands.

ICLARM/IFM (1996) *Analysis of Fisheries Co-management Arrangements: a Research Framework*. ICLARM, Manila, Philippines and IFM, North Sea Center, Hirtshals, Denmark.

Isaacs, M., Hara, M. and Raakjær Nielsen, J. (2007) Has reforming South African fisheries contributed to wealth redistribution and poverty alleviation? *Ocean and Coastal Management* 50, 301–313.

Jentoft, S. and Kristofferson, T. (1989) Fishermen's co-management: the case of the Loftoten fishery. *Human Organization* 48, 355–365.

Jul-Larsen, E., Kolding, J., Overå, R., Raakjær Nielsen, J. and van Zwieten, P.A.M. (2002) Management, co-management or no-management? Major dilemmas in the sustainable utilisation of SADC freshwater fisheries. Synthesis report, FAO Fisheries Technical Paper No. 426/1, FAO, Rome, Italy.

Khan, A.S., Mikkola, H. and Brummett, R. (2004) Feasibility of fisheries co-management in Africa. *NAGA, Worldfish Center Quarterly* 27, 60–64.

Low, B., Ostrom, E., Simon, C. and Wilson, J. (2002) Redundancy and diversity: do they influence optimal management? In: Berkes, F., Colding, J. and Folke, C. (eds) *Navigating Social-ecological Systems*. Cambridge University Press, Cambridge, UK, pp. 83–114.

McConney, P., Pomeroy, R. and Mahon, R. (2003) *Guidelines for Coastal Resource Co-management in the Caribbean: Communicating the Concepts and Conditions that Favor Success*. Caribbean Conservation Association, Bridgetown, Barbados.

Normann, A.K., Raakjær Nielsen, J. and Sverdrup-Jensen, S. (1998) *Fisheries co-management in Africa: proceedings from a regional workshop on fisheries co-management research*. Fisheries Co-management Research Project Research Report No. 12, Institute for Fisheries Management, Hirtshals, Denmark.

Nyikahadzoi, K. and Raakjær Nielsen, J. (2009) Policy evolution and dynamics of governance at the Lake Kariba kapenta fishery. *Development Southern Africa* 26, 639–648.

Nyikahadzoi, K, Hara, M. and Raakjær Nielsen, J. (2010) Transforming ownership and governance – lessons from capital intensive pelagic fisheries in South Africa and Zimbabwe. *International Journal of the Commons*. Available at: http://www.thecommonsjournal.org/index/php/ijc/article/viewArticle/227/157 (accessed 4 September 2010).

Olsson, P., Folke, C. and Berkes, F. (2004) Adaptive co-management for building resilience in socio-ecological systems. *Environmental Management* 34, 75–90.

Ostrom, E. (1990) *Governing the Commons: the Evolution of Institutions for Collective Action*. Cambridge University Press, Cambridge, UK.

Ostrom, E. (1992) *Crafting Institutions for Self-governing Irrigation Systems*. Institute for Contemporary Studies Press, San Francisco, California.

Ostrom, E. (1994) Institutional analysis, design principles and threats to sustainable community governance and management of commons. In: Pomeroy, R.S. (ed.) *Community Management and Common Property of Coastal Fisheries in Asia and the Pacific: Concepts, Methods and Experiences. ICLARM Conference Proceedings 45*, International Center for Living Aquatic Resources Management, Manila, Philippines, pp. 34–50.

Pinkerton, E. (1989) *Cooperative Management of Local Fisheries*. University of British Columbia Press, Vancouver, Canada.

Pinkerton, E. (1993) *Local Fisheries Co-management: a Review of International Experiences and their Implications for Salmon Management in British Columbia*. School of Community and Regional Planning, University of British Columbia, Vancouver, Canada.

Pinkerton, E. (1994) Summary and conclusions. In: Dyer, C.L. and McGoodwin, J.R. (eds) *Folk Management in the World's Fisheries*. University of Colorado Press, Boulder, Colorado, pp. 317–337.

Pomeroy, R.S. and Rivera-Guieb, R. (2006) *Fishery Co-management: a Practical Handbook*. CABI Publishing, Wallingford, UK and International Development Research Centre, Ottawa, Canada.

Pomeroy, R. and Viswanathan, K. (2003) Experiences with fisheries co-management in Southeast Asia and Bangladesh. In: Wilson, D.C., Raakjær Nielsen, J. and Degnbol, P. (eds) *The Fisheries Co-management Experience: Accomplishments, Challenges and Prospects*. Kluwer Academic Publishers, Dordrecht, The Netherlands, pp. 99–118.

Pomeroy, R., Katon, B. and Harkes, I. (2001) Conditions affecting the success of fisheries co-management: lessons from Asia. *Marine Policy* 25, 197–208.

Pomeroy, R., McConney, P. and Mahon, R. (2004) Comparative analysis of coastal resource co-management in the Caribbean. *Ocean and Coastal Management* 47, 429–447.

Raakjær Nielsen, J. (2009) *A Fisheries Management System in Crisis – the EU Common Fisheries Policy*. Aalborg University Press, Aalborg, Denmark.

Raakjær Nielsen, J., Degnbol, P., Viswanathan, K., Ahmed, M., Hara, M. and Abdullah, N.M.R. (2004) Fisheries co-management – an institutional innovation? Lessons from South East Asia and Southern Africa. *Marine Policy* 28, 151–160.

Raakjær Nielsen, J., Son, D.M., Stæhr, K., Hovgård, H., Thuy, N.T.D., Ellegaard, K., Riget, F., Thi, D.V. and Hai, P.G. (2007) Adaptive fisheries management in Vietnam: the use of indicators and introducing a multi-disciplinary Marine Fisheries Specialist Team to support implementation. *Marine Policy* 31, 143–152.

Sen, S. and Raakjær Nielsen, J. (1996) Fisheries co-management: a comparative analysis. *Marine Policy* 20, 405–418.

Steins, N.A., Edwards, V.M. and Roling, N. (2000) Re-designed principles for CPR Theory. *Common Property Resource Digest* 53, 1–3.

Sverdrup-Jensen, S. and Raakjær Nielsen, J. (1998) *Co-management in Small-scale Fisheries: a Synthesis of Southern and West African Experiences*. Institute for Fisheries Management and Coastal Community Development, The North Sea Center, Hirtshals, Denmark.

White, A.T., Hale, L.Z., Renard, Y. and Cortesi, L. (1994) *Collaborative and Community-based Management of Coral Reefs*. Kumarian Press, West Hartford, Connecticut.

Yandle, T. (2003) The challenge of building successful stakeholder organizations: New Zealand's experience in developing a fisheries co-management regime. *Marine Policy* 27, 179–192.

8 Climate Change and Other External Drivers in Small-scale Fisheries: Practical Steps for Responding

Stephen J. Hall

Introduction

Classically, management has concentrated on the fishery itself. Yet many of the challenges fisheries face are shaped by complex combinations of biophysical, social, political and economic forces. Many of these forces operate at scales beyond national level and outside the domain of fisheries. While there is usually limited scope for fisheries management to control these forces, policy makers and managers must understand them and plan for their impact.

Climate change is, perhaps, the driver that is receiving most attention at present, with governments increasingly calling for strategies to cope with the changes it will bring. Several other drivers, however, remain largely ignored by mainstream fisheries policy analysts; for example, finding an analysis of the likely impact of demographic, health and disease trends, or of wider development policy trends, is a challenge.

This chapter tries to meet two objectives. The first is to summarize the external drivers likely to affect small-scale fisheries over the next decade. The second is to offer some practical suggestions for researchers, managers and policy makers on how to develop responses to them. Given its prominence globally and its potential effect on fisheries systems, I focus in particular on climate change. However, the suggested approach to analysis and dialogue needed to identify the policy alternatives and management approaches to respond to climate change are equally applicable for developing responses to the other drivers that challenge fisheries systems.

Some readers may believe that worrying about the wider issues that impinge on fisheries is a diversion from the imperative of dealing with the more immediate internal problems that drive over-exploitation of fisheries. To some degree and for some fisheries, this is of course true, but ignoring these externalities is likely to prove short-sighted. These drivers have the potential to profoundly affect resource sustainability and the well-being of the people who depend on them. More positively, paying more attention to them might also help identify new arenas in which to also find solutions to more mainstream fisheries problems. Linking fisheries considerations into wider issues of climate change, migration, human rights or wider development policy, for example, might offer a more effective context for solving traditional fisheries management issues such as access rights, effort control or vessel decommissioning.

Key Concepts and Definitions

Adapting the definition used by Hazell and Wood (2008, p. 501) in their analysis of drivers of change in global agriculture, an external driver is defined here as '... any natural or human-induced factor that indirectly brings about change in a fishery system'. In this context, the term fishery system refers to a social-ecological system (SES) *sensu* Gallopin (1991, p. 707) i.e. ' ... a system that includes societal (human) and ecological (biophysical) subsystems in mutual interaction'.

Figure 8.1 summarizes a range of external drivers that fishery managers and policy makers should think about. Within each broad category several dimensions are highlighted.

Naturally, any categorization of drivers brings with it fuzzy boundaries and overlaps; it is arguable, for example, whether the growth of mobile phone technology should be included in a section on 'markets and trade' or 'technology'. I have simply grouped issues within the section that seemed most appropriate.

Along with a list of drivers, it is also helpful to have a framework for aspects of a fishery system where their impact is likely to be felt. Using a driver tree that categorizes the benefits that small-scale fisheries deliver and the network of conditions upon which they

	Driver	Key issues
	International trade and globalization of markets	• Economic integration • Trade liberalization • Export-led growth • Labour & capital mobility • Energy & food prices
	Technology	• Fish finding/navigation • Communications • Monitoring and compliance
	Climate and environment	• Extreme events • Water demand • Environmental services
	Health and disease	• Food safety • Infectious diseases • Health and safety
	Demography	• Population growth • Migration • Urbanization
	Governance	• Global governance • Democratization and decentralization • Political instability
	Development patterns	• Inequality & social change • Development policy • Rights-based approaches • Infrastructure and market access
	Aquaculture	• Competition for space • Changes in supply and demand • Demand for low-value fish

Fig. 8.1. Selected external drivers of change in small-scale fisheries.

are dependent, Fig. 8.2 offers such a framework.

The fundamental premise for this chapter is that by better understanding these external drivers, one will be in a better position to assess the vulnerability of fisheries to them. Since vulnerability varies greatly across production systems, households, communities, nations and regions, a second premise for this chapter is that systematic diagnosis in particular national, regional and local contexts is a prerequisite for sensible investment and action. The increase in understanding that a good diagnosis brings in turn allows one to identify options better to cope with, or adapt to, external threats, thereby helping to build resilience.

These four terms – vulnerability, coping, adaptation and resilience – are central to this chapter. Given their importance they warrant further amplification.

Vulnerability

The extent to which a particular fishery system or region is vulnerable to a particular driver results from a combination of three key factors: (i) the exposure of the system; (ii) the degree of sensitivity to the driver; and (iii) the adaptive capacity of the group or society experiencing those impacts.

Coping and adaptation

Coping refers to the actions people take to deal in some way with a stressor and to limit its effects. Although the roots of the concept of coping reside in the psychological literature, more recently development practitioners have thought about coping strategies in a broader livelihoods context. In this context,

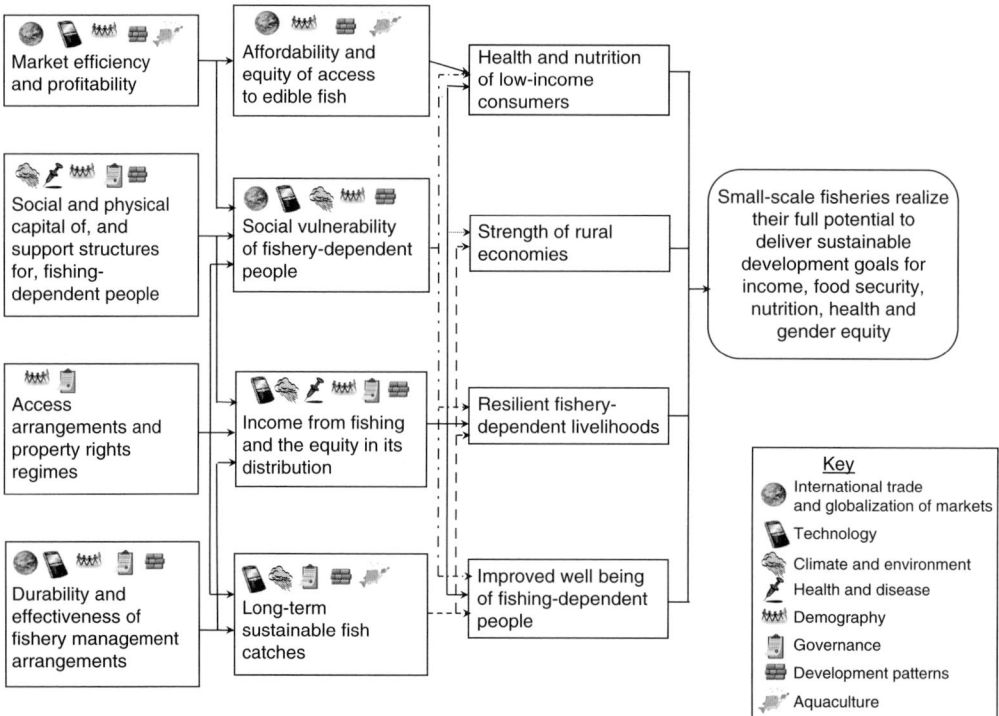

Fig. 8.2. A driver tree categorizing the benefits that small-scale fisheries deliver and the network of conditions upon which they are dependent.

the term is perhaps most usefully used to describe unpremeditated actions that are taken after a stress has occurred. Often, these actions erode long-term capacity to deal with future stresses. A fishing family that suffers property damage during a large storm might, for example, have to sell livestock to cover the costs of rebuilding. This sense of unplanned action after the event distinguishes coping from adapting, which involves anticipatory actions that reduce exposure to the stress or limit the damage caused. We can make a further distinction between reactive adaptation, where plans are enacted when a threat presents, versus anticipatory adaptations. An example of a reactive adaptation would be enacting cyclone evacuation plans, while an anticipatory adaptation would be permanently moving home away from the coastal zone to avoid storm damage, or physically reinforcing property against a storm's force. Of course, adaptation decisions might be triggered by experience and coping responses to past events and a desire to avoid them in future.

Resilience

Capacity to cope and the degree of adaptation contribute to the resilience of a social-ecological system. For our purposes, Walker *et al.* (2004, p. 2) provide a suitable definition of resilience as '... the capacity of a system to absorb disturbance and reorganize while undergoing change so as to still retain essentially the same function, structure, identity, and feedbacks'. This definition applies to both the ecological and human domains of the social-ecological system, but for the purposes of this chapter it is the human dimensions that are most central. For systems that are in a desirable state, we usually wish to increase the amount of change it can experience before crossing a threshold – this is a measure of its resilience.

Climate Change

Perhaps more than any other external driver, the issue of climate change has reached the consciousness of those involved in fisheries.

Current climate projections strongly suggest that the effects on coasts, lakes and rivers, and on the fisheries they support, will bring new challenges for these systems and to the people who depend upon them. The Intergovernmental Panel on Climate Change (IPCC) projects that global temperatures will increase by 1.8–4.0°C by 2100 (IPCC, 2007). This atmospheric warming will be accompanied by rising sea temperatures, changing sea levels, increasing ocean acidification, altered rainfall patterns and river flows, and increased incidence of extreme weather events. These effects will, in turn, affect the productivity, distribution and seasonality of fisheries, and the quality and availability of the habitats that support them. In addition, many fishing-dependent communities are located in regions of high physical exposure to climate change.

As with all change drivers, understanding the links between climate change, livelihoods and food security is critical for designing policies and management strategies for fishery-dependent communities, nations and regions. Within the context of the fishery system as described in Fig. 8.2, we can see that these climate-change impacts come into play through two pathways. The first is through effects on the social and physical capital of, and support structures for, fishing-dependent people (Table 8.1). The second is through biophysical impacts on fishery resources themselves (Table 8.2).

As a high-level generic summary of the issue, the above analysis is useful but somewhat limited. A fishery manager or policy maker with responsibility for a specific geographic domain requires a much more nuanced and context-specific picture that points towards clear action steps. Centred on the resource productivity of the fishery system, Fig. 8.3 provides an example framework for organizing thinking at this more detailed level.

Within such a framework we can build a picture of what really matters by asking the following three fundamental questions:

1. What precisely is the threat from climate change in my context and how does it compare with other threats?

Table 8.1. The effects of climate change on the social and physical capital of, and support structures for, fishing-dependent people (adapted from WorldFish, 2007, Table 1).

Drivers	Biophysical effects	Implications for social and physical capital and support structures
Changes in precipitation and water availability	Lower water availability for aquaculture; increased competition with other water users	Higher costs of maintaining aquaculture pond water levels and stock replacement Reduced production capacity; conflict with other water users
Drought	Changes in agricultural productivity	Increased adoption of fishing by farmers during periods of stress
	Lower water quality and availability for aquaculture Salinity changes	Increased production costs; loss of opportunity as production is limited
	Changes in lake water levels and river flows	Reduced wild fish stocks, leading to intensified competition for fishing areas and more migration by fisherfolk
Increase in frequency and/or intensity of storms	Large waves and storm surges; inland flooding from intense precipitation; salinity changes; introduction of disease or predators into aquaculture facilities during flooding episodes	Loss of aquaculture stock and damage to or loss of aquaculture facilities and fishing gear; higher direct physical risk to fishers during fishing operations; increased capital costs needed to design cage moorings, pond walls, jetties, etc. to withstand storms; higher insurance costs
Rising sea levels	Loss of land	Reduced area available for aquaculture
	Saltwater infusion into groundwater	Reduced freshwater availability for aquaculture
	Loss of coastal ecosystems such as mangrove forests	Worsened exposure to waves and storm surges and risk that inland aquaculture becomes inundated
Changes in sea surface temperature	More frequent and harmful algal blooms; less dissolved oxygen; increased incidence of disease and parasites; altered local ecosystems with changes in competitors, predators and invasive species; changes in plankton composition	For aquaculture, increased infrastructure and operating costs from aggravated infestations of fouling organisms, pests, nuisance species and/or predators

2. What does history tell me – how have people coped and adapted to past problems?
3. How might we respond and adapt for what may come?

To these three questions we must add a fourth, more pragmatic, one:

1. How can I obtain the financial and other support I need to make things happen?

In the following sections I will consider the first three of these questions in turn, followed by some thoughts on how to go about answering them; I will then consider question 4.

Understanding the threats

Developing policies and strategies to address climate change depends critically

Table 8.2. Ways in which climate change may directly affect fishery resources and aquaculture species (adapted from WorldFish, 2007, Table 1).

Drivers	Biophysical effects	Implications for fisheries and aquaculture species
Changes in sea surface temperature	More frequent harmful algal blooms; less dissolved oxygen; increased incidence of disease and parasites; altered local ecosystems with changes in competitors, predators and invasive species; changes in plankton composition	Worsened infestations of fouling organisms, pests, nuisance species and/or predators For capture-fisheries, impacts on the abundance and species composition of fish stocks
	Longer growing seasons; lower natural mortality in winter; enhanced metabolic and growth rates	Potential for increased production and profit, especially for aquaculture
	Enhanced primary productivity	Potential benefits for aquaculture and fisheries, but perhaps offset by altered species composition
	Changes in timing and success of migrations, spawning and peak abundance, as well as in sex ratios	Potential loss of species or shift in composition in capture fisheries; impacts on seed availability for aquaculture
	Changes in the location and size of suitable range for particular species	Aquaculture opportunities both lost and gained Potential species loss and altered species composition for capture-fisheries
	Damage to coral reefs that serve as breeding habitats and that may help protect the shore from wave action (exposure to which may rise along with sea levels)	Reduced recruitment of fishery species More severe wave damage to infrastructure or flooding from storm surges
El Niño-Southern Oscillation (ENSO)	Changed location and timing of ocean currents and upwelling alters nutrient supply in surface waters and, consequently, primary productivity	Changes in the distribution and productivity of open-sea fisheries
	Changed ocean temperature and bleached coral	Reduced productivity of reef fisheries
	Altered rainfall patterns bring flood and drought	See impacts from precipitation, drought and flooding above
Rising sea level	Loss of land	Loss of freshwater fisheries
	Changes to estuary systems	Shifts in species abundance, distribution and composition of fish stocks and aquaculture seed
	Salt water diffusion into groundwater	Damage to freshwater capture fisheries; for aquaculture, a shift to brackish water species
	Loss of coastal ecosystems such as mangrove forests	Reduced recruitment and stocks for capture fisheries and seed for aquaculture Worsened exposure to waves and storm surges and risk that inland fisheries become inundated

(Continued)

Table 8.2. Continued

Drivers	Biophysical effects	Implications for fisheries and aquaculture species
Higher inland water temperatures	Increased stratification and reduced mixing of water in lakes, reducing primary productivity and ultimately food supplies for fish species	Reduction in fish stocks
	Raised metabolic rates increase feeding rates and growth if water quality, dissolved oxygen levels and food supply are adequate, otherwise possibly reduction in feeding and growth; potential for enhanced primary productivity	Possibly enhanced fish stocks for capture fisheries or else reduced growth where the food supply does not increase sufficiently in line with temperature
		Possible benefits for aquaculture, especially intensive and semi-intensive pond systems
	Shift in location and size of potential range for a given species	Aquaculture opportunities both lost and gained
		Potential loss of species and alteration of species composition for capture-fisheries
	Reduced water quality, especially in terms of dissolved oxygen; changes in the range and abundance of pathogens, predators and competitors; invasive species introduced	Altered stocks and species composition in capture-fisheries For aquaculture, altered culture species and possibly more severe losses to disease (and so higher operating costs) and possibly higher capital costs for aeration equipment or deeper ponds
	Changes in timing and success of migrations, spawning and peak abundance	Potential loss of species or shift in composition for capture-fisheries; impacts on seed availability for aquaculture
Changes in precipitation and water availability	Changes in fish migration and recruitment patterns, and thus in recruitment success	Altered abundance and composition of wild stock; impacts on seed availability for aquaculture
	Lower water quality causing more disease	Change of culture species
	Altered and reduced freshwater supplies with greater risk of drought	
	Changes in lake and river levels and the overall extent and movement patterns of surface water	Altered distribution, composition and abundance of fish stocks
Increase in frequency and/or intensity of storms	Large waves and storm surges Inland flooding from intense precipitation Salinity changes Introduction of disease or predators into aquaculture facilities during flooding episodes	Impacts on wild fish recruitment and stocks
Drought	Lower water quality and availability for aquaculture Salinity changes	Loss of wild and cultured stock Increased production costs Loss of opportunity as production is limited
	Changes in lake water levels and river flows	Reduced wild fish stocks, intensified competition for fishing areas and more migration by fisherfolk

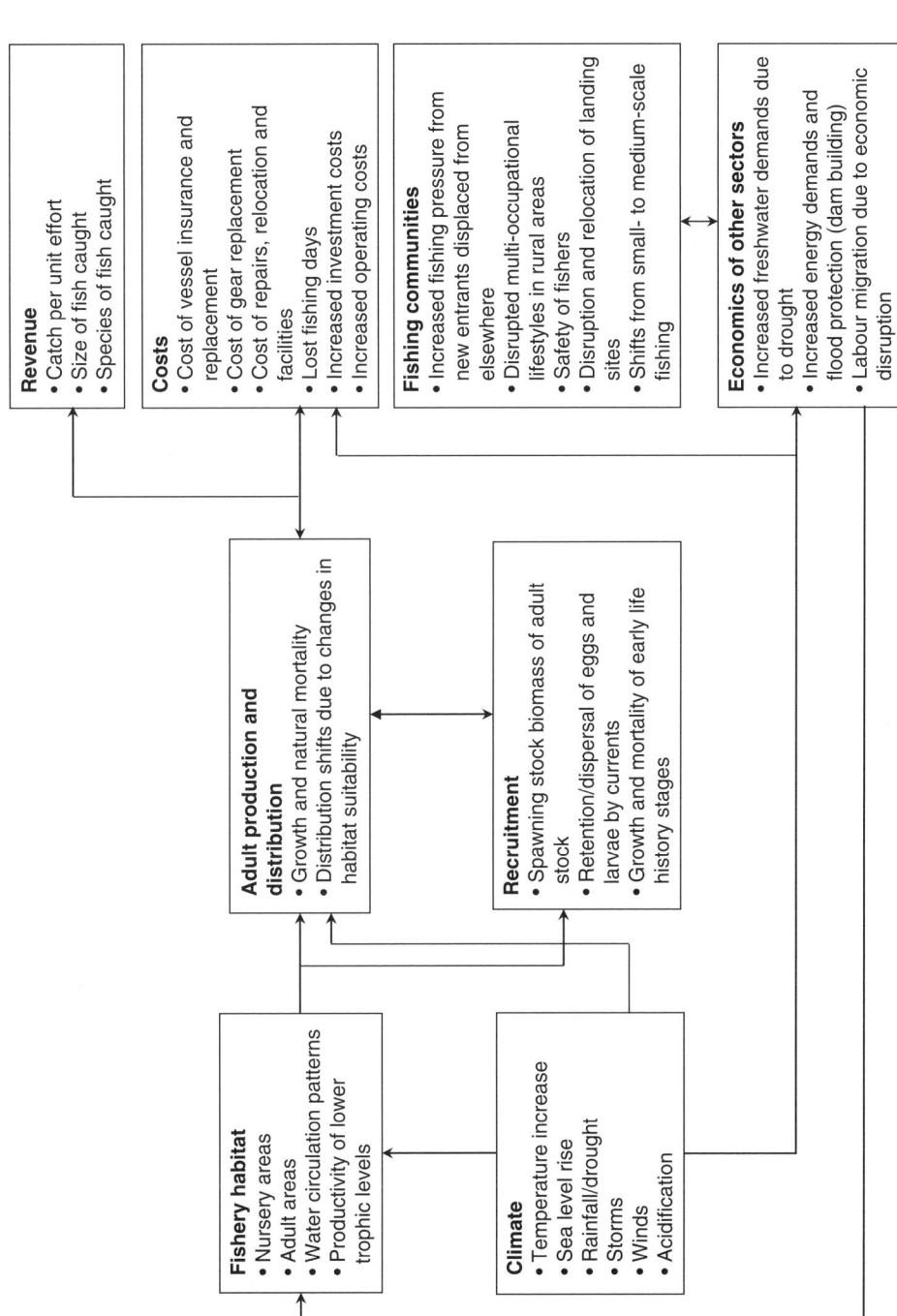

Fig. 8.3. An example framework for organizing thinking at a more detailed level, centred on the resource productivity of the fishery system (adapted from Mahon, 2002).

on identifying vulnerable places and people, and understanding what drives their vulnerability. This will require managers and policy makers to undertake, commission or draw upon vulnerability assessments at multiple scales, which take into account not only climate change but other interacting drivers. These two issues – multiple scales and multiple drivers – are key; we need to understand what is going on at scales from regional to local, and we must recognize that it will often be hard to separate the impacts or quantify the interactions between climate change and the other change drivers that will affect fisheries.

With regard to multiple drivers, consider river, lake or floodplain fisheries prosecuted in an agricultural landscape. Stimulated or not by impending droughts, increased water abstraction rates for irrigation are likely to affect fisheries productivity. Similarly, mangrove loss to aquaculture facilities in response to growing markets for prawns can affect coastal fisheries and render coastal landscapes more vulnerable to extreme weather (Handisyde et al., 2006). The point here is that the relative importance of the various change drivers and the pathways through which they might act must be weighed to help prioritize actions. Understanding climate vulnerability in the context of other drivers also helps to inform programmes aimed at mainstreaming climate change responses into other development policy and planning activities.

With regard to scales of analysis, there can be little doubt that any successful responses to the challenges external drivers present will require linkages between local, national and, quite possibly, regional levels. Recognizing that action at the local level, where vulnerability is actually felt, is especially important, Næss et al. (2006) examined how best to use vulnerability assessment as an instrument in aiding local adaptation. A key conclusion from their analysis is that it is important to link macro- and micro-level assessments through a dialectical process, rather than to try to integrate the information provided at these two levels. In other words, macro-level analytical and descriptive insights should be used to help raise awareness and frame local-level approaches that interpret perceived vulnerabilities in a locally relevant context (Table 8.3). A good example of such sub-national scale analysis in a fisheries context is provided by McClanahan et al. (2008).

A key step in making these connections is to find ways to communicate and raise awareness of threats at the scale of relevance to particular individuals. Visual presentation of both macro- and micro-level analysis in the form of maps can be especially powerful for this purpose. For example, national-scale maps showing differences across local government jurisdictions can often focus the minds of national policy makers, local government leaders and citizens in vulnerable areas and spark the dialogue needed to stimulate action. Such maps can also be an important guide for targeting local case studies that ground truth analyses and help to identify local- and state-level institutions and policies that influence coping and adaptation strategies (O'Brien et al., 2004).

Investing in the preparation of thoughtful maps is likely to be worth the effort, and preparing them may not be as difficult as one imagines. Efforts to map vulnerability at national level are now widespread, in both government departments and academic institutions. Such analyses can provide important elements of relevance to fisheries managers. Taking India as an example, one particularly instructive analysis is provided by O'Brien et al. (2004), who examined regional vulnerability of the agricultural sector to climate change in combination with the likely impacts of trade globalization. Here is a good example of a vulnerability assessment that seeks to examine the combined effects of two very different drivers. Data generated from this study would undoubtedly provide some important elements for an analysis to prioritize investment in fisheries. For many countries, there will be such studies and datasets to draw on. Patt et al. (2009) provide a good recent treatment of vulnerability assessment, particularly in relation to environmental

Table 8.3. Information needs and relative strengths of descriptive and interpretive vulnerability assessment approaches (adapted from Næss et al., 2006).

Level of government	Types of information needs	Assessment approach	
		Descriptive (national-level data)	Interpretive (local-level data)
National	Information for initiation of national discussions and awareness raising Information for the development of national adaptation/mitigation policies and priorities	National ranking/overview Identification of vulnerable sectors, social groups, geographical areas	Knowledge about causal mechanisms and interplay effects Identification of local needs for national policies
Local	Information legitimate for local discussions and awareness raising Fit between the scales of information and local institutions Integrated information across sectors and risk categories	Local vulnerability put into a national perspective	Transparent and inclusive assessment process Information in the form of local 'vulnerability maps'

change, and how such assessments can be used to help prioritize adaptation efforts and policy change.

Despite their intuitive appeal, however, a note of caution is warranted because the usefulness of vulnerability assessments for policy making is contested by some researchers. Most concerns centre around the relationship between researchers and stakeholders, the nature of the information contained in vulnerability assessments and the ability of stakeholders to make use of data they contain (Næss et al., 2006). Recommendations to improve their usefulness often include involving stakeholders as full participants in the assessment process from the beginning, combining both expert and lay knowledge and developing scenarios to facilitate learning (e.g. Schröter et al., 2005; Patt et al., 2009). I will return to the issue of how best to undertake needed diagnoses in a later section.

Although this is not the place to go into great detail about alternative approaches to vulnerability assessment, the importance of considering vulnerable groups within localities deserves special mention. In this context the issue of gender predominates. It is vital to remember that, as a general rule, women will be more vulnerable to the impacts of climate change and the impacts of other change drivers. Their social roles, inequalities in the access and control of resources, lower education, poorer health and their low level of participation in decision-making all contribute to this greater vulnerability. There is good evidence, for example, that gender differences in death rates following natural disasters are directly linked to women's economic and social rights. Analysis of disasters in 141 countries showed that in countries where women and men enjoyed equal rights, disasters caused the same number of deaths in both sexes, but where they didn't women suffered more (Neumayer and Plümper, 2007). This study also showed that discrepancies were the result of existing inequalities. For example, boys were given preferential treatment during rescue efforts and, following disasters, both women and girls suffered more from shortages of food and economic resources

(Neumayer and Plümper, 2007). The effects of these differences can be extreme. For example, during the cyclone disasters in Bangladesh in 1991, of the 140,000 people who died 90% were women (Ikeda, 1995).

Given the above, any vulnerability assessment must adopt a gendered perspective to analysis. This is particularly true for local-level assessment where understanding women's and men's resource use patterns, access and responsibilities and how these might change with climate change will be especially important.

Learning from history

There is much to be gained from understanding the role fisheries play in people's lives, and how fisheries and aquaculture systems have responded to past climate variability and other 'shocks'. Examining the responses of fishing communities to natural disasters or other change drivers, in particular the responses of women and the poor, can help us to understand which measures may reduce vulnerability and enhance resilience in the face of future climate impacts.

Although none are well studied, fishers' responses to declines or booms in fish resources are better documented than responses to other drivers. Responses to declines involve either movement elsewhere or *in situ* diversification into other activities, either within or outside of the fishery. In West Africa, for example, declines in coastal resources led fishers to diversify into hunting for bushmeat (Brashares *et al.*, 2004). Responses to El Niño events provide a particularly useful window into likely responses to large changes in fisheries productivity or distribution. The boom in scallop stocks during the 1982–1983 El Niño event, for example, led to migration from across Peru and shifts in fishing practices. In one location fishery statistics jumped from 250 families and 80 boats to 4500 crew, 3000 divers and 1500 boats (Morales, 1993, quoted in Badjeck *et al.*, 2009). Badjeck *et al.* (2009) provide an especially informative analysis of responses to change drivers in the Peruvian scallop fisheries. Understanding such responses to past events allows one better to anticipate possible futures.

Studying past events may also reveal unexpected barriers to adaptation. A good example of this is provided by Coulthard (2008), who showed how wealthier fishermen bound by their caste, specialized skills and status were unable to diversify their fishing techniques. As a result, they were less able to adapt to resource fluctuations than lower-caste fishers who enjoyed greater freedom in being able to use alternative fishing gears.

Better understanding the current livelihood strategies of fishing-dependent communities will also help to identify possible approaches for future change. Building on what people already do is likely to have more traction than novel alternatives, under many circumstances. Pomeroy *et al.* (2006), for example, recommend four key criteria need to be taken into account in assessing possible alternative livelihood options: social feasibility, technical feasibility, institutional feasibility, and supporting infrastructure and policy environment. By implication, finding ways to complement and extend current activities is more likely to satisfy these criteria, so understanding what these activities are is key.

One also needs to remember that change drivers may impinge on fisheries through indirect pathways. In the Caribbean, for example, damaged tourist infrastructure following hurricanes has led to increases in numbers of workers undertaking short-term work as fishers, thereby increasing overexploitation and conflict. Similarly, in Africa droughts have been shown to lead farmers into fishing (Conway *et al.*, 2005).

The objective when looking at the lessons of history, therefore, is to understand the extent to which past successful or failed responses to climate variability and other change drivers might confer or compromise resilience to future climate change. Lessons derived from failures to adapt can be especially informative. Examining ways in which societies might have coped better, and focusing on the political, cultural and socio-economic factors that inhibited them from doing so, will help to identify where intervention is most needed (Smit and Pilifosova, 2001; Thomas and Twyman, 2005). It is also important to ask whether probable short-term coping mechanisms might undermine long-term adaptive capacity. Research addressing such questions will provide governments, communities and their development partners with important lessons upon which to build.

Response options

There will probably be several alternative responses to each of the many threats that climate change poses. Indeed, long lists of measures are now proliferating in the literature (e.g. Smit and Pilifosova, 2001; USAID, 2009). For fisheries systems these measures include, among others, disaster response planning, mangrove rehabilitation, early warning systems, diversifying livelihood portfolios, improving fisheries management and governance systems and many others. Figure 8.4 illustrates some of the possible choices for dealing with climate impacts that may manifest themselves at the various points in the fisheries system described in Fig. 8.2. As an aide memoire, the lists by other authors, and diagrams such as Fig. 8.4, serve a purpose. The United Nations Framework Convention on Climate Change also provides access to a database of local adaptation measures that have been adopted (UNFCCC, undated). Deciding which measures are appropriate for a given location, however, can only be done by those who understand the prevailing circumstances.

In addition to debating which measures might be appropriate in a given circumstance, one must also be clear about three things: to what is the measure a response, who is the response designed to benefit and who is responsible for implementing it. In this context it is also important to remember that women are not only the primary victims of climate change, but they can also be effective change agents for adaptation. Women often have a unique perspective and extensive knowledge and expertise to help assess community risk, identify appropriate adaptation measures and mobilize communities to act (Enarson and Meyreles, 2004).

A central consideration in choosing any option is, of course, whether the societies and individuals have the capacity to exercise it. After examining the literature on hazard responses, resource management and sustainable development, Adger identified the main features of communities or regions that seem to determine their adaptive capacity (Adger, 2003). These features were: economic wealth, technology, information and skills, infrastructure, institutions and equity. As a general proposition, therefore, anything one can do to improve matters along these dimensions will pay dividends in improving the adaptive capacity of people and institutions.

Given the above list, it is unsurprising that capacity is usually related to poverty profiles of individuals and households (but see the study by Coulthard (2008) noted earlier). In Kenya, for example, fishers expressed greater willingness to exit the fishery in the face of a hypothetical 50% decline in catches when they had a higher material style of life and a greater number of occupations (Cinner et al., 2008). One must also ask, therefore, whether the extent to which, for developing countries in particular, the measures to respond to climate changes are really substantially different to those required to achieve broader development aspirations. With the possible exception of sea level rise, it can legitimately be argued that, under most circumstances, achieving broader development goals will contribute more than anything else to climate-proofing the small-scale fishery systems of developing countries. Key to this will be identifying generic policies and investment priorities that reduce recurrent vulnerability and increase resilience of fishery-dependent populations. Stimulating economic diversification, improving fisheries management and governance arrangements, providing social support for fishery-dependent populations and providing financial instruments that spread the risk of adverse consequences all bear close examination.

In the above context the issue of livelihood diversification often assumes particular prominence. Understanding and finding ways to support livelihood diversification that is sustainable and achieves intended outcomes, however, is a complicated and nuanced problem (Brugere et al., 2008).

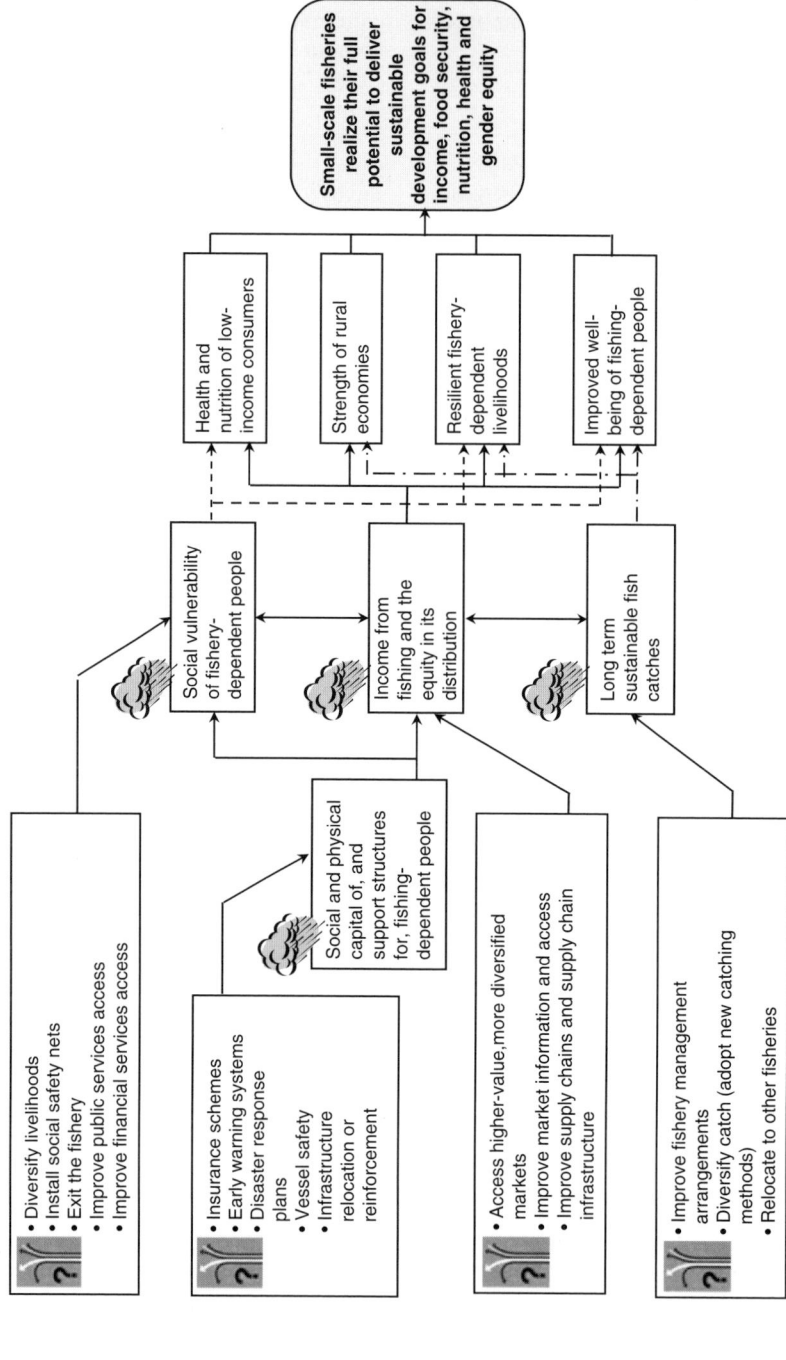

Fig. 8.4. Possible response options to the impacts of climate change at several key points in a fishery system.

Too often diversification is equated with job substitution, with attendant assumptions about fishers as predominantly specialists, who only fish. Where it is possible, however, livelihood diversification measures do provide a valuable means for insulating against resource declines. Livelihood diversification has a poor record, however, when the intended purpose has been to reduce fishing pressure and combating overfishing (Brugere *et al.*, 2008).

Answering the questions

In essence, obtaining answers to the three questions posed above constitutes a diagnosis and an options analysis. A key question, of course, is how does one best go about undertaking such a task? It will be no surprise to learn that there is no straightforward answer to this question. Indeed, the issue of effective diagnosis approaches for small-scale fisheries is one that has recently attracted growing research interest (see Evans and Andrew, Chapter 3, this volume). Here is not the place to repeat the advice offered in Chapter 3 or elsewhere, but it is worth emphasizing that whichever diagnostic tools are used, it is important that the diagnostic process has legitimacy and is 'owned' by the people carrying and managing risk.

The benefit of meaningful participation is a recurring one that managers and policy makers need to embrace thoroughly. I cannot put the argument for wide stakeholder engagement better than Neil Adger (2003, p. 401):

> Building trust and cooperation between actors in the state and civil society over adaptation has double benefits. First, from an instrumentalist perspective, synergistic social capital and inclusive decision-making institutions promote the sustainability and legitimacy of any adaptation strategy. Second, adaptation processes that are built from the bottom up and are based on social capital can alter the perceptions of climate change from a global to a local problem. When actors perceive adaptation to and the risk of climate change as being within their powers to alter, they will be more likely to make the connection to the causes of climate change, thereby enhancing their mitigative, as well as adaptive, capacity.

With this in mind, I offer a few additional thoughts concerning the softer (people) elements of required processes.

The first concerns the framing of the conversations needed to effect meaningful change. There are many frameworks for organizing your thinking to answer the questions posed above, and it probably doesn't matter much which one you choose. What is essential, however, is that you ensure that the required dialogues on key issues are framed in a clear and compelling way for stakeholders and that they are well facilitated. Getting high-quality professional help for these two tasks will usually be money well spent; the quality of the discussions and the process for arriving at decisions will be a key determinant of success, especially at the local level (Næss *et al.*, 2006).

As noted earlier, at all levels, pictures and maps can be enormously helpful. For local communities in particular, joint construction of pictures is an extremely powerful way of developing shared meaning and revealing barriers to change. Approaches that use influence diagrams to map out the actors and key issues look especially promising (see Hauck and Youkhana, 2008; Badjeck *et al.*, 2009 for fisheries examples).

Again, however, we should remember that gender inequalities are deeply ingrained and difficult to change. This often greatly affects the contribution that women can make, and overcoming this requires:

- ensuring that participatory planning methods are inclusive and motivate, support and encourage women and men to engage in the process;
- understanding practical barriers to women's participation in discussions, planning and decision-making, and in micro-enterprise;
- ensuring that issues identified and analysed are relevant and of interest to both men and women – this will help both genders formulate ideas and engage in the adaptation process;
- learning to recognize and handle conflict – personal attitudes and feelings about

equal participation and gender mainstreaming will vary and some may work against it; and
- establishing gender-focused and disaggregated monitoring (USAID, 2009).

The above issues of participation build logically into the wider consideration of partnerships, a topic that is only now starting to appear explicitly in the fisheries literature. Pinkerton (2009) provides a particularly useful review of the topic, which I draw on below.

Partnership arrangements differ in a very important way from advisory relationships because they involve the sharing of power. Because power-sharing facilitates conflict resolution and generates energy to solve critical problems, partnership arrangements have a particular and growing attraction (Pinkerton, 2009). Some of the most successful partnerships function at a regional scale around a common interest in a watershed or sub-basin. For local groups, being able to get beyond geographic isolation and cooperate on a regional scale can have a powerful effect on the parties. This is because, through partnership, people feel they have more power to effect change at a more meaningful scale, while simultaneously stabilizing or improving their own position. A good example is of groups working in partnership along the rivers in the Pacific North-west, combining data to obtain a better assessment of resource status (Ebbin, 2009).

Managers charged with helping fishery-dependent communities prepare and respond to the emerging challenges they face are well advised to step back and reflect on whether the partnerships among stakeholders in their fishery system are appropriate and functional. Figure 8.5 provides some key criteria for helping to answer this question.

Gaining funding

All managers have to work with limited resources, and the work required to deal with additional external drivers will stretch capacity even further. To deal with these added considerations it is worth thinking about additional avenues for garnering support.

Clearly, for a fishery manager or policy maker to argue for resources to help improve fisheries management per se, the costs must be commensurate with the benefits the fishery delivers (see Garcia and Cochrane, 2009, for a discussion of this topic). When seeking resources to help fishery-dependent communities cope with external threats, wider social welfare considerations come into play. Nevertheless, unless policy makers understand the contributions a fishery makes, or could make, and the losses that are likely to be realized by not acting, required investment will not be forthcoming.

A key challenge facing fisheries, especially those that are small-scale, is the indifference or neglect of governments. In a recent global review of 281 national policy papers, including 50 poverty reduction strategy papers, few countries included fishing and fish-farming communities among their target groups. Nor did they accord the fisheries sector an explicit role in poverty reduction and food security (see also Thorpe et al., 2007). Estimating the value of a fishery, however, to determine how much effort to invest is complicated – especially for small-scale fisheries. This is because the values of small-scale fisheries do not necessarily manifest themselves primarily in economic terms; issues of food security, social safety nets and other functions demand inclusion (Heck et al., 2007). Moreover, the realized value of a fishery may be low because fisheries management is failing, whereas its potential value under sound management may be much higher. Added to this are the challenges of including the value environmental externalities such as breeding habitats, or mangroves, that might be lost due to climate change or other forces (Barbier et al., 2002).

A key challenge for any manager, therefore, is to determine the current and potential 'value' of the fisheries under his or her jurisdiction, in all its dimensions, so that sound judgments on the appropriate levels of investment can be made. The technical skills to undertake such work are unlikely to reside within fisheries departments or other institutions that a manager or policy maker has direct access to. The best advice one can offer, therefore, is to reach out to appropriate

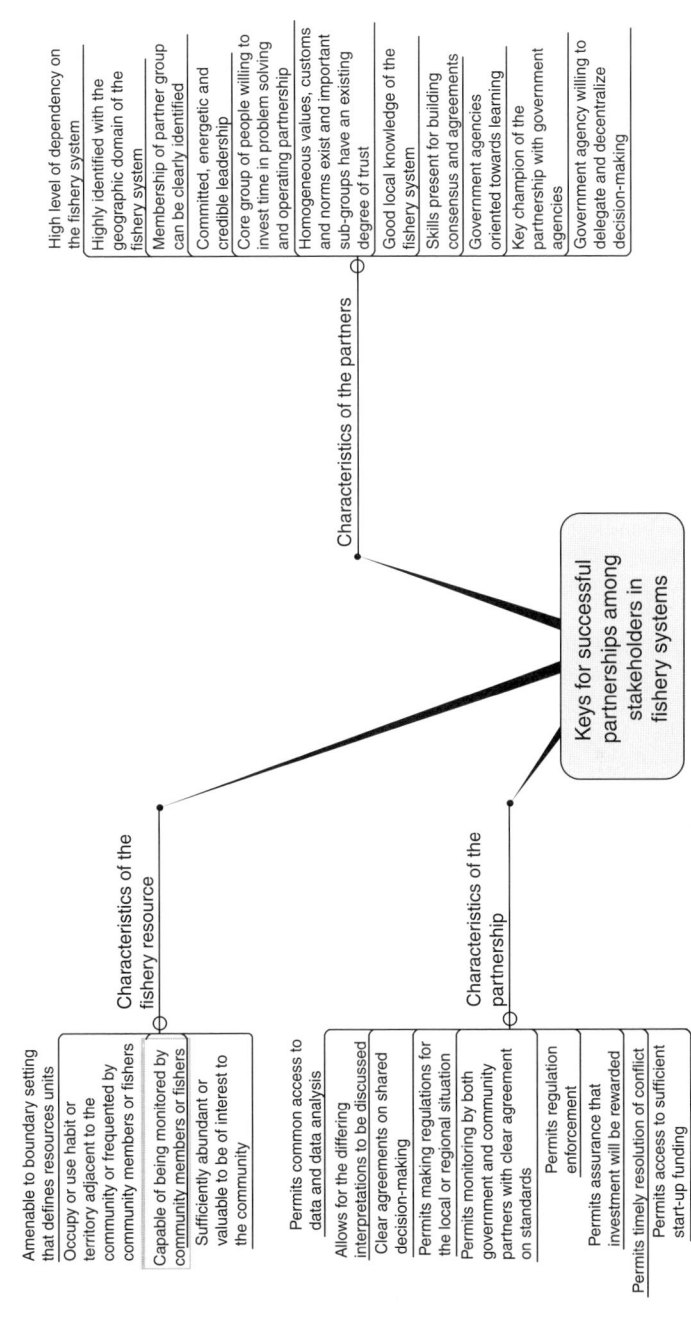

Fig. 8.5. Keys for successful partnerships for fishery systems (adapted from Pinkerton, 2009).

researchers in universities and other institutions and explore options for partnering to undertake such work.

It is also important to look beyond fisheries per se for support. In the context of climate change, a cornerstone of the response for countries that have been designated as Least Developed (LDCs) will be the National Adaptation Programme of Action (NAPA). The United Nations Framework Convention on Climate Change (UNFCC) provides support to the 50 LDCs, through NAPAs, which provide a process for identifying priority activities that respond to their urgent and immediate needs to adapt to climate change (UNFCCC, undated). As of September 2008, out of a total of 437 projects submitted 20 have fisheries in the title (Table 8.4). Excluding one very large project, the average project cost was about US$1.2 million and the median was US$462,000. There were also a total of 35 projects under the heading Coastal Zones and Marine Ecosystems, many of which, while they do not have fish or fisheries in the title, will no doubt deliver benefits for fishery-dependent coastal communities.

NAPAs place considerable emphasis on bottom-up strategies of consultation, empowerment and activity. Early analysis, however, suggested that the effectiveness of engagement was highly variable, with cursory consultation and wider national policy frameworks that made facilitating local empowerment and integrated programmes hard to achieve (Thomas and Twyman, 2005). Such concerns only serve to emphasize the importance of the points made earlier in this chapter about paying attention to process and quality of dialogue and participation.

While wider policy processes in support of climate change adaptation can be an important source of support for fisheries issues, it is also important to recognize that taking a wider cross-sectoral perspective can also be key to finding durable solutions to sectoral problems. In Malawi, for example, a NAPA project on the Lake Chilwa area recognizes the critical role of the fishery to the economy of the Basin. In consequence, it gives high priority to creating and maintaining the environmental conditions necessary to sustain the fishery's productive potential. Creating these conditions will require focus on managing the catchment, rather than the fish stocks directly (Kalk et al., 1979; Allison and Mvula, 2002). This will require policy dialogue and stakeholder consultation well outside the traditional boundaries of fisheries management.

A final area where there is the potential for support is climate change mitigation. Fishing burns 1.2% of the fossil fuel used globally each year (Tyedmers et al., 2005). While the potential benefit of investing in fishing energy efficiency and emission reduction is minor, the sector does provide opportunities to improve livelihoods and environmental and resource management in ways that mitigate climate change. Market instruments for financing mitigation, such as the United Nations Framework Convention on Climate Change (UNFCCC) Clean Development Mechanism (UNFCCC, undated) and voluntary carbon markets, may be used to fund work that contributes to the development of sustainable fisheries and aquaculture.

Mitigation strategies for fisheries include promoting the use of fuel-efficient fishing vessels and methods and removing such disincentives to energy efficiency as fuel subsidies. Similarly, conserving and restoring mangroves sequesters carbon, protects coastlines and enhances fisheries and livelihoods. Opportunities for funding adaptation through novel schemes that also contribute to mitigation, such as the Reduced Emissions from Deforestation and Degradation scheme for mangroves, could be promoted. Engaging with researchers to investigate the potential for fisheries to contribute to mitigation will provide governments, communities and their partners with a range of options for funding adaptation activities, as most mitigation initiatives are linked to markets or global funds. Reducing the carbon footprint of fisheries, as well as making a small contribution to halting climate change, can set an example to other food sectors in commitment to environmentally sustainable production.

Other Drivers

As noted above, the steps needed to help in coping with climate change will also help in

Table 8.4. Projects in NAPAs submitted (as of September 2008) that contain fish or fisheries in their title.

Country	Project ranking	Project title/description	Sector	Cost (USD, 000s)
Bangladesh	13/15	Adaptation to fisheries in areas prone to enhanced flooding in north-east and central regions through adaptive and diversified fish culture practices	Food security	4,550
Bangladesh	14/15	Promoting adaptation to coastal fisheries through culture of salt-tolerant fish especially in coastal areas of Bangladesh	Food security	4,050
Cambodia	3/3	Development and improvement of small-scale aquaculture ponds	Food security	4,000
Comoros	9/13	Introduction of fisheries concentration mechanisms (FCM)	Food security	132
Comoros	11/13	Short conservation of fish under ice	Food security	308
Comoros	12/13	Provender production	Food security	900
Gambia	10/10	Increasing fish production through aquaculture and conservation of postharvest fishery products	Food security	300
Guinea	9/25	Promoting adaptation-oriented technologies. 6. Promoting the use of solar energy for fish-drying to reduce pressure on mangroves	Energy	200
Guinea Bissau	9/14	Protection, Conservation and Enhancement of Fishing and Coastal Resources Project	Coastal/Marine	450
Maldives	9/11	Investigating alternative live bait management, catch, culture and holding techniques in the Maldives to reduce vulnerability of the tuna fishery sector to the predicted climate change and variability	Food security	1,027
Mali	4/19	Rehabilitation of aquaculture sites in Mali	Food security	25,760
Mauritania	14/28	Protection of the diversity of the fish population and prevention of overfishing with a view to sustainable development	Coastal/Marine	1,337
São Tomé and Principe	1/22	Training and readapt project of the new navigation technologies and fishing equipment for fishermen	Food security	350
São Tomé and Principe	4/22	Construction and installation of device for fish concentration (DFC) on coastal zone	Food security	250
Sierra Leone	14/24	Establishment of a permanent study programme of multi-species fisheries	Food security	395
Sierra Leone	16/24	Improving the quality of fisheries-related data and research	Food security	455
Solomon Islands	5/7	Fisheries and Marine Resources	Coastal/Marine	1,500
Tuvalu	7/7	Adaptation to Near-shore Coastal Shellfish Fisheries Resources and Coral Reef Ecosystem Productivity	Coastal/Marine	462
Yemen	11/12	Sustainable management of fisheries resources	Food security Fisheries	1,180
Zambia	6/10	Adaptation of land use practices (crops, fish and livestock) in light of climate change	Food security	1,200

building the resilience needed to cope with other drivers of change. That said, the more that fisheries managers and policy makers understand how these drivers might impact the fishery systems they serve, the better placed they will be to help develop well-targeted responses. To illustrate this, I offer a brief summary of several other change drivers and describe some of the research and policy responses needed to help in coping with them.

Demography

With the population of many developing countries expected to grow, and increasing movement of people both within and between countries, we can expect profound changes in fishery systems in the coming decades. Demographic trends are likely to have large implications for demand for fishery products and on fisheries as a source of employment. Unfortunately, the nature of those implications is hard to predict. This difficulty is increased because we have no reliable global picture of participation trends in the fisheries sectors of developing countries that might link to population growth and migration. Nor will one be easy to obtain; most national census data, for example, still do not distinguish between agricultural workers and fishers; accounting for the millions of people who fish part-time as part of a diversified livelihood strategy adds further complication.

Despite data deficiencies, however, some trends are apparent. Principal among these is that there is a general increase in the numbers who engage in small-scale fisheries in low- and middle-income countries. The case of Ghana illustrates this, with an increase from about 65,000 canoe fishers in 1959 to 123,000 in 2001 (Atta-Mills *et al.*, 2004). This contrasts with the developed world, where the number of fishers is falling. Hidden beneath the overall increase in participants in developing countries, however, is a fall in the number of full-time fishers while part-time fishers are rising. This seems to be especially true for Asia. In Indonesia, for example, part-time fishers increased by more than 50% to 1.1 million people between 1989 and 1998 (Indonesian Dept of Fisheries, 2000, quoted in Kura *et al.*, 2004).

Clearly, with current data deficiencies, the relative importance of various demographic and social processes that lead, or have led, to changes in fishing populations remains uncertain. Although explanation will undoubtedly vary with context, one rather seductive explanatory paradigm is Malthusian overfishing (Pauly, 1988). Under this model a combination of poverty and population growth leads to increasing demand for fish and fisheries-related income and push migration. This is then hypothesized, in turn, to lead to over-exploited stocks in both source and destination locations.

Although at first sight rather compelling, the evidence for widespread application of such a simple model appears scant. For West Africa, for example, there are very few indications that population pressures or over-exploitation of coastal resources drive the migration of fishers from Ghana to other countries in the region (Marquette *et al.*, 2002). Rather, migrants appear to be drawn instead by commercialization and urbanization. With the growth of commercial agriculture, ports and shipping and the concomitant improvements in road and rail infrastructure, migrants are attracted by the new markets for fish that have developed and the improved access to the markets inland (Marquette *et al.*, 2002). Other factors also play their part. The Côte d'Ivoire is attractive to Ghanaian fishers, for example, because it provides greater subsidies on fuel – the kind of attractor that may increase with continued upward pressure on fuel prices. These conclusions underline the importance of understanding markets for fish and the economics of fishing (Bremner and Perez, 2002).

Migrants can also affect fisheries indirectly, as can be seen in Indonesia, where national transmigration policies, settlement schemes and the promotion of aquaculture in coastal regions has seen the destruction of mangroves to make way for ponds given over to shrimp and fish culture. This has contributed to the increasing marginalization of coastal fishers and depletion of wild stocks (Armitage and Johnson, 2006). Given the continued growth of aquaculture driven by increasing demand for fish, we can, perhaps, expect such trends to increase.

Research needs and policy responses

Top of the list of needs is better census data on participation rates, socio-economic status and livelihood strategies of fishery-dependent people. We need to understand the sources of origin and demographic characteristics of migrant fishermen and the causes and courses of fisher migration in relation to their livelihood. It is especially important when collecting these data to disaggregate by gender, and also for residents and migrants. Development policies are more likely to succeed if they are founded on an understanding of gender dimensions and the multifaceted nature of migration, including temporary, circular and seasonal migration, within and between developing countries. Analysis of the impacts of labour mobility is also important, because women and migrants are often marginalized and vulnerable.

A further need is for us to develop scenarios for migration movement in fishery-dependent developing countries. What will happen if more dams are built on major river systems and fish catches fall? Where will people go when coastal areas are more frequently inundated due to climate change? These are the types of issues that nations need to think about and plan for now. We may not be able to predict the exact nature of the threat, but we can build plausible scenarios and explore them.

Markets, trade and the growth of aquaculture

In the past three decades global fish trade has risen more than fivefold, from US$15 billion in 1980 to US$78 billion in 2005, with developing countries accounting for more than 50% of the global export value. Asian developing countries are the largest fish producers, accounting for some 55% of global production, and aquaculture provides a major, and increasing, share.

Driven by increasing wealth and urbanization, it will be emerging economies that will show the strongest trend of increased demand for animal protein, including fish, over the next decade (Delgado *et al.*, 1997; Delgado and Courbois, 1999). For China and South-east Asia, it is likely that there will be a relatively higher growth in consumption of higher-value fish compared with those of low value. In contrast, for South Asia and India the opposite is true. For developed countries, while overall demand is unlikely to change, the value of purchases will rise through value addition and substitution of expensive for cheap fish products.

In the coming years we can expect demand-side processes such as seafood awareness, price, quality convenience, sustainability and ethics to become even more important. This will be driven not only by developed-country consumers but also by the growing middle class in the developing world. Wild-capture fisheries that currently supply wealthier urban markets will have to be particularly vigilant in adapting to these trends. With the rise in aquaculture, which is often better able to meet product quality standards and manage its supply chains, there is a real danger that the profitability of wild fisheries will suffer.

In the developed world consumer demand has evolved over decades, from a desire simply for increased convenience to the present inclusion of ethical and sustainability considerations. It is interesting to speculate on the pace at which attitudes of wealthier consumers in the developing world will evolve. Might one, for example, see a much more rapid increase in demand for value-added products, or adoption of ethical standards, rather in the way that the developing world jumped to mobile phones, largely missing out intermediate landline technologies?

The growth in aquaculture has far-reaching consequences for markets for wild fisheries, some of which are relatively easy to predict. As noted above, for example, aquaculture's greater ability to meet product quality standards and manage supply chains is likely to put downward pressure on wild-fish prices. Predicting the direction of change from other factors is more difficult. For example, expansion of total supply through aquaculture can put downward pressure on fish prices or increase demand for fish and help raise prices. Under the most likely scenarios developed by the Fish to 2020 study, however, high-value finfish will become about 15% more expensive relative to livestock-derived substitutes, whereas low-value food fish will increase in cost by about 6%. Fishmeal and fish oil prices are expected to increase by 18%.

Several other key market-related drivers will continue to affect the fish trade in the coming decades. Among these, the growing cost of fuel is likely to be particularly problematic, to which recent protests over rising prices by fishers in both developing and developed countries attest. The cost of fuel as a percentage of revenue rose from about 21% to 43% for developing countries between 2002/2003 and 2005 (FAO, 2006), and this despite the fact that diesel for fishing boats is either subsidized or government controlled in many countries. More recent dramatic increases in fuel costs will undoubtedly have eroded the profitability of developing-country fisheries further. Questions over the desirability and affordability of national fuel subsidies are also likely to grow, and it is inconceivable that fisheries will be unaffected.

Research needs and policy responses

The 2008 World Development Report emphasizes the critical role of trade in agricultural produce and services as a means of reducing poverty. An important question regarding the fish trade in developing countries is the extent to which it contributes to wider development objectives. This issue of how developing countries can best harness the benefits from the fish trade deserves further attention. A recent discussion of this topic by Béné et al. (2010), in the context of sub-Saharan Africa, concludes that the fish trade can indeed contribute substantially to development outcomes (see also Heck et al., 2007). They argue, however, that for this to occur the policy focus needs to shift away from large-scale operations focused on exports for the developed world towards small-scale operations and labour-intensive trading in low-cost fish among African nations. Such conclusions deserve further research and analysis to support improved policies and intervention.

In this context it is important to recognize that the trends of increasing fuel prices, changing markets, depleted stocks and product substitution from aquaculture are all leading to a restructuring of capital-intensive fisheries. In Gujarat, India, for example, a large proportion of the trawler fleet can now only afford to fish during the most productive three to four months of the fishing season. Wealthy fish processors are taking advantage of this situation to buy up vessels from those who are struggling financially (Armitage and Johnson, 2006). New global analyses and country-level case studies of the responses to these drivers by developing-country fisheries are badly needed.

Interestingly, in the Gujarat example cited above, it appears that the small-scale gillnet fishery is surviving better in the current crisis than the industrial fisheries. Such cases support the argument advanced by Béné et al., (2010) by suggesting that policy interventions to support smaller-scale operations may prove a more sustainable, equitable and employment-generating approach. This is a research topic that deserves further attention.

While cross-border and rural–urban trade bring new opportunities for small-scale producers, they also add to the pressure on aquatic resources. A key research question is how fish producers and traders can take advantage of the benefits, while avoiding the negative consequences of greater market integration. Answering this question may require working with producers to develop ways of critically assessing which markets to focus on to help them realize their own development goals. It may also require improved approaches for helping people understand and make trade-off decisions over risks and potential rewards in engaging with the highly segmented and differentiated markets for aquatic produce. For example, aggressive promotion of greater global market integration in the context of a small-scale capture fishery may be an inadvisable entry point for poverty reduction in situations where local nutritional dependency on fish is high, and where resources are poorly governed and thus likely to be rapidly depleted.

For developing countries sanitary and phytosanitary measures are viewed as the most important impediment to food exports to the developed world. This is partly a consequence of lack of scientific and technical expertise, information and finance. However, it also reflects a general mismatch between the prevailing small-scale production and marketing approach for fisheries in developing countries and the requirements for meeting export standards. Current implementation of, for example, EU hygiene standards threatens to badly weaken marketing opportunities for small-scale fishing communities. Several fisheries in East Africa have invested to improve catching and marketing operations, but are now finding it difficult to make the investments needed to meet international food health and safety standards. The result has been a monopoly of the export trade by a few larger companies able to process fish to the required standards. These now drive the exploitation and prices of the Lake Victoria Nile Perch fishery (Gitonga, 2007). In East African countries, for example, processors focus exclusively on exports, which account for less than 50% of fish landings. Means for establishing better links between the larger downstream enterprises and small-scale operators in the catching sector is a key requirement in rectifying this problem and improving the contributions to development from small-scale fishers.

Governance and development policy

I take the term 'fisheries governance' to mean the framework of social and economic systems and legal and political structures through which fisheries systems are managed. Fortunately, there have been some advances in both reforming governance systems and working to improve the technical basis for fisheries management, thereby making them more governable.

Developments from 1995 to 2010 in fisheries management, as with many other areas, have centred on ideas of decentralization, regionalization and participation (democratization; e.g. Adger, 2003; Satria and Matsida, 2004; Lebel et al., 2006). An important driver of this trend is increasing numbers of, and increasingly well-informed, stakeholders who are more determined to participate in fisheries management and governance arrangements. As a result, national agencies are becoming less able to control information flow and exclude other formal and informal institutions from influencing the management process (Gibbs, 2008). Gibbs (2008) describes this as a trend towards 'network governance', arguing that it is likely to erode the traditional command and control authority of many formal government-mandated agencies and, perhaps, increase the transaction costs in fisheries management processes. These cost increases, however, will be justified if they increase management effectiveness. Networked governance processes could well be an irresistible force in the next decade. It will require management agencies to be more transparent and open to scrutiny, and to resist elites seeking to develop exclusive arrangements. A growing commitment to local solutions seems the most positive reform in fisheries governance within countries.

Linking with decentralized government systems seems to be a promising way forward to out-scale previously isolated successes with community-based fisheries management (e.g. Satria and Matsida, 2004). Some observers, however, are expressing concern about the effectiveness of community-based or co-management approaches. First, the issue of defining an appropriate community is a difficult one that remains fundamental to the success of any such effort (see Johnson, 2004). Second, there is concern that important voices can be diluted by increased stakeholder participation (Suarez de Vivero et al., 2008).

The challenge of fisheries management is to deliver social and economic benefits in the long term, while maintaining fish stocks and their supporting social-ecological systems. This requires different solutions for different kinds of fisheries, and a willingness to invest in both local solutions and global and regional coordination. As a guiding framework to enable reform of national fisheries policies, the FAO Code of Conduct for Responsible Fisheries has played a key role. However, much of this internationally driven fisheries policy reform has not progressed to implementation of better policies at the national level. 'Responsible fishing' remains a paper policy only for many nations. The next wave of policy reform is probably going to take place at local and regional levels, and will involve working out how to make the principles of 'responsible fishing' workable.

Recent attention in this regard concerns illegal, unreported and unregulated (IUU) fishing and the introduction of port state measures (PSM). Nations will come under increasing pressure to ensure that foreign fishing vessels using their ports comply with agreed regulations. These might include prior notification of port entry, use of designated ports, restrictions on port entry and landing or trans-shipment of fish, restrictions on supplies and services, documentation requirements and port inspections. Many of these measures have, in recent years, seen their inclusion and development in international instruments.

Other new avenues for improving governance come from a growing realization that failing fisheries are a human security and human rights issue, particularly in developing countries with high fishery dependence. Fisheries collapse there can lead to displacement of coastal people, mass unemployment, increased pressure on land, the shift of large numbers of people to urban centres, and recourse of fishing people to other uses for their fishing vessels – such as smuggling of arms, people and drugs – or to engaging directly in piracy. These issues – and human rights abuses associated with poor working conditions in fisheries and fish processing, gender discrimination and widespread use of child labour – are leading to a broader rights-based agenda in fisheries (Kearney, 2007). The International Labour Organization has played an important role in highlighting these issues in the fisheries sector. This holds out the promise of fostering wider societal interest in a well-governed fishery and maritime sector.

A recent reorientation by the FAO to reassert its focus on hunger (the fundamental right to food) and the task of meeting the Millennium Goals has also led to a rethinking of the role and importance of small-scale fisheries. An FAO Global Conference, held in Bangkok in 2008 and focusing exclusively on small-scale fisheries, was the first on the topic in 25 years. The prevailing sentiment in the 1970s and 1980s was of a decline and replacement of small-scale fisheries by larger-scale industrialized production. The emerging interest in the efficiency and effectiveness of the small-scale fisheries sector in helping to meet development objectives is a trend that is likely to increase.

Research needs and policy responses

A key requirement is to improve understanding of key policy processes, particularly decentralization and democratization, and the opportunities and constraints they provide for fisheries. In this context, how best to link fisheries governance to local development issues becomes critical. To do this we must provide a more comprehensive understanding of the value of fisheries in relation to key development indicators. The earlier section of this chapter on attracting financial support for adaptation programmes touches on this issue.

As noted earlier, policies and investment to improve human and institutional capacities are also essential if governance is to improve. At present, in self-managed, rights-based approaches, fishers are being given greater responsibility for managing their fisheries without the corresponding transfer of skills and resources for information gathering, assessment, management and negotiation.

Human health and disease

The nexus between human health and wider fisheries issues is one that is rarely explored in a systematic manner. Certainly, some health researchers and practitioners focus on fishers as a target group, but rarely are health issues thought of as a driver of change for fisheries. Yet fishers and fishing communities are especially vulnerable to some communicable diseases, and any trends in the incidence of these diseases – or in health intervention approaches – could have profound implications for labour productivity, social structures and governance arrangements. These in turn can affect the contributions of wild-capture fisheries to human well-being. Using fishers and local fishery management institutions as an entry point for providing health care is also rarely discussed.

Among the infectious diseases, the one that has perhaps received most attention in a fisheries context is HIV/AIDS. Although we have seen promising developments in global efforts to address the AIDS epidemic, the number of people living with HIV and the number of deaths due to AIDS continues to grow. Data indicate that fishing communities in developing countries are among the groups with the highest levels of HIV/AIDS prevalence. A review of ten developing countries, for example, suggested that HIV prevalence rates in fishermen or fishing communities were up to fourteen times higher than national averages (Kissling et al., 2005). In many cases, both HIV prevalence and absolute numbers of infected people were also estimated to be higher among fishermen or members of fishing communities than in other sub-populations that are more usually considered to be 'populations at higher risk' by health researchers.

There are several ways in which the HIV/AIDS pandemic affects fisheries (Allison and Seeley, 2004). Most obvious are the direct impacts on the families of affected individuals. AIDS-affected families may, for example, sell their fishing assets (e.g. nets or boats) to meet immediate needs, which may deny the household more sustained income from lending the gear, or deny future opportunity for other household members to earn a living from fishing. Men who are no longer strong enough to fish may also take over roles traditionally occupied by women, such as fish trading or processing. The lack of other employment options for women in fishing communities may in turn lead them to undertake commercial sex work, thereby further increasing their own risk profile. Other direct effects can occur when, as is often the case, fishing forms part of a diversified livelihood strategy that includes farming. In these circumstances people often use cash from fishing to buy inputs such as fertilizer at key times in the production cycle. HIV/AIDS-related reductions in earning from fishing, therefore, can translate directly into reduced farm productivity. In common with other enterprises, effects can also be found in larger-scale fisheries and fish-processing sectors where loss of skilled labour and absenteeism through sickness or to attend funerals affects labour productivity. More broadly, HIV/AIDS threatens the ability of the fisheries sector to supply fish and fish products to consumers.

While there is now growing recognition of the high prevalence of HIV and AIDS in the fisheries sector, exposure to other diseases is also high. Not surprisingly, diseases that are water-borne are principal among these. Malaria infection rates in Ethiopia, for example, are greater and periods of transmission more intense for people living closer to the Koka reservoir. The presence of the reservoir, coupled with inter-annual climatic variations, explains more than half of the region's variability in malaria case rates (Lautsze et al., 2007); those who live 6–9 km from the water are 2.3 times less likely to contract the disease. Although not specifically analysed in Lautze et al.'s (2007) study, since fishers are more likely to live closer to the water, they bear this increased risk. Fishers are also likely to be at greater risk from the more neglected tropical diseases such as schistosomiasis and soil-transmitted helminths. These too are worthy of attention, as Parker et al. (2007) point out: '… it is likely that significant reductions in overall child and adult mortality and morbidity can be achieved if relatively small proportions of public health finance directed towards the control of HIV/AIDS, TB and malaria are re-directed towards the integrated control of some of the neglected, parasitic diseases'.

Research needs and policy responses

Although HIV/AIDS in the fisheries sector has not received much attention from policy makers, there are some notable exceptions. The Department of Fisheries Resources in Uganda, for example, has developed a strategy for reducing the impact of HIV/AIDS on fishing communities (Government of Uganda,

2005), which argues for the mobilization of a wide range of organizations to coordinate work at the community, district and national levels. It proposes that the poor quality of life in fishing communities contributes to risky behaviour and that part of the solution is to improve infrastructure, social cohesion and incomes in fishing communities. These conclusions underline the need to learn from successful experiences in other sectors, and argues for a rapid scaling up of community-level interventions (Gordon, 2005).

In particular, we need to understand much better the factors that contribute to high-risk behaviour (e.g. poor living conditions and low incomes) and the determinants of resistance to HIV infection and resilience to withstand the impacts of HIV/AIDS. This requires action research to identify and document emerging good practice, so that the lessons can in turn be taken up and applied in wider efforts to scale up the response (Gordon, 2005). What seems certain is that 'immediate and effective preventive efforts that consider the socio-cultural contexts are necessary to reduce the spread of the infection' (Yahya-Malima et al., 2007).

One good example of how better understanding fishers' origins and behaviours improves health interventions is described by Simonet (2004), who points to data showing that the more cash a fisherman has on shore leave, the more likely he is to engage in high-risk behaviours, such as patronizing brothels. In response, new arrangements in Thailand to reduce allowances and pay while on shore leave and pay a greater proportion of wage entitlements upon return home has decreased incentives to engage in high-risk behaviours during idle time. Simonet (2004) also describes how, when assumptions about fishers go untested, suboptimal choices are made. When AIDS prevention groups in Thailand, for example, realized that media campaigns featuring popular Thai figures did not resonate with the large number of Cambodian fishers working on Thai boats, they switched to using famous Cambodians, leading to much greater effectiveness. These and other examples show how information about the origin of the migrant flow helps to better target HIV/AIDS action programmes.

It is also important to understand the importance of fishers as vectors for other diseases. Siddique et al. (1994), cited in Colwell (1996), for example, describe how an epidemic of cholera started on temporary islands off the coast of the Sundarban area in the south-western coastal districts of Bagerhat, Bangladesh. Emerging at the end of the monsoon period, these islands are visited by migrant fishermen who arrive in October to fish in the Bay of Bengal. Because of the remoteness of the area the epidemic went unnoticed until December, when it was identified in the mainland of Bagerhat, presumably having arrived there by transmission through fishermen (a warranted assumption – or just the plankton?). Afterwards, it appeared in five neighbouring districts with an epidemic that lasted more than four months, involved a total of 46,965 cases and led to 846 deaths.

Finally, we must recognize that our understanding of the effects of overfishing on waterborne diseases such as schistosomiasis and cholera is in its infancy, yet the consequences for human populations of depleting fish assemblages in these systems may be profound (Allan et al., 2005). This is another area where further research is badly needed.

Conclusions

In this chapter, I have tried to move from a generic catalogue of the possible impacts of external drivers on fisheries systems to some practical approaches for thinking about and arriving at concrete responses in specific contexts. I have also tried to highlight those areas where further research can help people make better decisions.

The general conclusion from this effort is that meeting the fundamental challenge of improving fisheries management and supporting the wider development goals for fishery-dependent people will do more than anything else to build the resilience needed to cope with externally driven change. This will require fisheries management approaches and dialogue that is more outward looking and engages a wider range of actors across multiple scales. Ensuring such quality dialogue and analysis at multiple scales is essential for developing durable responses. Finally, we must recognize that no driver acts in isolation and we need, as far as practical, to consider all the externalities that might affect the fisheries system in question.

References

Adger, W.N. (2003) Social capital, collective action and adaptation to climate change. *Economic Geography* 79, 387–405.
Allan, D.J., Abell, R., Hogan, Z., Revenga, C., Taylor, B.W., Welcomme, R.L. and Winemiller, K. (2005) Overfishing of inland waters. *BioScience* 55, 1041–1051.
Allison, E.H. and Mvula, P. (2002) *Fishing Livelihoods and Fisheries Management in Malawi*. ODG, Norwich, UK. Available at: http://www.odg.uea.ac.uk/ladder/ (accessed 31 May 2010).
Allison, E.H. and Seeley, J.A. (2004) HIV and AIDS among fisherfolk: a threat to 'responsible fisheries'? *Fish and Fisheries* 5, 215–234.
Armitage, D. and Johnson, D. (2006) Can resilience be reconciled with globalization and the increasingly complex conditions of resource degradation in Asian coastal regions? *Ecology and Society* 11(2). Available at: http://www.ecologyandsociety.org/vol11/iss1/art2/ (accessed 31 May 2010).
Atta-Mills, J., Alder, J. and Sumaila, U.R. (2004) The decline of a regional fishing nation: the case of Ghana and West Africa. *Natural Resources Forum* 28, 13–21.
Badjeck, M.C., Mendo, J., Wolff, M. and Lange, H. (2009) Climate variability and the Peruvian scallop fishery: the role of formal institutions in resilience building. *Climate Change* 94, 211–232.
Barbier, E.B., Strand, I. and Sathirathai, S. (2002) Do open access conditions affect the valuation of an externality? Estimating the welfare effects of mangrove-fishery linkages in Thailand. *Environmental and Resource Economics* 21, 343–367.
Béné, C., Lawton, R. and Allison, E.H. (2010) Trade matters in the fight against poverty: narratives, perceptions, and (lack of) evidence in the case of fish trade in Africa. *World Development* 38, 933–954.
Brashares, J.A.P., Sam, N., Coppolillo, P., Sinclair, A. and Balmford, A. (2004) Bushmeat hunting, wildlife declines and fish supply in West Africa. *Science* 306, 1180–1183.
Bremner, J. and Perez, J. (2002) A case study of human migration and the sea cucumber crisis in the Galapagos Islands. *AMBIO* 31, 306–310.
Brugere, C., Holvoet, K. and Allison, E.H. (2008) *Livelihood Diversification in Coastal and Inland Fishing Communities: Misconceptions, Evidence and Implications for Fisheries Management*. FAO/DFID, Rome, Italy. Available at: ftp://ftp.fao.org/fi/DOCUMENT/sflp/wp/diversification_june2008.pdf (accessed 31 May 2010).
Cinner, J.E., Daw, T.M. and McClanahan, T.R. (2008) Socioeconomic factors that affect artisanal fishers' readiness to exit a declining fishery. *Conservation Biology* 23, 124–130.
Colwell, R.R. (1996) Global climate and infectious disease: the cholera paradigm. *Science* 274, 2025–2031.
Conway, D., Allison, E.H., Felstead, R. and Goulden M. (2005) Rainfall variability in East Africa: implications for natural resources management and livelihoods. *Philosophical Transactions of the Royal Society of London Series A* 363, 49–54.
Coulthard, S. (2008) Adapting to environmental change in artisanal fisheries: insights from a South Indian Lagoon. *Global Environmental Change* 18, 479–489.
Delgado, C.L. and Courbois, C. (1999) Changing fish trade and demand patterns in developing countries and their significance for policy research. In: Ahmed, M., Delgado, C., Sverdrup-Jensen, S. and Santos, R.A.V. (eds) *Fisheries Policy Research in Developing Countries: Issues, Priorities and Needs*. ICLARM Conference Proceeding 60, ICLARM, Manila, Philippines, pp. 21–32.
Delgado, C.L., Crosson, P. and Courbois, C. (1997) The impact of livestock and fisheries on food availability and demand in 2020. *American Journal of Agricultural Economics* 79, 1471–1475.
Ebbin, S.A. (2009) Institutional and ethical dimensions of resilience in fishing systems: perspectives from co-managed fisheries in the Pacific Northwest. *Marine Policy* 33, 264–270.
Enarson, E. and Meyreles, L. (2004) International perspectives on gender and disaster: differences and possibilities. *International Journal of Sociology and Social Policy* 24, 49–93.
FAO (2006) *The State of World Fisheries and Aquaculture 2006*. FAO, Rome, Italy.
Gallopin, G.C. (1991) Human dimensions of global change: linking the global and local processes. *International Social Science Journal* 130, 707–718.

Garcia, S. and Cochrane, K. (2009) From past management to future governance: a perspective view. In: Cochrane, K. and Garcia, S. (eds) *A Fishery Manager's Guidebook*. FAO and Blackwell, Rome, Italy, pp. 447–472.

Gibbs, M.T. (2008) Network governance in fisheries. *Marine Policy* 32, 113–119.

Gitonga, N. (2007) Fish safety and quality challenges by developing countries: the east African Nile Perch case. In: *Globalization and Fisheries: Proceedings of an OECD-FAO Workshop*, OECD, Paris, pp. 123–142.

Gordon, A. (2005) *HIV/AIDS in the Fisheries Sector in Africa*. WorldFish Center, Cairo, Egypt. Available at: http://www.worldfishcenter.org/resource_centre/HIV-AIDS-AfricaFisherySector.pdf (accessed 31 May 2010).

Government of Uganda (2005) *Uganda Strategy for Reducing the Impact of HIV and AIDS on Fishing Communities*. Ministry of Agriculture, Animal Industry and Fisheries, 12 pp. Available at http://www.aidsuganda.org/texbits/fisheries%20strategy-summary.pdf (accessed 31 May 2010).

Handisyde, N.T., Ross, L.G., Badjeck, M.C. and Allison, E.H. (2006) *The Effects of Climate Change on World Aquaculture: a Global Perspective*. Available at: www.aquaculture.stir.ac.uk/GISAP/gis-group/climate.php (accessed 31 May 2010).

Hauck, J. and Youkhana, E. (2008) *Histories of Water and Fisheries Management in Northern Ghana*. Center for Development Research, University of Bonn, Bonn, Germany. Available at: http://www.zef.de/fileadmin/webfiles/downloads/zef_wp/WP32_Hauck-Youkhana.pdf (accessed 31 May 2010).

Hazell, P. and Wood, S. (2008) Drivers of change in global agriculture. *Philosophical Transactions of the Royal Society of London Series B* 363, 495–515.

Heck, S., Béné, C. and Reyes-Gaskin, R. (2007) Investing in African fisheries: building links to the Millennium Development Goals. *Fish and Fisheries* 8, 211–226.

Ikeda, K. (1995) Gender differences in human loss and vulnerability in natural disasters: a case study from Bangladesh. *Indian Journal of Gender Studies* 2, 171–193.

IPCC (2007) Summary for policymakers. In: Parry, M.L., Canziani, O.F., Palutikof, J.P., van der Linden, P.J. and Hanson, C.E. (eds) *Climate Change 2007: Impacts, Adaptation and Vulnerability*. Contribution of Working Group II to the Fourth Assessment Report of the Intergovernmental Panel on Climate Change, Cambridge University Press, Cambridge, UK, pp. 7–22.

Johnson, C. (2004) Uncommon ground: the 'poverty of history' in common property discourse. *Development and Change* 35, 407–434.

Kalk, M., McLachlan, A.J. and Howard-Williams, C. (eds) (1979) *Lake Chilwa: Studies of Change in a Tropical Ecosystem*. Monographiae Biologicae vol. 35, Dr W. Junk, The Hague, Netherlands.

Kearney, J. (2007) Fulfilled, healthy, secure? *Samudra Report* 46, 18–21.

Kissling, E., Allison, E.H., Seeley, J.A., Russell, S., Bachmann, M., Musgrave, S.D. and Heck, S. (2005) Fisherfolk are among groups most at risk of HIV: cross-country analysis of prevalence and numbers infected. *AIDS* 19, 1939–1946.

Kura, Y., Revenga, C., Hoshino, G. and Mock, G. (2004) *Fishing for Answers: Making Sense of the Global Fish Crisis*. World Resources Institute, Washington, DC.

Lautsze, J., McCartney, M., Kirshen, P., Olana, D., Jayasinghe, G. and Spielman, A. (2007) Effect of a large dam on malaria risk: the Koka reservoir in Ethiopia. *Tropical Medicine and International Health* 12, 982–989.

Lebel, L., Anderies, J.M., Campbell, B., Folke, C., Hatfield-Dodds, S., Hughes, T.P. and Wilson, J. (2006) Governance and the capacity to manage resilience in regional social-ecological systems. *Ecology and Society* 11(1), art. 19. Available at: http://www.ecologyandsociety.org/vol11/iss1/art19/ (accessed 3 February 2010).

Mahon, R. (2002) *Adaptation of Fisheries and Fishing Communities to the Impacts of Climate Change in the CARICOM Region*. Caribbean Center for Climate Change (CCCC). Available at: http://homologa.ambiente.sp.gov.br/proclima/artigos_dissertacoes/artigos_ingles/fisheriesissues.pdf (accessed 31 May 2010).

Marquette, C.M., Koranteng, K.A., Overå, R. and Aryeetey, E.B.D. (2002) Small-scale fisheries, population dynamics, and resource use in Africa: the case of Moree, Ghana. *AMBIO* 31, 324–336.

McClanahan, T.R., Cinner, J.E., Maina, J., Graham, N.A.J., Daw, T.M., Stead, S.M., Wamukota, A., Brown, K., Ateweberhan, M., Venus, V. and Polunin, N.V.C. (2008) Conservation action in a changing climate. *Conservation Letters* 1, 53–59.

Næss, L.O., Norland, I.T., Lafferty, W.M. and Aall, C. (2006) Data and process linking vulnerability assessment to adaptation decision-making on climate change in Norway. *Global Environmental Change* 16, 221–233.

Neumayer, E. and Plümper, T. (2007) The gendered nature of natural disasters: the impact of catastrophic events on the gender gap in life expectancy, 1981–2002. *Annals of the Association of American Geographers* 97, 551–566.

O'Brien, K., Leichenko, R., Kelkar, U., Venema, H., Aandahl, G., Tompkins, H., Javed, A., Bhadwal, S., Barg, S., Nygaard, L. and West, J. (2004) Mapping vulnerability to multiple stressors: climate change and globalization in India. *Global Environmental Change* 14, 303–313.

Parker, M., Allen, T. and Hastings, J. (2007) Resisting control of neglected tropical diseases: dilemmas in the mass treatment of schistosomiasis and soil-transmitted helminths in north-west Uganda. *Journal of Biosocial Sciences* 40, 1–21.

Patt, A.G., Schröter, D., Klein, R.J.T. and de la Vega-Leinert, A.C. (2009) *Assessing Vulnerability to Global Environmental Change: Making Research Useful for Adaptation, Decision Making and Policy*. Earthscan, London.

Pauly, D. (1988) Some definitions of overfishing relevant to coastal zone management in South-east Asia. *Tropical Coastal Area Management* 3, 14–15.

Pinkerton, E. (2009) Partnerships in management. In: Cochrane, K. and Garcia, S (eds). *A Fishery Manager's Guidebook*. FAO and Blackwell, Rome, Italy, pp. 283–298.

Pomeroy, R., Ratner, B., Hall, S.J., Pimoljinda, J. and Vivekanandan, V. (2006) Coping with disaster: rehabilitating coastal livelihoods and communities. *Marine Policy* 30, 786–793.

Satria, A. and Matsida, Y. (2004) Decentralization policy: an opportunity for strengthening fisheries management system? *Journal of Environment and Development* 13, 179–196.

Schröter, D., Polsky, C. and Patt, A.G. (2005) Assessing vulnerabilities to the effects of global change: an eight-step approach. *Mitigation and Adaptation Strategies for Global Change* 10, 573–595.

Siddique, A.K., Zaman, K., Akram, K., Mutsuddy, P., Eusof, A. and Sack, R.B. (1994) Emergence of a new epidemic strain of *Vibrio cholerae* in Bangladesh. An epidemiological study. *Tropical Geographical Medicine* 46, 147–150.

Simonet, D. (2004) The AIDS epidemic and migrants in South Asia and South East Asia. *International Migration* 42, 35–67.

Smit, B. and Pilifosova, O. (2001) Adaptation to climate change in the context of sustainable development and equity. In: McCarthy, J.J., Canziani, O., Leary, N.A., Dokken, D.J. and White, K.S. (eds) *Climate Change 2001: Impacts, Adaptation and Vulnerability*. IPCC Working Group II, Cambridge University Press, Cambridge, UK, pp. 877–912.

Suarez de Vivero, J.L., Rodreguez Mateos, J.C. and Florida del Corral, D. (2008) The paradox of public participation in fisheries governance. The rising number of actors and the devolution process. *Marine Policy* 32, 319–325.

Thomas, D.S.G. and Twyman, C. (2005) Equity and justice in climate change adaptation amongst natural resource-dependent societies. *Global Environmental Change* 15, 115–124.

Thorpe, A., Andrew, N.L. and Allison, E.H. (2007) Fisheries and poverty reduction. *CAB Reviews: Perspectives in Agriculture, Veterinary Science, Nutrition and Natural Resources* 2, 1–12.

Tyedmers, P., Watson, R. and Pauly, D. (2005) Fueling global fishing fleets. *AMBIO* 34, 635–638.

UNFCCC (undated) *Clean Development Mechanism*. Available at: http://cdm.unfccc.int (accessed 31 May 2010).

USAID (2009) *Adapting to Coastal Climate Change: a Guidebook for Development Planners*. Available at: http://www.usaid.gov/our_work/cross-cutting_programs/water/docs/coastal_adaptation/adapting_to_coastal_climate_change.pdf (accessed 31 May 2010).

Walker, B., Holling, C.S., Carpenter, S.R. and Kinzig, A. (2004) Resilience, adaptability and transformability in social-ecological systems. *Ecology and Society* 9. Available at: http://www.ecologyandsociety.org/vol9/iss2/art5 (accessed 31 May 2010).

WorldFish (2007) The threat to fisheries and aquaculture from climate change. WorldFish Policy Brief No. 1, The WorldFish Center, Penang, Malaysia. Available at: http://www.worldfishcenter.org/resource_centre/ClimateChange2.pdf (accessed 31 May 2010).

Yahya-Malima, K.I., Matee, M., Evjen-Olsen, B. and Fylkesnes, K. (2007) High potential of escalating HIV transmission in a low prevalence setting in rural Tanzania. *BMC Public Health* 7. Available at: http://www.biomedcentral.com/1471-2458/7/103 (accessed 31 May 2010).

9 Developing Markets for Small-scale Fisheries: Utilizing the Value Chain Approach

Eusebio R. Jacinto and Robert S. Pomeroy

Market Development: Perspective of Small-scale Fishers

There is a growing body of literature on how markets can be developed to benefit small agriculture producers, with cross-fertilization among development approaches employed in diverse settings (Kanji and Barrientos, 2002; Dorward *et al.*, 2003; MacFayden *et al.*, 2003; Kanji *et al.*, 2005; Albu, 2008). This literature is principally aimed at development professionals who work with these small producers at the community level, as well as small producer organizations themselves, in crafting means towards economic growth, enhanced human development and the realization of sustainable livelihoods in the context of the accelerated pace of international economic integration.

This, however, does not seem to hold true for the small-scale fisheries (SSF) sector, which, despite many development programmes throughout the last few decades, remains largely impoverished and lacks access to markets, both local and international.

There are an estimated 36 million capture-fishers in the world today, with an estimated 97% in developing countries. The vast majority of these fishers are considered to be small-scale. While there is debate on the exact definition of the small-scale fisher given the varied settings from least developed and developing to developed countries, they are characterized as capture-fishers who use small boats, typically non-motorized, and who employ rudimentary fishing gears such as hand-lines and simple nets. In addition to those directly engaged in catching fish, there are also an estimated 82 million employed in postharvest activities (processing, distribution and marketing) within the ambit of the global fishing industry. It is estimated that almost half (47%) of those engaged in small-scale fisheries are women (see Mills *et al.*, Chapter 1, this volume).

One of the binding constraints for small-scale fisher engagement in markets is the fragmentation of the sector – they largely operate individually in both production and marketing of fishery products. Despite agreement among social scientists that the establishment of associations and cooperatives among small-scale fishers would be beneficial over the long term, many small-scale fishers still find it difficult to maintain functional organizations. The issue of fragmentation is fundamental to addressing problems arising from the nature of fisheries as common pool resources. Being such, fisheries are vulnerable to depletion in situations wherein property rights are not well articulated and rules that would ensure sustainable use are either not in place or not effectively enforced – an open-access situation sometimes referred to as the 'tragedy of the commons'.

© CAB International 2011. *Small-scale Fisheries Management*
(eds R.S. Pomeroy and N.L. Andrew)

This situation can therefore be viewed as a collective action problem wherein an increased level of organization among resource users and stakeholders could serve as the key solution. The socio-economic fragmentation that inhibits the sector from effectively participating in realizing sustainable utilization of fishery and coastal resources also prevents them from accessing and effectively engaging markets. The challenge for fishery managers, then, is to pursue sustainable resource conservation while keeping in mind local, national and international market development for fisheries products. That is, to match the productive capacity of fisheries with the demands of the market.

Parallel with efforts to upscale and ensure the long-term viability of community-based coastal resources management, market development should also be seen as being linked from micro to the macro levels, the success of which depends on taking advantage of the interplay of market forces at these different levels.

While due consideration must be given to the need for food availability at the local level, horizontal and vertical integration of specific fishery value chains can be an important part of a comprehensive approach to developing markets for small fishery producers. This can enable them in two important areas of value chain engagement – increasing economic benefits (improving/equalizing benefits from the value chain) and enhancing their roles in decision-making in the value chain (increasing bargaining power).

Stabilizing Fish Supply: a Prerequisite for Market Development

It is essential in fisheries to ensure a stable supply before one can successfully engage in market development activities. Similar to agriculture, under which it belongs as an economic subsector, there are also serious constraints to achieving this in capture fisheries.

Globally, it has been estimated that wild fisheries have reached their limit, i.e. the maximum sustainable yield and additional effort exerted in the form of more efficient extraction technologies (overcapitalization) or increased number of fishers (overparticipation) will only lead to, on the whole, lower production. This trend also applies to specific fish stocks such as North Atlantic cod, specific fishing grounds such as Tonle Sap in Cambodia and specific countries such as the Philippines.

A large part of overfishing globally has been attributed to industrial fisheries that are heavily subsidized, fishing out territorial waters of their home countries, and eventually seeking out fishing grounds often of developing countries through access agreements worked out by developed-country governments under pressure from their own industrial fishing fleets.

Even as this continues, coastal fisheries are also under increasing pressure from small-scale fishers, many of whom are poor and sometimes resort to illegal, unregulated and unreported (IUU) fishing that damages fish stocks and the habitats where they live and spawn (i.e. coral reefs, mangrove stands and seagrass beds).

Amidst this, stabilizing fish supply from small-scale fisheries sufficiently to allow for favourable conditions under which to undertake market development initiatives is a difficult task. An increasing number of governments and private sector groups are now realizing the need to rationalize the level of capture-fishing effort to ensure the long-term sustainability of the fishing industry in their countries, by formulating the appropriate policies and programmes at the national level (van Mulekom *et al.*, 2004).

International legal instruments such as the United Nations Convention on the Law of the Sea (UNCLOS) and the Code of Conduct for Responsible Fisheries (CCRF) are invoked as the basis for enacting policies that are envisioned as leading to more sustainable fisheries. These efforts employ a variety of methodologies, approaches and measures, such as development of specific community tenure systems, to remedy the open-access problem, installation of marine protected areas (MPAs) over coastal ecosystems and the institution of open/closed seasons that may be fishing ground- or species-specific.

Given the continuing predominance of small-scale fisheries in terms of their sheer number in developing countries, the emergence of community-based coastal resources management (CBCRM) is considered by many development organizations as a firm foundation upon which efforts towards sustainable resource utilization of fishery resources can be based and, ultimately, substantially reduce poverty among small-scale fishers.

Overfishing having being brought about by heavily subsidized industrial fisheries and a largely fragmented small-scale fisheries sector, points to the probability that future increases in fisheries production would come from aquaculture, a trend already apparent in fishery production statistics over the past few decades. In this scenario, CBCRM would need to be involved with regard not only to the sustainable utilization of fish stocks and protection of fish habitats, but also to the use of coastal waters for farming of various fishery products that will not compromise the health and productivity of ecosystems.

In the end, the aim is to ensure non-interruption of benefit streams – economic gain for small-scale fishers and communities and continued provision of ecosystem services for the benefit of society across scales. Doing this can produce virtuous cycles wherein economic benefits from sustainable fisheries result in incentives to conserve resources that would lead, in turn, to increased economic benefits, and so on.

Organizing Groups for Market Development

To address the issue of fragmentation, it is important for small-scale capture-fishery producers to ensure access and control over resources. This is often a difficult task to achieve in many countries, given the prevalent open-access nature of fisheries wherein property rights are often not clearly articulated due to weak governments and rent capture by narrow private interests. This situation brings about resource depletion and ecosystem destruction as it becomes difficult to allocate use rights among different competing resource users.

Fragmentation of the small-scale fishers sector has been historically addressed through the setting up of fisherfolk associations and cooperatives. Studies point to the efficacy of this approach in helping to address the problems of overfishing and resource degradation.

Cooperatives

Fishers have often complained about having limited bargaining power in selling product or purchasing inputs or services. Cooperatives have been proposed as one possible solution to these problems. By organizing into a cooperative, fishers are able to have greater control over their product, obtain a wide variety of services and have greater bargaining power than an individual fisher would have (Shaw, 1986; Kohls and Uhl, 2001).

What is a cooperative? A cooperative is a group of fishers who act together to achieve some common business objective. They can do this informally, but more often they will have a written agreement specifying the terms of their cooperation. Two aspects of cooperatives are that they: (i) are a legal, institutionalized device that permits group action that can compete within the framework of other types of business organization; and (ii) are voluntary organizations set up to serve and benefit those who are going to use them. Each country around the world has laws to support cooperatives, and these laws should be examined and understood before developing a cooperative.

There are three fundamental concepts that help differentiate a cooperative from other forms of business enterprises:

1. Ownership and control of the cooperative must be by those who utilize its services.
2. Business operations shall be concluded to approach a cost basis. Returns above cost will be returned to members on an equitable basis.
3. Return on the owner's invested capital shall be limited.

Thus, a cooperative is managed and controlled by its members who have either a vote per head or in proportion to the amount of trading conducted through the cooperative. Larger cooperatives may have a manager and other staff. The profits return to the owners through their use of the cooperative rather than to owners as investors.

There can be several different kinds of cooperative business when classified according to the tasks performed:

1. Marketing cooperatives are those through which farmers sell their products.
2. Purchasing cooperatives are those through which members buy the inputs or supplies they need.
3. Service cooperatives are those that provide their members with improved services or with services they could not otherwise obtain, such as credit and insurance.
4. Processing cooperatives are those used for packing or processing of the members' products.

Cooperatives can also be classified according to how they are organized, membership affiliation, control and geographic area:

1. Independent local cooperatives in which people hold direct membership and are able to participate in the affairs of the cooperative.
2. Federated cooperatives are those composed of several local cooperatives that operate together as an integrated unit.
3. Centralized cooperatives are those in which the patron is a direct member of the central organization and exercises control through delegates sent from the various areas to the annual meeting of the central organization.
4. Mixed cooperatives are a mix of centralized and totally federated associations. These federated associations often undertake new operations that are organized on a centralized basis.

Advantages of cooperatives:

- Savings through economies of scale and size;
- Sharing of risks; and
- Opportunities to increase bargaining power through better information, price and supply.

Difficulties of cooperatives:

- Developing joint responsibility means working with others to achieve the same objectives;
- Inefficient management resulting from either lack of experience or working with others;
- Inadequate membership support and relations;
- Lack of sufficient capital and financing; and
- Size and complexity of operations can result in a breakdown in direct membership control.

Successful cooperatives must accomplish the following things:

- Increase returns from sales of products of its members; and/or
- Reduce the price or improve the quality of the purchases of its members; and/or
- Render new or improved service or give more equitable treatment to its members.

Several factors are necessary for success:

- Will the fishers make more profits with the cooperative than if they stay independent of each other?
- Are the interests of members similar enough for them to be able to work together?
- Can an adequate volume of business be secured and maintained?
- Can adequate and reasonable financing be secured?
- Is efficient management available and will the cooperative pay its price?
- Is the membership prepared to meet competitive trouble?

If these questions can be answered positively, then the next steps include:

1. Having specific objectives stating exactly what the cooperative aims to achieve year by year.
2. Having a set of rules and responsibilities written down and understood by all members, such as how profits will be divided, how much of production is to be sold to the cooperative, quality of product and decisions to be made by members and staff.

3. Having open and transparent communication; there is a need for regular meetings and discussion between members and staff.
4. Having clear definition of staff responsibilities.
5. Ensuring that all members know what is going on, what the rules are and what is expected of them.

This has been the experience of Fishery Cooperative Associations (FCAs) in Japan, where concrete gains were achieved not only in resource management but also in addressing the issues of lack of bargaining power of these fishers in markets. This also can be said for the fisheries marketing system in Norway that revolves around the operations of fish sales organizations, considered by fishers as their own means to mitigate risk and benefit from the market.

While fisheries cooperatives, associations and other organizational forms have an uneven record of success in developing countries, there are potentially beneficial roles that these organizations can play such as: (i) increasing the resilience and stability of fishing communities; (ii) helping stabilize markets by managing supply; (iii) enhancing the negotiating position of small-scale fishers in relation to traders; (iv) improving product quality and value added; (v) developing postharvest facilities and practices; (vi) improving logistics and access to market information; and (vii) managing risk through collective action.

A higher level of organization among small-scale fishers would also provide the basis for increased market competitiveness through: (i) the setting up of auction systems; (ii) establishment of cooperative ventures such as ice plants and fish-processing facilities; (iii) bulk purchases of fishing input such as gears, engines, equipment and fuel; and (iv) facilitation of credit schemes that would provide increased access to necessary financing.

Not least, being better organized would give small-scale fishers more political clout and negotiating power with local authorities and government agencies.

However, the experience of many development workers who have worked with small-scale fishers and fishing communities points to specific characteristics that set them apart from other marginalized socio-economic sectors and render organizational development programmes problematic.

Many fish stocks are migratory and do not observe human-determined boundaries. This affects the behaviour and lifestyles of small-scale fishers who also tend to move around in a similar way to the fisheries resources upon which they depend. Apart from this, there is also seasonality of fish stocks that further influences their livelihood strategies, especially for those who also engaged in other occupations such as farming during the lean fishing season.

This spatial and temporal mobility affects the optimal way in which to develop small-scale fisher organizations, including practical matters such as deciding upon the best time to set meetings and other organizational activities on a daily, monthly and yearly basis. These patterns by which small-scale fishers conduct their lives also result ultimately in irregular income and cash availability, which would affect organizational concerns such as schedules for loan repayment and capital mobilization.

Small-scale fishers and their households also tend to be isolated from society at large. Fishing as an activity occurs at a considerable distance from settlement areas and at night or very early in the morning, when most people are resting or asleep. Their residences are commonly located on a narrow strip along the coast with many of them not having formal ownership of the land on which their houses stand. These factors combine and result in the spatial, temporal and, ultimately, social isolation of small-scale fishing households. This isolation may be a determinant of their low social standing and marginalization in many societies is manifested by their lack of access to education, health and other social services.

Another crucial factor, already mentioned, is the nature of fisheries as common-pool resources and its potential implications on the building and strengthening of small-scale fisher organizations. These include conflicts that could arise when collective fishing and fish farming methods and techniques are introduced in the context of competing operations and the absence of individual fishing tenure. A possible model to address this is the Japanese FCA system that is based on collective tenure within which both cooperative and individual fish-capture and culture practices are exercised.

When setting up and/or strengthening small-scale fisher organizations for resource management, capacity building for market engagement must be pursued on a parallel track, which may be simultaneous or sequential with the former. There are expected to be sometimes marked, sometimes subtle, differences between organizing for coastal resource management and organizing for market engagement in terms of organizational form (associations or cooperatives), needed skills sets (environmental management or business development) and success indicators.

In addition to their crucial role as stewards of fishery resources, small-scale fishers also need to play a growth or development role since they comprise the majority (in terms of number of participants) of the fisheries sector in developing countries. Specifically, they have important roles in market engagement, which manifests in the formal economy as economic productivity (i.e. value added) and in making social contributions by providing subsistence and safety nets to small-scale fisher households.

As small-scale fishers often belong to the 'poorest of the poor' in many developing countries, they usually lack the capacity to link to integrated value chains and find it difficult to meet the desired quantity and quality of fishery products required by these markets. These issues have been addressed by employing such strategies as aggregation through marketing cooperatives, technical assistance for the improvement of fishery product quality and safety and involvement in group certification.

It is important for non-governmental organizations (NGOs) and business service providers (BSPs) who aim to facilitate market access for small-scale fishery producers to possess the internal capacity to engage in business and enterprise development, i.e. they themselves must be of proven competence in their particular fuel and/or commodity.

These organizations should have extensive experience in facilitating multi-stakeholder arrangements, i.e. those involving government (as policy makers/executors), business (as organizations engaged in production, marketing and other functions in the value chain) and other civil society organizations (as facilitators of development at the community and sectoral levels).

The linking organization (NGO and/or BSP) would be in the best position to assist the community/sectoral organization in meaningful and gainful engagement in markets, while also remaining true to their core principles and values that are oriented towards social and environmental goals as well as beneficial economic outcomes.

Strengthening Relations in the Value Chain

As small-scale fishers strengthen their organizations for resource management and market development, they need to identify and examine the types of markets with which they can engage and benefit from. In most cases meeting the demand for fish of the immediate community is the primary option, especially in cases when there is a lack of postharvest facilities and fish, whether from the wild or from aquafarms, needs to be disposed of quickly. There is thus a trade-off between lower buying prices and reduced risk of spoilage. Marketing is often done by women from fishing households – wives, daughters, mothers or sisters of the predominantly male fishers – a set-up typical in many communities.

Higher prices are often obtained if small-scale fishers cater to other than their own local market, and the role of traders then expands because of the need for tradeable volumes, methods of preservation and means of transport. Traders become providers of services essential to the functioning of local markets, in many cases also becoming the principal providers of credit for the often cash-strapped small-scale fishers. Production for large urban centres and international markets further increases the role of other chain actors such as large traders (those not based in fishing communities), processors and exporters.

The further the markets targeted by small-scale fisher organizations, the greater the need for thorough analysis of the relevant value chains (Jacinto, 2004). Small-scale fisher organizations are often not sufficiently familiar with the tools needed to do this and do not have access to the information needed to make an accurate analysis. This is a current need that can be filled by development organizations that provide services to the sector, such as non-government organizations, academic institutions and even business groups who are interested in developing partnerships with communities.

Value Chain Analysis

The value chain describes the full range of activities required to bring a product or service from conception, through the different phases of production (involving a combination of physical transformation and the input of various producer services), delivery to final consumers and final disposal after use (Kaplinsky and Morris, 2001). Capture-fisheries feed into diverse and spatially extensive networks of supply and trade that connect production with consumers, adding significant value and generating important levels of employment (the value chain). To some extent, this system can be used to provide an important mediation and buffering function to increasing variability in supply and source location, but direct impacts will also affect its ability to do so. This system can also be used to reduce vulnerability and increase adaptive capacities of fishers and fishing households.

The value chain encompasses many economic agents (individuals, companies, government). From the perspective of the value chain, it is relatively unimportant how impacts are distributed among the economic agents that comprise it. As a result, value chain analyses do not have to address many of the difficult policy decisions that determine how impacts are distributed. The value chain perspective is important because it offers insights that would not surface in studies focused on individual economic agents or particular policy frameworks. A value chain analysis can also uncover insights into the challenges that face the sector as a result of different drivers of change (see Hall, Chapter 8, this volume), such as climate change, including small firms' and fishers' competitiveness in changing markets. A value chain perspective of the small-scale fisheries sector can reveal response strategies that enhance the sustainability and competitiveness of the entire value chain and the economic agents that comprise it. Value chain analysis helps effectively to isolate the binding constraints that affect the sector in a systematic manner. The set of issues that emerge from such a detailed analysis at a sector level has implications for both the public and private sectors alike. Some of the issues are sector-specific, and others are relevant across an economy and apply to many sectors and firms in a country. It also provides an opportunity to find policy positions that can be supported by the sector's different economic agents and important stakeholders.

The idea of a value chain is quite intuitive. A value chain refers to the full range of activities that are required to bring a product (or a service) from conception, through the different phases of production, to delivery to final consumers and disposal after use (Kaplinsky, 2000; Kaplinsky and Morris, 2001). Furthermore, a value chain exists when all the stakeholders in the chain operate in such a way as to maximize the generation of value along the chain. This definition can be interpreted in either a narrow or a broad sense. In the former, a value chain includes the range of activities performed within a firm to produce a certain output. The 'broad' approach to a value chain looks at the complex range of activities implemented by various economic agents (primary producers, processors, traders, service providers, etc.) to bring a raw material to the retailing of the final product. The 'broad' value chain starts from the production system of the raw materials and will move along the linkages, with other enterprises engaged in trading, assembling, processing, etc. The concept of the value chain encompasses the issues of organization and coordination, the strategies and the power relationship of the different economic agents in the chain. The idea of a value chain is associated with the concept of governance, which is of key importance for fisheries because fisheries value chains crucially depend on the utilization of natural and environmental resources. The value chain framework can also be used to understand social ties and traditional norms, which can be used to draw conclusions on the participation of the poor and the potential impact of value chain development on poverty reduction and food security.

The world of production and exchange is complex and heterogeneous. Not only do value chains differ (both within and between sectors), but so, too, do national and local contexts. There is therefore no mechanistic way of applying value chain methodology; each chain will have particular characteristics, whose distinctiveness and wider relevance can only be effectively captured and analysed though an understanding of the broader issues involved.

Three main research streams may be recognized in the value chain literature: (i) the filière approach; (ii) the conceptual framework elaborated by Porter (1985); and (iii) the global approach proposed by Kaplinsky (2000), Gereffi (1994, 1999) and Gereffi and Korzeniewicz (1994).

Value chains are complex, and particularly in the middle tiers, individual firms may feed into a variety of chains. Which chain – or chains – is/are the subject of enquiry therefore very much depends on the point of entry for the research inquiry. In each case, the point of entry will define which links and which activities in the chain are the subject of special enquiry. The entry point and the concentration of the value chain analysis are directly related to the desired development outcome from supporting the value chain. For example, if the focal point of the enquiry is in the design and branding activities in the chain, then the point of entry might be on design houses, or the branding function in key global marketing companies. This will require the research to go backwards into a number of value chains that feed into a common brand name (for example, the different suppliers to Nestlé). At the other end of the scale, a concern with small and medium-sized firms that feed into a number of value chains might require the research to focus on final markets, buyers and their buyers in a number of sectors, and on a variety of input providers. The key entry point that will be used in this proposal is the impact of the development and operation of the small-scale fisheries value chain on food security and poverty resulting from climate change.

The value chain approach is mainly a descriptive tool to look at the interactions between different economic agents. As a descriptive tool it has various advantages in so far as it forces the analyst to consider both the micro and macro aspects involved in the production and exchange activities. Commodity-based analysis can provide better insights into the organizational structures and strategies of different actors and an understanding of economic processes often studied only at the global level (often ignoring local differentiation of processes) or at the national/local level (often downplaying the larger forces that shape socio-economic change and policy making).

The methodology should address the following issues, and begin with an understanding of the nature of final markets, which are increasingly the driver in many value chains:

- the point of entry for value chain analysis;
- mapping value chains;
- product segments and critical success factors in final markets;
- how producers access final markets;
- benchmarking production efficiency;
- governance of value chains;
- upgrading in value chains; and
- distributional issues.

Kaplinsky and Morris (2001) stress that there is no 'correct' way to conduct a value chain analysis: rather, the approach taken fundamentally rests upon the research question that is being answered. None the less, four aspects of value chain analysis as applied to agriculture are particularly noteworthy.

1. At its most basic level, a value-chain analysis *systematically maps the economic agents* participating in the production, distribution, marketing and sales of a particular product (or products). This mapping assesses the characteristics of economic agents, profit and cost structures, flows of goods throughout the chain, employment characteristics and the destination and volumes of domestic and foreign sales (Kaplinsky and Morris, 2001). Such details can be gathered from a combination of primary survey work, focus groups, participatory rural assessments (PRAs), informal interviews and secondary data.

2. Value chain analysis can play a key role in *identifying the distribution of benefits of economic agents in the chain*. That is, through the analysis of margins and profits within the chain, one can determine who benefits from participation in the chain and which economic agents could benefit from increased support or organization. This is particularly important in the context of developing countries (and agriculture in particular), given concerns that the poor in particular are vulnerable to the process of globalization (Kaplinsky and Morris, 2001). One can supplement this analysis by determining the nature of participation within the chain to understand the characteristics of its participants.

3. Value chain analysis can be used to *examine the role of upgrading within the chain*. Upgrading can involve improvements in quality and product design that enable producers to gain enhanced value or through diversification in the product lines served. An analysis of the upgrading process includes an assessment of the profitability of actors within the chain, as well as information on constraints currently present. Governance issues play a key role in defining how such upgrading occurs. In addition, the structure of regulations, entry barriers, trade restrictions and standards can further shape and influence the environment in which upgrading can take place.

4. Value chain analysis can *highlight the role of governance* in the value chain. Governance in a value chain refers the structure of relationships and coordination mechanisms that exist between economic agents in that value chain. Governance is important from a policy perspective through identification of the institutional arrangements that may need to be targeted to improve capabilities in the value chain, remedy distributional distortions and increase value added in the sector.

At the heart of the analysis is the mapping of sectors and key linkages. The value added of the value chain approach, however, comes from assessing these intra- and interactor linkages through the lens of issues of governance, upgrading and distributional considerations. By systematically understanding these linkages within a network, one can better prescribe policy recommendations and, moreover, further understand their reverberations throughout the chain.

Building mutually beneficial relations among the various actors in the value chain while maintaining priority on improving the livelihoods of small-scale fishers can start from the hypothesis, on the part of small-scale fishers and their support organizations, that traders can be potential partners rather than being the adversaries in a zero-sum game.

Traders play necessary roles in the functioning of value chains, such as helping to develop consumer markets, providing financial services and adding value to fishery products. On occasion they bear risks even more so than do primary producers – spoilage, low prices in consumer markers, non-payment of loans – and in the course of trading operations devise means to manage and mitigate such events. The margins that they obtain in the markets should be appraised in the light of these risks, as well as the costs they incur and the services they provide.

The Role of Fish Traders

Fish traders (market intermediaries, middlemen) are important actors in any market system. They provide small-scale producers with incentives and access to markets, but they also provide a variety of services to the producers. Traders can be differentiated according to the services they perform, such as transport, processing, money lending, risk bearing and market information. They can be distinguished by the function they perform such as primary buyer, wholesaler or retailer. Fish traders may also buy and sell other products. Frequently, they provide a social insurance mechanism to individual fishers through credit arrangements. Yet, the productive role of the trader in providing services advantageous to the fisher and in reducing the fisher's market risks is often overlooked. Such close relations between producers and local traders are well documented in small agricultural production systems, and reciprocal agreement and credit arrangements between the two have been examined for small-scale fisheries (Smith, 1979; Smith et al., 1980; Scheid and Sutinen, 1981; Ishak, 1988; Pomeroy, 1989).

The *suki* relationship in the Philippines, a credit/marketing linkage, is often assumed to be exploitative of the fisher. In its simplest form, the it provides the fisher with a guaranteed outlet for his fish and access to capital, while providing the trader with a steady supply of fish. When a fisher enters into a *suki* relationship, he must sell his fish exclusively to that trader, the purchase price being established by the trader. The trader provides the fisher with a wide range of services and the majority of the fishers are in debt to the trader. It has been argued by some that the *suki* relationship is exploitive of fishers. In cases where credit is extended and a lower ex-vessel price is given, it is felt that oligopolistic control (an imperfect competitive market situation where relatively few buyers handle a large percentage of the fish landed by and purchased from fishers and thus can influence the price paid to fishers) over the fisher exists. Others feel, however, that the potentially large number of traders with whom a fisher could establish a *suki* relationship and social and kinship ties within the community exert a modifying influence over oligopolistic tendencies. In a study in the Philippines, Pomeroy (1989) found that traders did not exploit *suki* fishers, and that the lower price paid to the *suki* fishers reflected a competitive charge for the services provided. Factors that were found to inhibit or reduce the level of fisher exploitation included social and kinship ties, the beneficial nature of the relationship to both parties, fear of entry of new traders and the existence of a relatively large number of non-*suki* fishers in the area.

The market both provides for, and restricts, livelihood opportunities for small-scale fishers and market traders. The constraints to market access include weak bargaining power and poor marketing strategies, monopolies among traders, poor product-holding infrastructure, difficulties meeting quality standards and lack of market information. With specialized traders, fishers often have little, if any, control over marketing outlets and the prices that they receive. Women producers and traders face additional gender-related barriers including lack of access to credit and technology, increased dependence as well as a lack of representation in local decision-making related to fisheries and other livelihood opportunities. Low incomes create a situation of potential dependence that influences decisions about production and marketing by the fisher. This dependency may become a motive for excessive exploitation of open-access resources, or it may undermine compliance with formal resource governance institutions. Relations and potential inequalities between fishers and traders point to the need to find ways to address these issues in order to increase the return received by fishers and to better sustain fisheries resources. This requires a better understanding of fisher–trader relations and how these relations affect decisions about resource use and ecological outcomes.

This is not to say that there are no unscrupulous traders, for many studies have shown there are those who profit disproportionately and unjustly from the disadvantaged position of small-scale fishers in value chains. A more nuanced approach needs to be applied in terms of analysing the benefits and costs of confronting or collaborating with particular traders.

One way of differentiating between 'good' and 'bad' traders is to look at their behaviour in relation to other chain actors (especially small producers) – are they working towards the development of long-term relationships with both suppliers and buyers? Do they refrain from short-term speculative activities that tend to 'degrade' value chains, i.e. reduce stability and profitability over the long term? Do they facilitate the flow and sharing of market information to the benefit of their partners in the value chain?

One of the main challenges in value chain intervention is to facilitate the transformation of 'bad' traders into 'good' traders by generating respect among chain actors sufficient for the emergence of mutually beneficial chain partnerships.

Fish Market Analysis

In small-scale fishing communities, combinations of economic, technical and informational factors appear in varying degrees. These factors imply that sufficient conditions may exist for non-competitive exploitation of fishers by traders. Market analysis studies can be undertaken to identify the existence of non-competitive market conditions in the fishery (Pomeroy and Trinidad, 1995). In developing countries, conditions do not always allow studies of fish marketing systems to be neatly dealt with. Since so little is usually known about the marketing system and its operation, the first need is to describe accurately the system that exists.

Market analysis studies can be classified using descriptive, price efficiency and organizational criteria. The descriptive approach usually contains little statistical analysis, but reaches conclusions based on the investigator's subjective analysis of the situation. This approach has been extensively used as a basis for studying commodity flows and marketing techniques. The price efficiency approach analyses marketing in its dimensions of space, time and form. Examination of the efficiency with which the marketing system transmits information among the different producers, wholesale and retail markets is achieved through the application of various pricing criteria. The industrial organization approach is a standard methodology for industry analysis. The theory tells us that the market structure (the market environment) determines market conduct (the behaviour of economic actors with the environment) and thereby sets the level of market performance (how close the industry comes to meeting the norm or standard of reference of social welfare). If we can uncover a reliable link between elements of structure, conduct and performance, we have a powerful tool for economic analysis.

The descriptive analysis of the seafood marketing system focuses on a functional (exchange, physical, facilitating) and institutional (traders) description of existing and historical arrangements. It includes a description of socio-economic characteristics of traders, market channels, trade flows and distribution routes, and the marketing process, including fisher–trader relationships. It also provides a description of the seafood products currently traded, as well as those available in the ecosystem but currently not traded. Assessing species trade will provide insights into what ecosystem processes are being traded. Through descriptive analysis, it will also allow for identification of the various marketing arrangements that exist in the study area, including the relationship between fishers and traders and between traders and traders. The details of these relationships will be studied in more depth through both surveys and interviews with fishers (using different gears and targeting different species) and various types of trader (traders at different levels in the marketing chain). The outcomes of these relationships will also be examined, including the impacts on prices, resource exploitation, livelihoods and trade.

The description of the marketing system will lead to a hypothesis about the competitive nature of the market. This hypothesis can then be tested through a systematic analysis of the market using a methodology from the field of industrial organization, which traces a causal flow between market structure, conduct and performance. The industrial organization analysis will allow for an understanding of whether or not non-competitive conditions exist for traders in the seafood marketing system. Market structure consists of characteristics of the organization of a market that appear to influence the nature of competition and pricing behaviour within the market. Through direct observation of the functioning of the market and its participants, analysis can be made of variables such as buyer concentration, barriers to entry, vertical integration and product differentiation. Market conduct refers to the pattern of behaviour that traders follow in adapting or adjusting to the markets in which they buy and sell. Two aspects of market conduct that are examined are buying and selling practices and pricing behaviour. Market performance refers to the impact of structure and conduct as measured in terms of variables such as prices, costs and volume of output. By analysing the level of marketing margins and their cost components, it is possible to evaluate the impact of the structure and conduct characteristics on market performance. Price and cost differentials are examined in their dimensions of space, time and form.

The limitations of the industrial organization approach include the dynamic rather than causal relationship between market structure, conduct and performance. The approach also implicitly assumes a fairly substantial amount of data, preferably collected over a fairly long time. Also critical to this approach is the participation and astute observation of the researcher when it becomes time to analyse market conduct.

Development of chain partnerships is a means to integrate value chains that can foster innovation and good practices. Value chain integration in the industrial sector is typically driven by large manufacturing combines around which other chain actors gravitate because of their production efficiency and the technology available. While this model can also be seen in the fisheries sector, there are many value chains, especially local ones, that are not dominated by such big chain actors and can be the subject of value chain intervention aimed at fostering chain partnerships focused on small-scale fishers.

Lately, public–private partnerships (PPPs) have been developed for commodities such as coffee and palm oil, aimed at the incorporation of social and environmental criteria into product standards, certification systems, labelling schemes and codes of conduct across the length and breadth of the value chain involving producers, processors, traders, retailers and consumers.

Another type of partnership is that between communities and companies that features the buying of contracts for specific products that involve the application of standards and practices agreed upon and that aim to give a price premium to the members of that community.

These partnerships can evolve from initial disrespect, to continuing dialogue to mutual understanding of the specific roles that each actor plays in the value chain. It can even lead to an active appreciation of the specialization that occurs in the development of value chains. In a highly evolved, fairly governed value chain that distributes benefits equitably among the different actors, net social benefits can be greater when small-scale fishers specialize in production while leaving traders to perform their marketing role.

Building Market Institutions

Parallel with strengthening relations among different actors in fisheries value chains is the building of requisite market institutions that will serve to facilitate and regulate the functioning of these chains. Specifically, these institutions pertain to: (i) market information systems; (ii) product and process standards; (iii) rules for enforcing contracts; (iv) financial services provision; and (v) an appropriate set of policies for formal chain governance.

Market information

Transparent market information is founded on strong relations between and among fisheries value chain actors, with the most crucial being open communication between small-scale fishers and community-based traders that they deal with on a daily basis. This applies as well to the relations among traders operating across different scales (from small traders based in fishing communities to large traders based in urban areas), and the more powerful chain actors such as large-scale processors, big retailers and exporters.

Free flow of market information contributes to a working consensus on buying and selling prices, and profit margins that are acceptable to the different actors in the value chain and that are generally perceived as just and non-exploitative. Good information flow also speeds up reaction time to changes in the value chain (especially on the demand side) and facilitates quantitative and qualitative adjustments in production.

But, perhaps more importantly, transparent market information can help accelerate upgrading of value chains, especially in situations where mechanisms for dialogue are in place and there is active and participatory governance from the different chain actors.

Product and process standards

Product and process standards that are generally accepted and have a basis in formal regulations is a firm foundation for the continuous development of fisheries value chains. In accord with good information flow through the chain, actors on the demand side such as retailers and consumers would have a rationale for providing premium to small-scale fisheries production, especially that which adheres to social and environmental criteria.

However, setting standards is just an initial phase and these standards must be effectively verified by credible actors. In this way, standards in place will necessitate the development of certification and accreditation systems and would need to include, especially in schemes requiring third-party involvement, actors that are not embedded in the value chains in question.

Some of these eventually culminate in labelling schemes that attempt to communicate to final consumers that fisheries products from specific value chains have adhered to production and processing criteria that ensure food safety, environmental sustainability and social responsibility.

Rules for enforcing contracts

While appropriate and workable rules for enforcing contracts is an indicator of robust market institutions, these rules, and the processes through which they are operationalized, are embedded in the larger field of social capital, i.e. the level of trust between market actors that makes repeated transactions possible without the need for formal sanctions in instances of deviation from actions and procedures previously agreed upon.

It is necessary to differentiate between traditional trust, developed among small-scale fishers and traders involving numerous transactions over time, and modern trust mediated by formal contracts and often involving legally constituted business firms. Traditional trust usually operates with a lot of flexibility, while modern trust is often more rigid.

In terms of facilitating the effective interaction and/or eventual integration of small-scale fishers with 'modern' markets wherein relations are consummated through the fulfilment of contractual obligations, mechanisms for flexibility can be worked into these contracts with the aim of managing/mitigating risks, specifically those associated with fishing as an economic activity, coastal communities as a locale and fishers as a social sector. These contracts could include: (i) the provision of inputs such as seeds, feeds, fuel and credit for start-up and operations to small-scale fishers (a role played by traditional traders); (ii) renegotiation provisions in cases of extreme price volatility (possibly with a variance price trigger); and (iii) resolution mechanisms (in cases of contract disputes involving 'small sums') where value chain associations composed of representatives from the different market players (including small-scale fishers) serve as arbitration boards.

Ultimately, the goal is for enforceable contracts among small-scale fishery producers, traders, processors and other market players that can run parallel to or even supplant traditional/informal market relations. These contracts can be replicated throughout the small-scale fishers sector and would be expected eventually to build trust and foster cooperation (i.e. development of social capital) among value chain participants, with preferential consideration for small-scale fishers in relation to other market players such as traders and buyers, input suppliers, processors and exporters.

Financial services provision

Lack of finance capital is one of the major constraints faced by small-scale fishers across the world. They are generally considered as a credit risk because of the unpredictability of fish catches due to resource variability, the seasonality of fishing as source of income, the perishability of fishery products and the price volatility resulting from the combination of these factors.

Because of the uncertainties of fishing, small-scale fishers often find it difficult to get financing for day-to-day fishing operations. Their needs are not met by micro-finance schemes that cater mainly to micro-enterprises or by banks that are geared to provide services to medium- to large-scale enterprises. This gap in formal financing is commonly filled by what can be termed a chain credit, i.e. credit provided by traders to whom the small-scale fishers sell their catch.

Credit provided by traders for small-scale fishing operations has been criticized in fisheries social science literature as a means for the unscrupulous to exploit impoverished fishing households by using their position to drive down buying prices for fishery products.

While this is true in many instances, it is also true that small-scale fishers do not have other options in cases where cooperatives and other collective efforts to address their financing problems have failed. It seems the case that these traders, many of whom are neighbours and even relatives of fishers themselves, are the only ones willing to take the risk of providing credit to small-scale fishers. Furthermore, it can be asserted that the purpose of chain credit is not primarily for earning interest but rather for maintaining good trade relations and a continuous flow of products.

In building the capacity of small-scale fishers to engage in markets, access to formal finance is a crucial concern. This includes access to both formal credit for capital expenses and financing for fishing operations.

With regard to the former, facilities and financial products can be developed in partnership with development and rural banks, wherein small-scale fishers will be able to access financing for medium- to long-term investment in small-scale fisheries production, both in the capture and aquaculture sectors.

With regard to the latter, what is currently termed as chain credit can be transformed into formal chain financing schemes with a large role to play for traders as they relate to small-scale fishers in specific value chains. Formal chain financing can be described as a system wherein product flow in the value chain becomes the carrier for providing financial services. Small-scale fisheries can be transformed from a credit risk to a performance risk in the course of developing formal chain finance, by devising mechanisms whereby the risk of non-willingness to pay becomes the lesser risk of production shortfall (on the part of producers) or stoppage of trading activities (on the part of traders). This can be in the form of production contracts, collateral, storage receipts or other means to mitigate risks.

Potentially, this can be advantageous in situations wherein buyers (including processors) are willing to extend credit to ensure that they have sufficient supply. On the other hand, buyers willing to provide capital/financing for production/operations are sometimes dissuaded from doing so in instances of extra-contractual marketing wherein producers violate contracts (formal or informal) to sell to other parties who offer higher/better prices.

Enabling policy environment

Fostering an enabling policy environment for market development for small fishers can be facilitated by bringing together multiple stakeholders, i.e. public and private. This environment would not only encompass the body of laws governing the fisheries sector but also policies such as taxation, market regulation and provision of public goods and support services. To develop robust and beneficial linkages between small fishery producers and local, urban and export markets, it is necessary to have a bare minimum of policies in place that would enable market actors to act freely in pursuing economic gains.

In order to push forward this policy environment, small-scale fishers and their advocates should focus not only on the formulation of policies but also on their implementation and practical application. Furthermore, even as it is good for policy advocacy to work towards solving problems and resolving conflicts, the positive intent of policies to stimulate development and business should not be overlooked. Studying the value chain should help small-scale fishers evaluate the playing field, i.e. an analysis of how power is distributed across the value chain.

Developing the enabling environment for successful resource management and market development for small-scale fisheries would need to be guided by two basic precepts – 'establish the basics' and 'kick-start the markets'. The first precept pertains to investment in public goods to develop technologies that will raise the potential productivity of small-scale fishery producers, while the second refers to coordinated complementary investments to improve the access of small-scale fishers to both financial services (including insurance as risk management) and the input and output markets necessary for technology adoption.

From these precepts, we can envisage this raft of policies and programmes as consisting of: (i) extending the 'right' subsidies; (ii) facilitating contract enforcement; (iii) protecting and promoting property rights; (iv) providing public goods and infrastructure; (v) developing market information and intelligence; (vi) conducting market-oriented research and product innovation; (vii) developing applicable fishery product standards and certification systems; (viii) instituting export quality controls as necessary; and (ix) promoting PPPs.

'Right' subsidies would be effective grants for activities such as assessing local markets, preparing business plans, piloting of specific products and strengthening of skills in organizational development, financial management and postharvest handling.

As an integral part of developing market institutions, policies should facilitate the enforcement of contracts with consideration for necessary flexibilities in response to uncertainties associated with small-scale fisheries. In the end, this should result in the emergence of strong relations of trust between small-scale fishers and other actors in the value chain.

There must be policies that lead to secure tenure/property rights over coastal/fishery resources. This would afford small producers the foundation upon which they can enter into market relations. Specifically, such tenure would serve as formal/informal collateral for small producers so that they can access the resources they need to increase resource productivity and economic benefits (Jacinto, 2004).

There must be policies in place that would lead to the provision of basic infrastructure, social services and protection to enable small fishery producers to emerge from destitution through their own efforts. This would enhance productivity at the household and individual levels by maintaining human resource capacity at its optimum.

Ensuring conditions for the continuous flow of market information is another area of policy development. Small-scale fishers are often on the losing end of market asymmetries wherein traders and processors use market information that they possess exclusively as leverage in setting prices. By making vital market information available through traditional and modern means, policy makers and implementers will contribute to levelling the playing field in fisheries markets.

Related to market information flow is the need for creating and disseminating new knowledge that would lead to the innovation and development of products from small-scale fisheries. While large- and even medium-scale fishery producers have the means to engage in what is often referred to as 'research and development', proactive policies are needed to ensure that there is appropriate attention and resources given to product development in small-scale fisheries.

As discussed in relation to the development of market institutions, policies should facilitate the adoption and implementation of fishery product standards and corresponding certification systems. This leads to greater added value as well as increased market access for products from small-scale fisheries. It can develop to the point that there will be an eventual need for export quality controls in cases where these products are able to fill demand in the international market.

A potentially effective means in the provision of extension services for small-scale fisheries development is through the development of public–private partnerships involving small fishery producers, traders, processors, exporters and other market players. While PPPs are seemingly the result of an extended development of the value chain, often in a developed-country setting, some initiatives focused on small producers have been successful in certain least-developed countries. A proactive approach to developing PPPs can be a crucial component in a comprehensive strategy for market development in small-scale fisheries.

Summary

Even though small-scale fishers do not derive optimal economic benefits from markets across different scales (especially national and global), the whole range of values that they contribute to society and ecosystems has not yet been accorded proper recognition.

Mentioned above is the contribution of small-scale fisheries to employment (livelihood), nutrition (food security) and income (including export earnings), but there are also the services provided by fisher- and community-managed coastal ecosystems, such as coral reefs, mangrove forests and seagrass beds. These include not only provisioning services (the values of which are realized in the market as food, water, fuel and fibre) but also the regulation, support and cultural services that are more difficult to quantify and for which there are few existing markets. While there are already valuation methods and systems for these services, policy frameworks – especially in least-developed and developing countries – often do not take this into account in decision-making with regard to fisheries and coastal resources.

In relation to large- and medium-scale capture-fisheries and most types of aquaculture, small-scale fisheries also appear to be more efficient when certain economic and ecological factors are taken into consideration. On aggregate, small-scale fisheries employ more people, consume less fossil fuels (both in absolute terms and in relation to fish catch), produce less by-catch and fishmeal/fish oil and require less government subsidy while producing approximately the same amount of fish for human consumption as large-scale capture-fisheries.

On this basis, it has been argued that instead of international NGOs allocating resources in attempting to reform large-scale fisheries into being sustainable through eco-labelling and certification schemes, work and funds should be put into reducing current levels of government subsidies to large-scale, industrial fishing fleets in order substantially to reduce effort and allow fisheries to recover on the global scale.

From the value chain perspective, it would appear that the non-recognition of the values and services that SSF produce, and the efficiency with which they bring this about, is inversely related to the 'length' of the value chains in question. At the local level, it seems more likely that communities, and even local governments and the private sector, are able to arrive at an appropriate valuation of small-scale fisheries in relation to other economic activities and other uses of resource in coastal areas.

Thus, it is in the interest of the small-scale fishers and their communities to keep fisheries value chains 'as short as possible', not only for them to capture more economic benefits from these chains but also to maintain them as efficient creators of both market and non-market values for the local community, as well as for society and ecology as a whole.

In cases where it will be counterproductive to dismantle functioning fisheries values chains that have a global character, it would be ideal to employ all means available to fisheries managers and development workers, as discussed above, to ensure that resource management and market engagement are integrated into approaches that are environmentally sustainable and socially responsible.

Small-scale fisheries function within the larger context of the national and global economy. The current international economic crisis has reminded us that while markets can be effective and efficient in distributing goods and services among the various individuals and institutions across the different countries of the world, this does not necessarily happen: markets can and do fail.

Even as debate rages with regard to the proper role of government in relation to markets (and market actors such as trans-national corporations) in specific instances in specific places, current thinking points to the need for judicious intervention and regulation by the state not only to address but also to prevent market failure.

As it is apt for capital and property markets, so it is for the markets for products and services derived from fisheries and coastal resources and ecosystems. While a long list of possible policies and programmes that can be formulated and implemented by government in partnership with small-scale fishers and the private sector institutions has been enumerated above, it is in the actual application of these partnerships that we will find out which ones will work for which communities.

References

Albu, M. (2008) *Making Markets Work for the Poor: Comparing M4P and SLA Frameworks: Complementarities, Divergences and Synergies.* FAUNO Consortium, Bern, Switzerland.

Dorward, A., Poole, N., Morrison, J., Kydd, J. and Urey, I. (2003) Critical linkages: livelihoods, markets and institutions. ADU Working Paper 02/03, Imperial College Wye, Wye, UK.

Gereffi, G. (1994) The Organization of buyer-driven global commodity chains: how US retailers shape overseas production networks. In: Gereffi, G. and Korzeniewicz, M. (eds) *Commodity Chains and Global Capitalism.* Praeger, London.

Gereffi, G. (1999) A commodity chains framework for analysing global industries. In: *Institute of Development Studies, 1999, Background Notes for Workshop on Spreading the Gains from Globalisation.* Available at: http://www.ids.ac.uk/ids/global/conf/wkscf.html (accessed 6 July 2010).

Gereffi, G. and Korzeniewicz, M. (eds) (1994) *Commodity Chains and Global Capitalism.* Praeger, London.

Ishak, H.O. (1988) Market power, vertical linkages and government policy: the Malaysian fish industry. PhD dissertation, University of East Anglia, Norwich, UK.

Jacinto, E.R. (2004) A research framework for value-chain analysis in small-scale fisheries. *Lundayan Journal* 5, 41–61.

Kanji, N. and Barrientos, S. (2002) Trade liberalization, poverty and livelihoods: understanding the linkages. A review for the Africa Policy and Economics Department. Department for International Development, UK, IDS Working Paper 159, Institute of Development Studies, London.

Kanji, N., MacGregor, J. and Tacoli, C. (2005) *Understanding Market-based Livelihoods in a Globalizing World: Combining Approaches and Methods.* International Institute for Environment and Development, London.

Kaplinsky, R. (2000) Spreading the gains from globalisation: what can be learned from value market chain analysis? IDS Working Paper 110, Institute of Development Studies, University of Sussex, Brighton, UK.

Kaplinsky, R. and Morris, M. (2001) *A Handbook for Value Chain Research.* International Development Research Center, Ottawa, Canada.

Kohls, R.L. and Uhl, J.N. (2001) *Marketing of Agricultural Products,* 9th edn. Prentice Hall Publishing, Upper Saddle River, New Jersey.

Macfadyen, G., Banks, R., Phillips, M., Haylor, G., Mazaudier, L. and Salz, P. (2003) *Background Paper on the International Seafood Trade and Poverty.* Network of Aquaculture Centres in Asia-Pacific (NACA), Bangkok, Thailand.

Pomeroy, R.S. (1989) The economics of production and marketing in a small-scale fishery: Matalom, Leyte, Philippines. PhD dissertation, Cornell University, New York.

Pomeroy, R.S. and Trinidad, A. (1995) Industrial organization and market analysis: fish marketing. In: Scott, G.J. (ed.) *Prices, Products and People: Analyzing Agricultural Markets in Developing Countries.* Lynne Rienner Publishers Inc., Boulder, Colorado.

Porter, M.E. (1985) *Competitive Advantage: Creating and Sustaining Superior Performance.* The Free Press, New York.

Scheid, A.C. and Sutinen, J. (1981) The structure and performance of wholesale marketing of finfish in Costa Rica. In: Sutinen, J. and Pollnac, R. (eds) *Small-scale Fisheries in Central America: Acquiring Information for Decision-making.* International Center for Marine Resource Development, University of Rhode Island, Rhode Island.

Shaw, S.A. (1986) Marketing the products of aquaculture. FAO Fisheries technical paper 276, UN Food and Agriculture Organization, Rome, Italy.
Smith, I.R. (1979) A research framework for traditional fisheries. In: *ICLARM Studies and Reviews 2*, International Center for Living Aquatic Resources Management, Manila, Philippines.
Smith, I.R., Puzon, M.Y. and Vidal-Libuano, C.N. (1980) Philippine municipal fisheries: a review of resources, technology and socioeconomics. In: *ICLARM Studies and Reviews 4*, International Center for Living Aquatic Resources Management, Manila, Philippines.
van Mulekom, L., Axelson, A., Batungbacal, E.P., Batker, D., Siregar, R. and De la Torre, I. (2004) Trade and export orientation of fisheries in Southeast Asia: underpriced exports at the expense of domestic food security and local economies. *Lundayan Journal* special issue, 19–40.

10 Communication

Patrick McConney and Carmel Haynes

Issues and Challenges

Communication is an essential aspect of fisheries management. It is especially challenging in small-scale fisheries (SSF) due to their diversity and complexity (Berkes *et al.*, 2001). When there are many stakeholders, often with differences in formal education, knowledge bases, cultures, interests and ways of learning, it is easy to be misunderstood. Since we all communicate every day it is normal to think that communication is just plain common sense – but then we are surprised when our communication is not effective. At the other extreme, some stakeholders may decide that no communication at all is preferable to difficult communication. This non-participatory stance may escalate conflicts or erect additional barriers. Understanding communication is important.

However, if we look at the typical fisheries authority's capacity to communicate, we find that few fisheries officers are exposed to training in communication in their academic or professional careers. Well-intentioned fisheries managers may also find themselves unable to communicate as they would like to due to restrictive information policies and controls in government agencies. Fisheries resource users and non-governmental organizations (NGOs) in civil society may see communication primarily as a tool of advocacy, and interest- or pressure-group tactics, but be equally ill-equipped. The private sector and donor agencies may view communication mainly as a marketing tool to sell their goods and services to customers and beneficiaries by advertising and announcements, respectively, but their messages may not reach intended audiences. What can we do to improve communication in various SSF situations so that each of these stakeholder groups and others can engage in meaningful exchanges to achieve shared aims? Tackling this requires understanding that communication is not only about what you share but how, when, where, why and with whom. A bit of insight and some tips can pave the way for improvement.

This chapter first sets communication in context, describes what it is and the basics of how it can be made more effective. The next sections feature communication strategies, communication-based processes and specific media guidelines. The chapter draws upon a wide literature, particularly development communication (e.g. Ramírez and Quarry, 2004), sprinkled with examples and experiences from SSF.

Readers will not become communication experts upon perusing the chapter. They can, however, use the practical guidelines offered to improve their communications and develop their expertise from learning-by-doing accompanied by adequate monitoring and evaluation. Communication toolkits are available (e.g. Gauthier, 2005; Hovland, 2005; OECS, 2007; ESRC, 2008) to further facilitate this process. As with many aspects of SSF, the aim is to experiment carefully and responsibly to learn and adapt in order to be able continuously to make improvements to communication over time.

Fisheries Data, Information and Messages

A common saying is that 'the information speaks for itself'. This is seldom true. Fisheries data (fish length and weight measurements, catch and effort records, trade statistics, consumption figures, socio-economic and demographic profiling numbers, etc.) need to be analysed and interpreted into useful information (fisheries model outputs, summary charts with text, layers in a geographic information system, technical report, etc.). But these outputs usually need to be further encapsulated in messages (given meaning or 'speak') in order to reach target audiences and have the intended effect (Fig. 10.1).

Although all of these steps involve some aspects of communication, the last, messaging stage, is often overlooked. This stage needs careful attention in this so-called information age if fisheries communications are to cut through the increasing competing noise. This noise may depend in part upon the scale or level of the communication. Scale can be temporal, spatial, institutional or other, and level is a measurement on the scale such as daily or annual if temporal, settlement or country if spatial, individual or community if institutional, and so forth. Typically, noise increases as the level increases on any particular scale. Many factors contribute to this, such as greater numbers of stakeholders, more economic sectors, diversity of norms and values, etc. So, although care must be taken with all messages, those at higher levels require additional attention.

Understanding Messages

Definitions of communication abound, but the central theme is information exchange through some system of messaging. From the beach to the boardroom, within any grouping, one of the most highly prized skills that an individual can possess is the power to communicate effectively. Experience within and beyond fisheries has shown that the ability to communicate effectively can make a profound difference when it comes to securing physical or human resources, gaining funding or rallying parties to reach decisions. Sometimes, providing information is the most powerful strategy available. Not only is information a tool for empowering people to help themselves, information sharing is also an important mechanism for encouraging accountability, transparency and participation in the decision-making process – and these are necessary ingredients for good fisheries governance.

In order to understand effective communication, one must understand the communication process (GTZ, 2006). This is not a very difficult task, as each of us engages in forms of verbal or non-verbal communication every day. We all carry out the six stages of the communication process in any conversation,

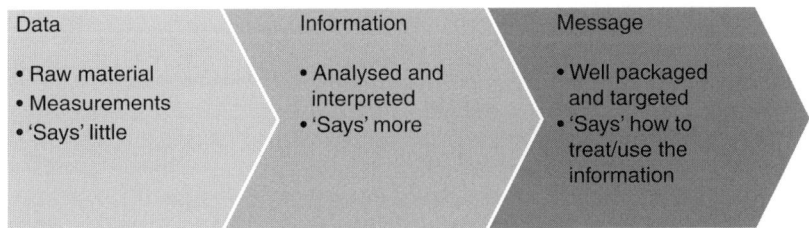

Fig. 10.1. The messaging sequence.

when a captain instructs a crew member to carry out a task, or when you try to influence a top policy maker to take a fisheries planning decision. has been effective. Although communication can be an individual process, fisheries organizations may be the Senders. Organizational perspectives and processes have then to be taken into account.

Stages of communication

Communication is a process carried out over six stages in which both parties (the Sender and the Receiver) engage in dialogue (Fig. 10.2) using the following continuous process:

- A Sender decides to initiate a message with a specific set of intended meanings.
- The Sender then encodes the intended meanings by selecting specific words, gestures or images, which the Receiver is expected to understand.
- The message is then transmitted in visual, spoken and/or written form across the gap between Sender and Receiver.
- The Receiver perceives the incoming message as a specific pattern of symbols.
- The Receiver then decodes the message through his or her own interpretation of the symbols.
- The Receiver is then influenced in some way (whether aware of this influence or not).

It is mainly the final stage of the communication process – the ability to influence the Receiver of the message in the way intended – that is the hallmark of effective communication. When the Sender has prompted the desired response from the Receiver – aroused an emotion, triggered action or simply taught him or her new facts – then the communication

Barriers to effective communication

If the Receiver does not interpret the words, gestures or images in the way in which the Sender intended, then the message is distorted or lost and the intended response cannot be achieved. The factors that can create barriers to communication include the following and many more:

- lack of shared cultural meanings for the same words (e.g. use of fishing terms);
- unfamiliarity with the language (e.g. official language versus creole or dialect);
- a physical condition that interferes with the transmission and reception of the information (e.g. either person might have difficulty hearing, seeing or speaking);
- memory failure or lack of recall (e.g. forgetting a major fishing event;
- faulty perception or interpretation (e.g. in understanding fisheries science statistics);
- background noise or other distractions (e.g. when holding a meeting in a public fish market); or
- poor lighting and other environmental faults (e.g. posters mounted in dimly lit areas).

The Sender and Receiver of the message can do two things in order to remove or reduce

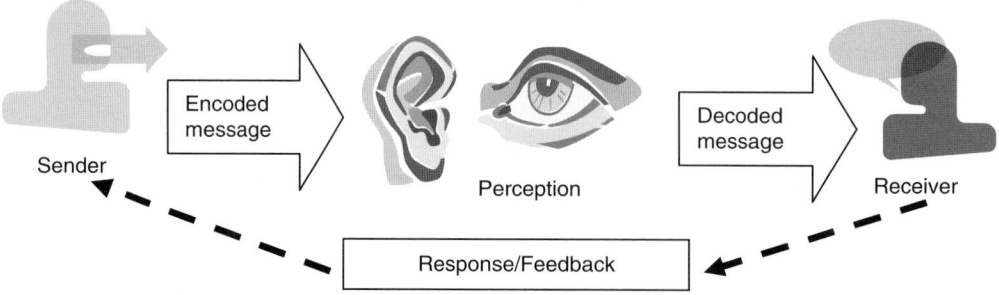

Fig. 10.2. The communication process.

barriers, and so increase the accuracy of the message and effectiveness of communication:

- The Receiver can provide feedback – essentially reverse communication that indicates whether the message is getting through in the way intended.
- The Sender can engage in sensitive role-taking – in other words, the Sender tries to take the point of view of the Receiver in constructing and transmitting the message.

When Senders and Receivers are in a collaborative relationship to address shared interests or achieve common goals, they both use mechanisms to increase communication effectiveness.

Communication flows

Good communication is key to fishery organizations because their success is based on beneficial relationships among people. People cannot interact with each other without communication. In its absence, transactions in any fisheries organization would grind to a halt. Four main types of communication flow occur in fisheries authorities, private firms and NGOs:

1. Downward, or enabling, communication that moves instructions and other directive information down or through the organization structure.
2. Upward, or compliance, communication that provides feedback to the people who originate the downward communication.
3. Lateral, or coordinating, communication that moves between members of the organization at the same levels.
4. The grapevine, often a social network, that fills in gaps in official communication and provides answers to unaddressed questions.

The importance of communication downward (instructions) and upward (reporting) is easily understood. However, we may overlook the significance to members of any organization of being able to communicate laterally and formally (and through the informal 'grapevine') to be able to carry out their instructions, often by conferring with colleagues. This makes use of the social networks that exist within all organizations or groups, regardless of size, and whose lines of communication may be very different from what organizational charts show (Fig. 10.3).

Figure 10.3 shows that, compared with formal channels, one middle-level worker is especially well connected in the informal network structure and may be an effective communicator and leader. It also shows that the lower-level workers are much better linked, and perhaps self-organized, than top management. The notion and use of social

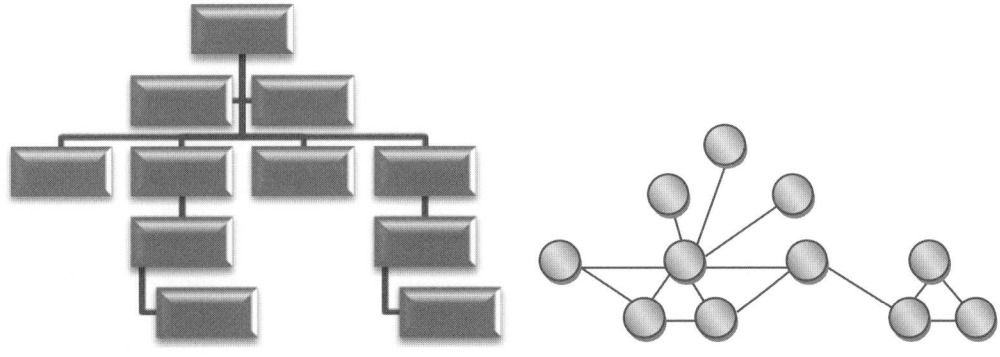

(a) Formal organizational structure   (b) Informal social network structure

Fig. 10.3. Formal and informal communication in organizations: (a) formal organizational structure; and (b) informal social network structure.

networks can be especially useful in small, closely knit organizations or communities (place-based, of practice or interest, etc.). Key 'brokers' can help communications between stakeholder organizations and informal groups. Targeting these key communicators can be quite effective. If there is conflict between parties, then the use of brokers becomes even more important. In the coordination of government agencies, and among some non-governmental bodies and in the private sector, hierarchy (and consequently power) adds yet another dimension.

Fisheries organizations cope with very dynamic and uncertain circumstances that often require swift action. They may sometimes get things done faster by informal means, ignoring formal communication flows. If you by-pass the proper information channels within your organization too often this can undermine authority and arouse suspicion. It can also reduce transparency if there are no formal records of decisions, and this damages institutional memory (the ability of groups and institutions to maintain collective means of recall, often through well-maintained written records useful for learning). Real or imagined financial problems in fisherfolk groups are classic examples. The lesson here is to ensure that intra- and inter-organizational communication is effective in all directions, but not to try to stifle informal communication that gets things done. Organizational and network assessments can be useful tools for investigating and designing communication webs and knowledge networks (Lusthaus et al., 2002; Creech and Ramji, 2004).

Communication Strategies

Your fisheries organization's communication strategy should be a holistic approach to planning how it is going to engage the people and agencies that matter most to its success. Your strategy is a combination of: (i) creating the message you want to get across; with (ii) the most appropriate medium (or product) for moving that message; and (iii) the channel (or pathway) to deliver it. A good communication strategy will ensure that you are communicating effectively and increasing your chances of creating a successful fisheries communications campaign (a set of coordinated communication initiatives) (Norrish et al., 2001; Government Communications Unit, 2004; Media Trust, undated). Bear in mind these following eight steps for planning a communications campaign:

- Identify the issue that is the subject of the campaign.
- Know the audience to whom you will communicate the information.
- Set objectives or list the aims of the campaign.
- Create a communication strategy or determine the methods you will use to communicate.
- Design the messages to suit the media to be used and the audience to be reached.
- Make a plan to manage the campaign so that at all times you will know exactly what is being done and by whom, what is next and who is responsible.
- Develop methods to evaluate the effectiveness of the campaign or answer the question: has the communication plan worked?
- Consider and plan how to sustain the campaign until it has achieved all of its agreed aims.

The following eight key points when formulating and implementing a fisheries communication strategy and plan are:

- Before starting, find out what those outside of your organization, especially the potential target audiences (e.g. fishers, donor agencies, policy makers), think about your organization. This will help you to decide upon priority areas for the communication strategy.
- Start by thinking about how your organization is communicating now. Are you saying what you want to say, to the people you want to say it to, with the desired impact?
- Be clear about the target audiences and user groups you wish to reach. Prioritize them according to importance and

influence relative to the communication objectives. Do not just think about the 'usual suspects' (e.g. fishers). Consider the entire market for your messages.
- Think about what communications your target audiences prefer (e.g. audio-visual rather than text) and plan to use the right ones for maximum impact. Market research can guide choices.
- Keep it manageable. Do not underestimate the time involved in communication. Include key deadlines, milestones and review points to guide your progress and keep on track.
- Ensure value for money by targeting communication effectively: prioritizing the audiences and channels and focusing on high-impact/low-cost activities.
- Keep the communication strategic plan simple. It will be easier to evaluate and update a simple plan quickly and appropriately as various aspects change over time.
- Develop a communication action plan to go along with your strategic plan to explain:
  - what you're going to do;
  - when you're going to do it;
  - who is going to do it; and
  - how much it will cost.

In addition, Creech (2006) poses ten questions that researchers should ask when planning communications, especially when they wish their research project to influence policy. She asks:

1. What do you anticipate may be the mid- to long-term outcome(s) of the project?
2. Who are the individuals you most want to engage with or influence the project?
3. Are you working with one or more partners (other organizations) on the project?
4. How do you want to work together to achieve the outcome?
5. What internal communications support tools will you need to work with your partner(s)?
6. What are all of the possible products/services that could be delivered throughout the project cycle?
7. What are the principal information-gathering points and media contact points for the target group?
8. What are the key messages from this project that you want to communicate to your target audience?
9. How do you want to promote and distribute the final deliverables to the target group?
10. How do you intend to inform broader audiences of the work?

Stakeholders and Audiences

Clearly identify audiences or key stakeholders with whom you need to communicate to achieve your aims and objectives. Make sure you dedicate the resources necessary to reach all of them. Remember, the best audiences to target in order to achieve an objective may not always be the most obvious ones, and targeting audiences such as the media may not always help achieve your objectives. It is likely that only a small group of people can actually change things to achieve the effect that you desire. These people who can actually make the changes that you want are your *primary audience*. A primary audience can be considered as the group of persons you must reach if you want to solve 80% of your problem. For example, it may be better to target 20 boat owners than their 60 fishers if the communication is to improve safety equipment aboard vessels.

However, this is not to say that you do not need to engage with your *secondary audience* as well, especially if that audience can influence your primary targets to help make the change in attitude or behaviour that you want. The secondary audience essentially facilitates your targeting. Therefore, also communicating with fishers may encourage boat owners to invest in safety equipment.

In devising a communication strategy it is critical to identify the primary and secondary target audiences early. This will help decide who to collaborate with to achieve the goals, who to involve in your planning process, who to defend your organization and its messages against, and whose activities to monitor to see what impact they could have or do have on your organization.

The critical process described above is a stakeholder analysis (Pido *et al.*, 1996; Renard, 2004). A stakeholder is any person, group of people or institution that has something to gain or lose through the process and outcomes of implementing your communication strategy. Your level of success is higher if you can win the support of your stakeholders. A stakeholder analysis equips you with information needed to determine who the parties are that have to be won over, or otherwise addressed, and cannot be ignored. Once you identify the stakeholders and the communication strategy that is appropriate to each stakeholder, then you can move from reacting to issues and instead put communication tools to use ahead of time to answer questions or head off crises before they occur. In fisheries situations it usually pays to be proactive, especially if capacity for conflict management is low.

For example, the challenge of establishing a Marine Protected Area (MPA) can be made easier by first identifying the different interest groups who have a stake in the area – be it the fishers, their families, the tourists who visit the area or the developer who wants to build a new hotel on the coast. All of these groups (see Fig. 10.4) have a stake in the MPA but all do not necessarily share the same level of interest. It is important to know which of these groups will help you achieve your objectives, which ones you have to sway to your cause and which ones will stand in your way and for whom you need to develop early counter-arguments to their negative stance.

Stakeholder analysis allows you to identify early the kind of positive or negative influence the stakeholders will have on the outcome of your project. This can be achieved by employing a stakeholder matrix to plot those stakeholders who are in positions of relative importance and whose strong influence is critical to the project's success. For example, a small project such as improving the sanitary conditions within a fish market would involve several actors within and

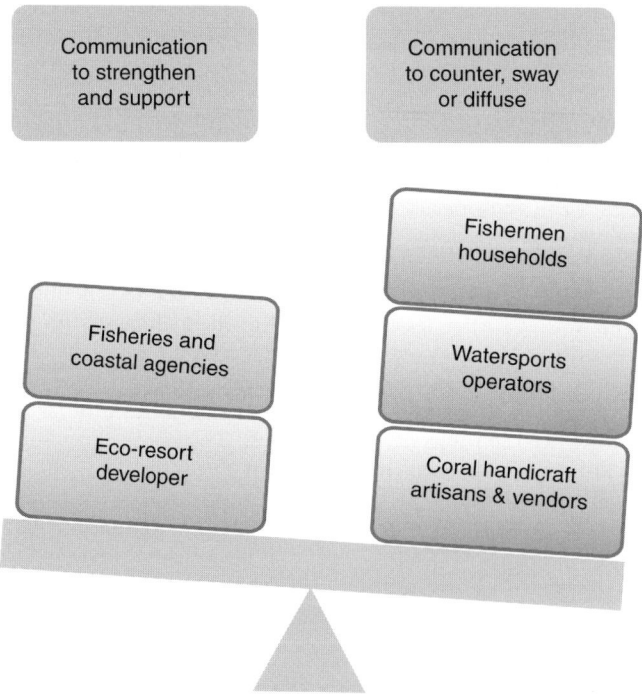

Fig. 10.4. Communication target audiences in the balance.

outside of government, such as the market authorities, public health inspectors, the sanitary service authority, a supplier of garbage receptacles and the market users – vendors, fishers and the general public. These groups are stakeholders who need to be involved and drawn in at various stages of the project, and therefore communication with each of these stakeholders is very necessary.

The stakeholder matrix should have four headings: (i) Allies (those with whom you collaborate closely); (ii) Opposers (those against whom you have to defend the interests and objectives of the project); (iii) Interested Parties (those with whom you do not collaborate closely but that you must keep informed); and (iv) Indirect Associates (those whose activities you must monitor in order to measure what impact their activities could have on the success or failure of your project).

This classification of stakeholders can further be refined by creating a grid on which the interest and power of each stakeholder in relation to the success of the project can be graded. 'Interest' measures to what degree they are likely to affect or be affected by your goal, and what degree of interest or concern they have in or about it. 'Power' is the influence they have over the project, and to what degree they can help achieve, or block, the desired change (Table 10.1).

Once you understand who has the most, or least, influence and power in relation to your project or activity, you can then determine the level of information that should be passed on to them in order to keep them involved or to prevent any negative feedback. Once you determine who the most powerful stakeholders are, then you can consult their opinions in shaping the project, which guarantees their support and input, thus improving the quality of the project. Their support also leads to more resources, which would assist the project's success. Stakeholder input allows you to anticipate the general reaction to the project and develop communication to reduce obstacles to the successful implementation of the project.

Delivering Messages

Delivering the right message to your fisheries audiences is not simply a case of restating your goals. Your message needs to make a case that will be compelling to your target audience; it should be designed towards your target audience – not based on your own knowledge and beliefs. The knowledge systems, cultures, beliefs and means of sharing information will vary considerably as you go from fisher to buyer to fisheries manager or scientist to donor.

The best messages are short and simple. Strategic targeting and consistency are key to your organization's messages. Create a comprehensive case covering all the key messages, and emphasize the different elements of the case for different audiences. To maximize impact you should summarize the case in three key points which can be constantly repeated. Create a message that your audience will understand and that is:

- simple and clear;
- up to date with facts;
- attention grabbing; and
- reveals information that is little known or poorly understood.

Make sure your message is being delivered by a source that the audience finds credible. Deliver a consistent message to an audience through a variety of channels over

Table 10.1. Stakeholder matrix.

Allies		Opposers		Interested Parties		Indirect Associates	
Stakeholder group		Stakeholder group		Stakeholder group		Stakeholder group	
Level of interest	Level of power	Level of interest	Level of power	Level of interest	Level of power	Level of interest	Level of power

an extended period of time. Keep saying it over and over! Learn from how effective civil society organizations communicate to engage and influence policy through carefully crafted campaigns, rigorous context assessment, critical-stage policy interventions, relevant evidence and use of networks (Court et al., 2006).

Communication products and pathways

In order to deliver your messages effectively you need to determine the best product or container for the message, and the best pathway or channel for delivering the product. Identify the tools and activities that are most appropriate to communicating the key messages within the time and human and financial resources available (Creech, 2006). For example, Caribbean fisherfolk prefer having fisheries officers in the field for one-on-one exchanges as their culturally preferred mode of communication (McConney et al., 2003), but this can be costly (in terms of time and human resources if not direct finances). Choosing the right pathway (Table 10.2) is determined by:

- What are the costs associated with delivery of the product?
- Do the target groups have access to the chosen medium (e.g. do they have televisions)?
- Is the medium simple to use (e.g. is your target group Internet-savvy)?
- Is the medium credible (e.g. is it a tabloid-style newspaper known for embellishing the truth)?
- Does it encourage participation (i.e. does it encourage user interaction or feedback)?
- Does it allow for long-term dissemination (i.e. is the pathway sustainable)?
- Is it consistent with your objectives (i.e. does the pathway allow you to meet your goal)?

Table 10.2 does not address the many small-group and interpersonal communication pathways, such as face-to-face and workshop settings, that may reach fewer audience members but be very powerful for delivery in interactive settings. Most fisheries stakeholders are familiar with these, and a large number of resources are available via the Internet for both planning and execution.

Monitoring outputs and outcomes

After implementing your communication strategy, you need ways of answering the following questions:

- Did we obtain the stakeholder support needed to create the change we wanted?
- Did we succeed in creating a positive change that helped to achieve our goal?
- Is the change that was created sustainable over the period required for change?

Sometimes you can use simple methods for evaluating whether a communication strategy is working. For example, as an output measure, count and compare the number of people who attended your meetings before and after you implemented your strategy to indicate whether messages to persuade people to come out to meetings were successful. To measure outcomes, you can visit local fishing communities and see whether techniques on which you communicated are actually being used.

Policy influence is a particularly difficult, but important, area in which to measure the outcomes of communication. For example, you may wish to see what impact a research finding has had upon fisheries management policy in order to confirm uptake of the message.

Communication-based Processes

Several participatory processes and approaches used in fisheries management, planning and policy are based upon communication. Some of these processes are explained in this section.

Conflict management and negotiation

Conflicts are not necessarily negative. They may cause more equitable power relationships to emerge, correct bad practices or improve policy. The issue is how to manage conflicts in order to reach solutions (at least

Table 10.2. The pros and cons of some communication pathways for large audiences.

Pathway	Pros – favour pathway	Cons – do not favour pathway
Newspapers	Relatively inexpensive; graphics and tables feasible; longer shelf life than other mass media; can contain more information than other mass media	Limited readership; no guarantee that your message will be placed unless you pay; little control over where your message will be placed in the newspaper
Television	Reaches viewers when they are most attentive; can convey message with images, sound and motion; allows for greater creativity in delivering your message	Can be most expensive mass medium; not as easy or cheap to update or adapt messages as with other media; because audience is compartmentalized you might not reach all targets within the same time slot
Leaflets and pamphlets	Easy to print, easy to hand out; can be passed from one person to another easily; generally cost-effective	Easily disposable; might not be read if distributed with other informational material; needs literate audiences
Newsletters	Easy to start (desktop publishing); easy to deliver electronically and store for reference; creates regular contact with your stakeholder base	Content can be difficult to create on a regular basis; locked into a schedule for delivering information; more limited space than a newspaper for your articles
Radio	Most popular medium, especially in rural areas; can grab attention using a catchy jingle; cheaper to produce than other broadcast media messages; can be repeated more often over a time period; call-in allows dialogue	Little opportunity for retention by audience once the message is delivered; most popular slots can become full quickly; can be difficult to reach some audience members who do not normally listen to the radio
E-mail	Use anywhere once connected to the Internet; can send from a computer or any e-mail-enabled device; fast and cheap relative to other forms of direct contact; messages are easily stored, retrieved and can be mass-circulated	It can be difficult to convey emotions accurately; can be impersonal; not always a secure method of communication; can find its way into other hands than the intended recipient; requires technology
E-mail lists	Allows those focused on a particular area of interest to communicate with others of like mind; reaches a large number of people; creates a vast pool of people with varying depths of experience and knowledge	Lots of redundant messages; recipients bombarded with messages on topics in which they have no interest; recipients' mailboxes can become cluttered with e-mail; response time can be slow; only those who sign up can respond
Websites	Can be interactive, entertaining; may host a variety of supporting media; can take advantage of links; popular among many audiences	Web hosting can be costly; requires specialized knowledge to create and update; updating can be time-consuming; active engagement necessary; needs technology
Internet chat rooms	In-depth discussion or debate can take place in real time between a group of people; can be a good pathway for bringing together like-minded people; means of refining and improving communication skills	Meetings need to be scheduled for specific times, not always convenient; many people might not be familiar with how Internet chat rooms work; difficult to verify the identity or authority of the person to whom you are speaking; can be time-consuming
Wiki technology (such as Wikipedia)	Allows for voluntary collaboration in creating online information; people with new knowledge on a subject can update or edit other people's entries to increase timeliness and accuracy	Edits can be made freely and without restriction; multiple authors can create conflicting goals with the information; Wiki entries can be vulnerable to corruption and destruction; fairly sophisticated audiences required

(Continued)

Table 10.2. Continued

Pathway	Pros – favour pathway	Cons – do not favour pathway
Blogs	Keep websites fresh with new content; easy to post new content; build a tight-knit community of interest among those who respond to a particular thread	Maintenance and moderating can be time-consuming; constant editorial oversight needed; require active participation for best results; technology required
Theatre and the arts	May be culturally most appropriate in some places; novelty can attract large audiences; may appeal more to children and young people	May seem culturally inappropriate; actors and production process can be expensive and time-consuming; may need facilities; time- and place-based unless broadcast
Corporate communications	Access to a specific customer base; not discarded as quickly as other print media; creates a brand identity for your organization	Can be expensive; requires planning well in advance; can become dated if your area is very dynamic; may be perceived as too self-promotional or biased

temporarily) in the most appropriate and least disruptive or harmful manner (Krishnarayan, 2005). The goal of conflict management and negotiation is not to avoid conflict, but to employ communication skills that can help people to express their differences and solve their problems for win-win, or mutually beneficial, outcomes. Conflict management is facilitated negotiation.

Not all disputes are candidates for conflict management. Conflicts cannot be managed well, or a negotiated solution achieved, without adequate information exchange. Correct identification of the nature and source of the conflict requires communication amongst the parties involved to get past symptoms until root causes are known. There are several stages in conflict management:

1. Initiation: a stakeholder or outsider invites help to manage the conflict.
2. Preparation: conflict analysis, information sharing, rules, participant selection.
3. Negotiation: articulating interests, creating win-win options, packaging preferred options.
4. Agreement: concluding jointly on best option package, recording final decisions.
5. Implementation: publicizing outcomes, signed agreement (optional), monitoring.

One of the most difficult activities, but sometimes also a liberating one, is fact-finding and information sharing. Seeing the dispute from the other side is vitally important. However, in highly technical situations there may be serious disparities in the capacities of stakeholder groups to interpret and use the information provided. In such situations it may be necessary, as part of the process, to identify means of communication that suit diverse audiences. Jointly examining visual information, such as that from geographic information systems (GIS), may assist in the occurrence of information exchange from different perspectives more than numerical or other analyses.

Facilitation

Facilitation is a process that helps meetings or decision-making processes run smoothly and reach desirable ends. A trained facilitator communicates in order to work with diverse groups of stakeholders and under sometimes difficult circumstances, such as when there is conflict as described above (Rees, 1998). Communicating is the lifeblood of facilitation. A facilitator achieves the following:

- distinguishes process from content;
- manages the client relationship;
- prepares thoroughly for planning;
- uses time and space intentionally;
- evokes participation and creativity;
- maintains objectivity at all times;
- reads underlying group dynamics;
- releases blocks to the process;

- adapts to the changing situation;
- shares responsibility for process;
- demonstrates professionalism;
- shows confidence and authenticity;
- maintains personal integrity; and
- learns from each facilitation experience.

In addition to a sound process of communication, good facilitated outputs and outcomes are usually based on good information. Informed stakeholders can provide valuable inputs, ranging from local or traditional ecological knowledge to technical and scientific knowledge. However, communicating across knowledge systems can be a difficult and daunting task for stakeholders (Berkes et al., 2001).

Leadership

Good leadership, especially among fisherfolk groups, has a lot to do with communication. Boat captains are leaders of fishing enterprises, and many are exceptionally knowledgeable about their working environment, communicating effectively to their crew members. The crews follow instructions from captains at sea, but these same captains may be out of their depth when ashore communicating with the fisheries authority or fishing cooperative members. Different skills of communication are required in different settings, and leaders may not be equally proficient in all.

Almerigi (2000) lists some of the most important characteristics and personal qualities that Caribbean fishers look for in their organizational leaders, and many of these are communication-based:

- embraces, and is committed to pursuing, the group's goals;
- identifies the needs, and respects the values, of members;
- knows the problems and aspirations of the membership;
- values consensus decision-making and every contribution;
- treats the members fairly, transparently and equitably;
- encourages flexibility, creativity, tolerance and self-discipline; and
- learns from mistakes and motivates others to excellence.

Policy and planning

Policy development and fisheries planning, when approached correctly, are usually participatory to some extent (Berkes et al., 2001). The extent of participation, coupled with the number and nature of participants, can be the determinants of communication requirements. Learning by doing collaborative activities successfully builds capacity, trust, respect and legitimacy of both content (the plan) and process (the planning). The nature of the participation needs to be decided early on, since bottom-up is not always feasible or affordable, although it is usually desirable. If stakeholders are not well informed, or do not have the capacity or time, it is not appropriate to start at the bottom (McConney et al., 2003). This usually means that resource users will communicate their input after there is a first draft, or at least an outline of policy or plan content, communicated to them along with criteria for organizing their contributions. For example, an MPA or fisheries measure may have scientifically determined aspects, but leave implementation features partly up to the stakeholders' preferences.

In these cases the science will need to be communicated to lay persons and political choices be communicated to scientists or managers. However, the process must genuinely consider and use the input of stakeholders in order to be credible. Since fisheries policies and plans often have to be endorsed at a political or legal level in preparation for implementation, other audiences are also involved. Prior to implementation the plan should be widely publicized and disseminated for it to be actively adopted. Even though stakeholders should have bought into the plan, it may be ignored unless it is well known and becomes standard operating procedure (McConney et al., 2003). This type of promotional communication helps to institutionalize fisheries policies and plans. A specific sequence of stages is followed in order to progress logically, but within the overall sequence there may be feedback loops of communication that allow plans to be evaluated and revised.

Good governance

In the Caribbean, coastal resource co-management partners reported the need for considerable improvement in communication, cooperation and coordination (McConney et al., 2003). These terms are closely related, but communication is the basis for the other two. Cooperation follows communication if the parties decide to work with each other, but this does not necessarily result in coordination, unless there is closer communication. Coordination requires communication and leadership for synchronization of activities. All three concepts contribute to good governance.

Transparency and accountability are two of the more communication-oriented principles of good governance. Transparency, particularly in government agencies, tends to have follow-on effects that facilitate positive developments in fisheries governance. These include fostering more trust in access to information, and this is often reciprocated via information flows from fisherfolk.

Managers and others need to be aware of the functional details of communication. There must be adequate attention paid to issues of language and literacy. Dialects are spoken in most countries, and it is often assumed that resource-users such as fishers are less literate than average citizens. Factors such as these determine the most appropriate products and media. In co-management it is especially important to ensure that stakeholders can receive information, and also present it, in the manner that is most suitable for them to share equitably. This reflects respect in governance.

Mass Media Communication

Mass media communication deserves special attention as a process for distributing messages widely, rapidly and continuously to large, diverse audiences in an attempt to influence them (DeFleur and Dennis, 1996). While not having the benefit of immediate feedback associated with face-to-face communication, it has the alternative benefit of reaching a wider audience than those directly involved in your fisheries activity or project. Considering the typical suite of stakeholders and others that you have to reach in most fisheries situations (Fig. 10.5), mass media communication has benefits.

Generally, when people think of media for mass communication, they identify the traditionally popular media of print, radio and television. However, these examples are only the conventional 'news media'. There are a variety of other tools that can be used to get your messages across, and the category of mass media is continuously growing thanks to advances in information and communication technology (ICT) (recall Table 10.2). Detailed practical guidelines are given below for a few types of media and processes often employed by fisheries stakeholders.

Press releases

The best way to ensure accurate and consistent coverage of your fisheries-related activities is to be helpful to the media. One way of helping the media help you is to accompany all announcements with press releases. A press release is a short summary of a piece of news. Use it to publicize the key elements of your story to journalists. Its most important feature is that it needs to be topical – it should make clear what is new. You can also use press releases as part of a marketing strategy to publicize a forthcoming fisheries-related event. The structure of a basic press release (Martineau, 2008) is as follows.

Title

Needs to be brief, contain major keywords and say exactly what the story is about.

The five Ws

The opening paragraph should answer the following questions (try for no more than 30 words):

- Who (is involved)?
- Where (is the activity)?
- Why (is it new)?
- What (is new about it)?
- When (does it happen)?

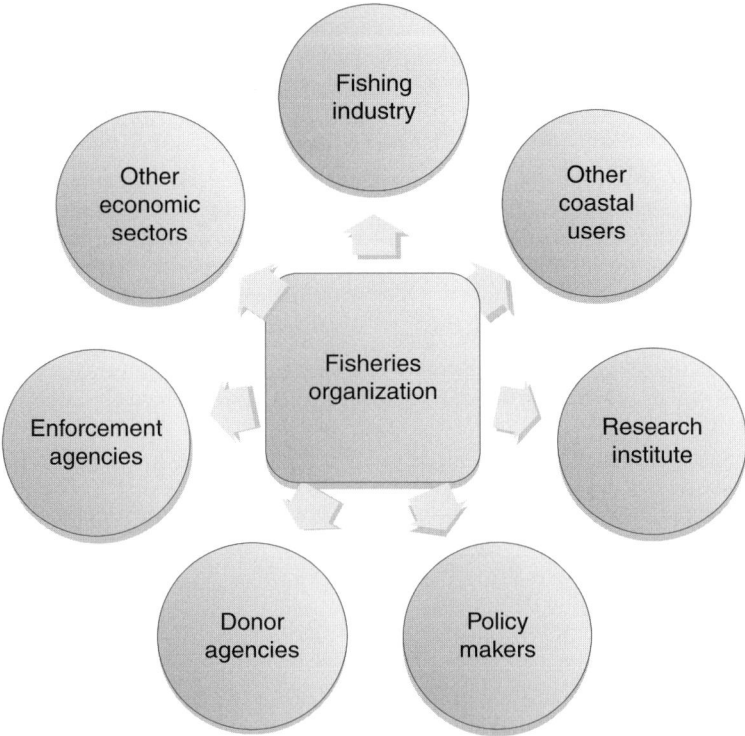

Fig. 10.5. Typical target audiences.

Body of the release

The main body contains further information about your story and sets it in a familiar context. Stick to key points that support your message, and do not get too technical (as fisheries scientists often do).

Quotes

Include a brief quote from someone directly involved with the story. It should sound like something someone would normally say, not what he or she might have specially prepared.

Contacts

It is crucial to provide regular and after-hours (international) telephone and e-mail information for the press office and other individuals involved with the story (with their agreement).

Content

Here are some points to remember in creating an effective press release on a fisheries topic:

- Write in the active voice ('The association has decided', not: 'It has been decided').
- Use everyday language and avoid (or explain) all jargon, technical terms and acronyms or difficult words.
- Put the most interesting things at the start in order to 'hook' the reader.
- Determine (from your strategy) what is the main news angle you wish to communicate.
- Connect your information to something topical, something that has been in the news.
- Check deadlines for local publications/television/radio bulletins to ensure the media release is received in time to be published before the event.

- Use a first paragraph of no more than 30 words telling briefly who, what, where, when and why about the event, issue or project.
- Use short sentences; each sentence should be a separate paragraph; keep it simple.
- Make numbers (statistics and measurements) more meaningful by making comparisons or breaking them down into familiar units.
- If using quotes in the body of the release, quote credible spokespeople and identify them by their position in the organization.
- Keep information clear and unambiguous.
- Keep releases short, no longer than one page. If the media want more information, they will contact you.
- Include in the media release the date the release was written, and a contact name and phone number for someone who is easily available during office hours. Put information on your website and include the link in any press release.
- If later offering interviews, make it clear whether this is an exclusive for one media outlet.

Distributing a press release also requires special attention. Deciding on whom you are targeting will help you decide on how you want to send out your release. Journalists often receive far more press releases than they have space or time to cover, so sending out your release intelligently is as important as writing it in the first place.

Before you send it out

Always make sure all parties mentioned in the press release agree to it (especially for quotes and contact details). Inform any interested parties, people or organizations about your release before you send it out to journalists.

To whom should I send it?

Identify a named journalist (on a relevant news desk – science/news/local/national), and send your release straight to them. Your choice of journalist should be informed by the message you want to get across.

When should I send it?

Depends on the source of the story, and also which journalists you want to cover it. Think about embargoes (you will need to keep to deadlines of any associated journal). Some days of the week (for example, towards the beginning) are better than others (for example, when competing stories come out). Avoid competing against significant national or international events unless your story has relevance to them.

How should I send it?

Obvious channels of communication include post, fax, e-mail and websites. Make sure all letters and faxes are clearly addressed to a named journalist; possibly follow up with a phone call (mornings better). Put a brief title in the e-mail subject line that summarizes the story; include the release in the body of the message, not as an attachment. Putting your release on your own website is a good reference point (you will need to tell journalists it is there) and enables links to relevant sources of further information and pictures. You can also post on other web-based news services regularly visited by journalists. When distributing pictures, always include a caption and photo credit; put it on a website and include a direct link to the page in an e-mail. Make sure any hard copies are of good quality. After the press release is sent:

- Ensure that you (or an informed colleague) are available for interview, or to provide further background information and explanation once your release reaches a journalist.
- Familiarize yourself with how journalists work and what information they will need; be prepared for any question or approach.
- Provide prompt and informative answers; equip yourself with basic facts and figures.
- Be aware that journalists from abroad may approach you outside of normal hours.
- Leave reliable contact details with your colleagues.

- Sort out organizational logistics and permission for photographic, sound recording or filming opportunities in advance.
- Evaluate your experience: note down questions you find hard to answer.
- Keep a record of the press coverage you generate, and inform colleagues, press offices, bosses, fund-providers, etc. of what you have achieved.
- If you do not like the coverage you receive, think about how you could have presented your work in a different way to communicate the message you wanted to get across.
- Note the names and contacts of journalists you enjoyed working with, and keep in touch.
- Make an archive of your news releases available on your website.

Broadcast media

Outlined below are tips on creating the best image when dealing with the broadcast media, especially when handling difficult questions posed by them.

- Be prepared before being interviewed. Research your topic in detail and be comfortable with it. Jot notes before the interview and refer to them when the need arises.
- If put on the spot, avoid saying 'No comment'. This raises suspicion and may cast you in a negative light. Rather, say, 'I am unable to shed any light on the matter at this time; however, we will issue a statement on the matter shortly' ... or something to that effect.
- Avoid getting into arguments, hostile discussion or confrontation with media workers.
- Remember, the media should be an ally, never an adversary, in a communication strategy.

Speaking in public

Everyone who has ever faced a journalist for an interview, given a presentation or hosted a press conference wishes to get rid of their nervousness and project an aura of confidence and authority. Here are two guides for assisting you in having a flawless public speaking moment.

When speaking at press conferences:

- Always, always, be thoroughly prepared to speak on your topic. Avoid speaking at a press conference on a topic with which you are uncomfortable.
- If you have to make an impromptu comment or statement, keep it brief and promise to expand on it later. Avoid nervous chit-chat or rambling.
- If you are leading a press conference, prepare an agenda and ensure that every member of the team and the media personnel receive a copy. Sticking to the agenda will ensure that the press conference goes smoothly, with your desired outcome.
- Always insist that the question-and-answer segment be held at the end of the session. This keeps order and helps maintain control of the session.

When making a presentation:

- Ensure that you devote adequate time for research and knowledge gathering.
- Use props, graphs, flip-charts and/or a slide show to enhance your presentation.
- Actively engage your audience either during or after your presentation. This ensures that they get the most out of your presentation.
- K.I.S.S. (Keep It Simple, Stupid!) is an industry standard acronym to remind you never to forget your audience. Ensure that your language, sentence structure and overall presentation are clear, concise and easily understood.

What's Next?

The main message of this chapter is that communication is an important part of fisheries management to which all stakeholders need to pay more attention in order to be successful. In the past few years approaches to fisheries have become more comprehensive and participatory. There is now more emphasis upon governance, ecosystem-based management (EBM) and the ecosystem approach to fisheries (EAF). This is good. However, it means that fisheries stakeholders will have to both compete and collaborate with other actors in this larger arena. In order to ensure that their voices are clearly heard, and not misunderstood, an increasing allocation of resources to communication is anticipated. Already some agencies include communication in the design of projects and programmes, not as an afterthought or confined to the limited dissemination of highly technical reports, but as comprehensive communication in order to encourage wider audiences to contribute towards their goals and outcomes. If the message of this chapter has reached you, then you may wish to improve your knowledge and skills even further. In order to facilitate this, some resources available via the Internet are listed in Appendix 10.1.

References

Almerigi, S. (2000) Leadership for fisherfolk. *CARICOM Fisheries Research Document* 24, 1–63.
Berkes, F., Mahon, R., McConney, P., Pollnac, R. and Pomeroy, R. (2001) *Managing Small-scale Fisheries: Alternative Directions and Methods*. International Development Research Centre, Ottawa, Canada.
Court, J., Mendizabal, E., Osborne, D. and Young, J. (2006) *Policy Engagement: How Can Civil Society be More Effective?* Overseas Development Institute, London.
Creech, H. (2006) *Ten Questions to Guide the Development of Communications Tactics for Research Projects*. An IISD Knowledge Communications Practice Note, IISD, Winnipeg, Canada.
Creech, H. and Ramji, A. (2004) Knowledge networks: guidelines for assessment. International Institute for Sustainable Development Working Paper, IISD, Winnipeg, Canada.
DeFleur, M. and Dennis, E. (1996) *Understanding Mass Communication: a Liberal Arts Perspective*. Houghton Mifflin Company, Boston, Massachusetts.
ESRC (2008) Communications Toolkit. Economic & Social Research Council. Available at: www.esrcsociety-today.ac.uk/ESRCInfoCentre/CTK/default.aspx?ComponentId=25076&SourcePageId=19467 (accessed 10 February 2008).
Gauthier, J. (2005) *Popularize, Produce, Disseminate! Putting Information to Work for Research Projects*. Reference sheets for field researchers, International Development Research Centre, Ottawa, Canada.
Government Communications Unit (2004) *How to Write a Communication Strategy for an Australian Government Campaign*. Commonwealth Government of Australia, Melbourne, Australia.
GTZ (2006) *Strategic Communication for Sustainable Development: a Conceptual Overview*. Deutsche Gesellschaft für Technische Zusammenarbeit, Eschborn, Germany.
Hovland, I. (2005) *Successful Communication: a Toolkit for Researchers and Civil Society Organisations*. Overseas Development Institute, London.
Krishnarayan, V. (2005) *Understanding and Managing Natural Resource Conflicts*. Caribbean Natural Resource Institute (CANARI) Guideline Series No. 6, Caribbean Natural Resources Institute (CANARI), Laventille, Trinidad, West Indies.
Lusthaus, C., Adrien, M.H., Anderson, G., Carden, F. and Montalván, G.P. (2002) *Organisational Assessment. A Framework for Improving Performance*. International Development Research Centre, Inter-American Development Bank, Washington, DC.
Martineau, N. (2008) *How Do I Write a Press Release? Science and Development Network*. Available at: http://www.scidev.net/en/science-communication/practical-guides/ (accessed 10 February 2008).
McConney, P., Pomeroy, R. and Mahon, R. (2003) *Guidelines for Coastal Resource Co-management in the Caribbean: Communicating the Concepts and Conditions that Favour Success*. Caribbean Coastal Co-management Guidelines Project, Caribbean Conservation Association, Barbados, West Indies.
Media Trust (undated) *Developing a Communications Strategy*. Available at: http://www.mediatrust.org/training-events/training-resources/online-guides-1/guide_developing-a-communications-strategy (accessed 10 February 2008).
Norrish, P., Lloyd Morgan, K. and Myers, M. (2001) *Improved Communication Strategies for Renewable Natural Resource Research Outputs*. Socioeconomic Methodologies for Natural Resources Research, Best Practice Guidelines, Natural Resources Institute, Chatham, UK.

OECS (2007) *Toolkit for Communication Planning*. Organisation of Eastern Caribbean States (OECS), Environment and Sustainable Development Unit (ESDU), Castries, St. Lucia, West Indies.

Pido, M.D., Pomeroy, R.S., Carlos, M.B. and Garces, L.R. (1996) *A Handbook for Rapid Appraisal of Fisheries Management Systems (Version 1)*. ICLARM Education Series 16, ICLARM, Manila, Philippines.

Ramírez, R. and Quarry, W. (2004) *Communication for Development: a Medium for Innovation in Natural Resource Management*. IDRC, Ottawa, Canada.

Rees, F. (1998) *The Facilitator Excellence Handbook: Helping People Work Creatively and Productively Together*. Jossey-Bass/Pfeiffer, San Francisco, California.

Renard, Y. (2004) *Guidelines for Stakeholder Identification and Analysis: a Manual for Caribbean Natural Resource Managers and Planners*. CANARI Guidelines Series No. 5, Caribbean Natural Resources Institute (CANARI), Laventille, Trinidad, West Indies.

Appendix 10.1

Communication resources are plentiful on the Internet, although few are specific to fisheries. Be prepared to adapt them to your circumstances. Consider establishing learning groups to share your lessons.

Resource and/or organization	Internet address (accessed March 2009)
FAO Fisheries and Aquaculture Department	www.fao.org/fishery/
Vocational Information Center, Communication Skills	www.khake.com/page66.html
Science and Development Network	www.scidev.net/en/
ESRC Communications Toolkit	www.esrcsocietytoday.ac.uk/ESRCInfoCentre/
W.K. Kellogg Foundation communications toolkit	www.wkkf.org/
Connecting with Communities: communications toolkit	www.idea.gov.uk/idk/core/page.do?pageId=7816073
Successful Communication: a Toolkit for Researchers and Civil Society Organizations	http://www.odi.org.uk/resources/odi-publications/toolkits/rapid-successful-communications.pdf

11 Small-scale Fisheries Compliance: Integrating Social Justice, Legitimacy and Deterrence

Maria Hauck

Introduction

Fisheries management, worldwide, struggles to find a balance between protecting resources, ensuring equitable access to resources and promoting economic efficiency and stability (Hanna, 2003). To try to achieve this balance, central governments have intervened by formulating policies and establishing rules and regulations, with the aim of ensuring compliance. However, it has been widely recognized that non-compliance in fisheries is widespread, and researchers and management authorities continue to grapple with the factors that lead to non-compliance (Hemming and Pierce, 1997; Kuperan and Sutinen, 1998; Hatcher *et al.*, 2000; Hønneland, 2000; McKinlay and Millington, 2000; Gezelius, 2003; Raakjær Nielsen, 2003; Hauck and Kroese, 2006). From a theoretical perspective, the past two decades have seen a shift taking place in understanding fisheries compliance and conceptualizing appropriate responses to non-compliant behaviour.

This chapter provides a brief overview of the development of fisheries compliance thinking. In particular, it highlights the applicability of this theory to the small-scale fisheries sector, in which the context of fisher compliance behaviour may need to be understood differently. Research conducted in different parts of the world will be drawn on to highlight the need to challenge and develop compliance thinking further in relation to small-scale fisheries. A conceptual framework for understanding small-scale fisheries compliance will then be outlined. This framework, which originally emanated from empirical research conducted in South Africa (Hauck, 2008, 2009a, b), is presented here by drawing on other studies, arguing that the framework could be applied more broadly. Emphasis will be placed on the need to understand fisher behaviour within the context of a fishery system as a whole, and to recognize the role of law and power in influencing the way in which fishers comply with rules and regulations. Further, what are argued to be the preconditions for enhancing small-scale fisheries compliance – social justice, legitimacy and deterrence – will be introduced and discussed in detail. This chapter, therefore, calls for a more innovative and integrated approach to understanding and addressing the challenge of fisheries compliance in the small-scale sector.

Understanding Compliance

Compliance is generally understood as the behaviour of people to conform to rules that have been developed to influence actions (Tyler, 1990). These rules may exist as formal laws or as informal norms[1], thus being monitored and enforced through either formal or informal mechanisms, and sometimes by both. Compliance research has largely been rooted within two schools of thought. One explores the 'rationalist' models of deterrence and law enforcement that assume rational actors calculate costs and benefits of their actions, and the second explores the 'normative' models that investigate norms, morality, legitimacy and social and cultural influences on individuals' decisions to comply with rules and laws. These different perspectives to understanding compliance have also translated into different methods and strategies for regulating behaviour.

Traditionally, the fisheries compliance literature was built upon Becker's neoclassical model of rational criminality (Becker, 1968), arguing that non-compliance was determined by the balance of expected gains and losses from illegal activities (Sutinen and Andersen, 1985; Anderson and Lee, 1986; Charles et al., 1999). This rationalist approach argues that external influences (such as rewards and punishment) prompt individual fishers to act in their own immediate self-interest. Based on this perspective of rational choice, fishers will choose to comply (or not) based on economic gains, the likelihood of detection and the severity of sanctions. Fisheries management systems worldwide have embraced this approach, and governments often respond to non-compliance by increasing law enforcement efforts in order to increase the probability of detection and conviction (Sutinen et al., 1990; Hatcher et al., 2000; Raakjær Nielsen, 2003).

However, over the years, there has been an increasing realization that compliance can be achieved even when formal law enforcement is weak. This led to an interest in the normative approach to compliance, which recognizes that norms and morals, as well as the legitimacy of law and governance, are important factors that influence fisher decision-making (Kuperan and Sutinen, 1994, 1998; Sutinen and Kuperan, 1999; Hatcher et al., 2000; Jentoft, 2000; Gezelius, 2002, 2003, 2004; Raakjær Nielsen, 2003). This, therefore, led researchers to argue that reliance on traditional law enforcement as a primary means to enhance compliance ignores the complexity of the socio-economic and political context of fishers and coastal communities. Thus, there was a call for '... a radically different approach to enforcement and compliance' (Berkes et al., 2001, p. 162). Increased policing and punishment for non-compliant fishers often led to further conflict and violent confrontation between fishers and authorities (Hauck, 1999; Gupta and Sharma, 2004; van Ginkel, 2005). Furthermore, in many developing countries, law enforcement capacity is weak and therefore largely ineffective (Berkes et al., 2001; Christie et al., 2007; Pomeroy et al., 2009). Thus, there is arguably a need to understand compliance, and fisheries management more broadly, by tackling the social, economic, political and institutional challenges of the fishery in question. Although many governments continue to rely on law enforcement strategies to enhance compliance, research indicates the importance of combining the approaches and strategies of both rational and normative action theories to improve compliance outcomes (Kuperan and Sutinen, 1998; Sutinen and Kuperan, 1999; Gezelius, 2003; Raakjær Nielsen, 2003).

While combining these schools of thought may be necessary, it has been argued that the compliance analysis needs to be taken even further to question the role of law and power in influencing fisher behaviour (Hauck, 2008). The marginalization of small-scale fisheries, particularly when social and economic inequities are rife, needs also to be acknowledged in compliance discourse. Thus, before embarking on a revised conceptual framework for understanding small-scale fisheries compliance, it is necessary to provide the context of compliance discourse, and how it has evolved from the perspective of industrial fisheries. In particular, it is important to highlight how this perspective

---

[1] Informal norms may also be embedded in customary (or traditional or folk) law, in which 'pre-existing systems' of governance influence human behaviour outside of state law (Bavinck, 2005). Local community institutions, therefore, can establish their own legal order, including their own property regimes (Meinzen-Dick and Pradhan, 2001).

has influenced compliance strategies aimed at the small-scale sector.

The compliance discourse and its impact on small-scale fisheries

Although there are a handful of empirical studies aimed at understanding compliance in the small-scale fisheries sector (such as in Norway and Newfoundland (Canada) by Gezelius (2002, 2003, 2004); in Malaysia, Indonesia and the Philippines by Kuperan *et al.* (1997), Kuperan and Sutinen (1994, 1998) and Sutinen and Kuperan (1999); and in South Africa by Hauck (1999, 2007, 2008, 2009a, b)), most compliance research and theory has emerged from a focus on industrial fisheries. As a result, compliance strategies, which are often developed in the context of the industrial sector, are then transferred and assumed to be applicable in the small-scale fisheries context. The transfer of fisheries management tools from one context (usually from a developed nation perspective) to another has been identified as highly problematic (Christie *et al.*, 2007; Ruddle and Hickey, 2008; McClanahan *et al.*, 2009). In fact, Ruddle and Hickey (2008, p. 566) argue that the underlying cause for the mismanagement of tropical nearshore fisheries '… is the projection of Western policies and programs based on Western models and approaches into areas for which they are inherently unsuitable'.

The discourse on illegal, unregulated and unreported (IUU) fishing is one particular example that will be highlighted here in relation to compliance. Concerns about fisheries non-compliance grew in importance through the efforts of the United Nations (UN), which identified IUU fishing as a major contributor to fisheries collapse worldwide (FAO, 2001; UN, 2006). In fact, the Environmental Justice Foundation (EJF) in the UK states that IUU fishing represents '… one of the most serious threats to the future of world fisheries' (EJF, 2005, p. 4). IUU fishing is defined largely on the basis of fishing activity that contravenes national, regional and/or international laws and regulations (FAO, 2001). It evolved primarily from the international community's concern over illegal fishing in the high seas, and still largely focuses on industrial and international fleets (Rigg *et al.*, 2003; EJF, 2005; MRAG, 2005). Certainly in terms of the impact of IUU fishing, there is reason to be concerned. In addition to the resource and ecosystem impacts, economic loss to developing countries due to illegal fishing is considered to be in the region of US$2–15 billion per year (EJF, 2005). Developing countries with a high dependence on coastal resources for poverty reduction and livelihoods are severely impacted by foreign vessels fishing illegally in their waters (EJF, 2005; CEC, 2007).

The drivers of IUU fishing are largely attributed to economic incentives and inadequate laws and enforcement strategies to conserve marine resources (EJF, 2005; CEC, 2007). Thus, a focus on monitoring, control and surveillance (MCS), particularly in terms of enhancing enforcement, has been a key focus of international organizations with an interest in eliminating IUU fishing (FAO, 2003; EJF, 2005; CEC, 2007). Although MCS is considered a critical component of fisheries management (Flewelling *et al.*, 2003), the problem is that the discourse of IUU fishing is being transferred to small-scale inshore fisheries, where the context of 'illegal fishing' in the coastal zone is very different. Thus, although the focus of IUU fishing is on the large-scale industrial sector, concepts, approaches and interventions aimed at this problem are being incorporated into an understanding of fisheries non-compliance in the small-scale sector.

As a result, even in recent documentation, where poverty and marginalization are recognized as possible drivers of IUU fishing (SIF, 2008), the overall message is that MCS capacity needs to be strengthened. For example, in a Stop Illegal Fishing report (SIF, 2008, p. 3), while 'individual starvation' is identified as a possible cause of IUU fishing that needs to be addressed, the following paragraph states: 'The mindset that motivates crime is not too different to that which motivates IUU activities. Seeking personal advantage to the disadvantage of others is a fundamental motive for many IUU fishery operators.'

Thus, the focus on increasing the costs of IUU fishing through enforcement and sanctions, which may be appropriate in some cases, ignores the complexity of the drivers of IUU fishing, which are not adequately acknowledged or understood in the small-scale sector. This is of great concern due to the fact that empirical research has highlighted the diverse factors that are influencing the compliance behaviour of small-scale fishers (Kuperan and Sutinen, 1998; Sutinen and Kuperan, 1999; Gezelius, 2002, 2003, 2004; Hauck, 2009a).

The trend is for developed countries to assist developing countries with IUU fishing challenges (CEC, 2007; SIF, 2008), but the different scales of IUU fishing are not sufficiently differentiated in order to develop appropriate strategies. As a result, there are significant challenges in transferring compliance approaches and strategies from the developed world to nation states with very different social, economic and political contexts (Gezelius and Hauck <in press>). Government authorities participating in these regional and international developments, which are heavily funded, are being encouraged to enhance MCS capacity as an effective means with which to address IUU fishing – broadly. Small-scale fisheries, however, should rather be understood in the context of customary fishing practices, fragile livelihoods, market dynamics, institutional arrangements and legitimate laws, all of which will be highlighted further below.

Re-conceptualizing Small-scale Fisheries Compliance

As highlighted above, the response by governments is often to increase law enforcement as a means of mitigating non-compliance, without adequately considering the history, or circumstances, of small-scale fishers (Berkes et al., 2001). There is wide recognition, for example, that small-scale fishers are often the poorest members of society (Berkes et al., 2001; Béné, 2003), requiring an understanding of the social and economic factors motivating people to fish. Further, the political, institutional and biophysical dynamics that influence fisheries management decision-making also need to be considered when understanding fishers' behaviour.

Understanding compliance within a fishery system

If one refers to a number of small-scale fisheries case studies around the world (e.g. Durrenberger and King, 2000; Hauck and Sowman, 2003, 2005; Wilson et al., 2003; McClanahan and Castilla, 2007a), it is clear that an overall understanding of the nature of compliance, as well as the strategies to achieve compliance in different contexts, is different between researchers, managers and fishers. In some cases, law enforcement has been significantly strengthened (Hauck and Kroese, 2006) and, in other cases, formal law enforcement is weak because of a lack of government capacity (Berkes et al., 2001; Christie et al., 2007; Pomeroy et al., 2009). In both cases, non-compliance remains an ongoing challenge for the fisheries authority. This highlights the need to understand the factors that are driving fishers to behave in the way that they do, and to develop appropriate fisheries management arrangements that reflect these.

Thus, a key approach to understanding small-scale fisheries compliance is to understand small-scale fisheries as being embedded in a 'system', which encapsulates the ecological, social, economic, political and institutional aspects of the fishery, and how they interrelate (Charles, 2001). In the area of natural resource management, there is an increasing realization that traditional, natural science-based methods of addressing problems are no longer appropriate, and there is a need to look for broader approaches and solutions (Berkes et al., 2003). This is certainly the case in fisheries management, where there is broad consensus that many of the world's fisheries are in crisis, and there is a need to move beyond a primary focus on the biophysical aspects of a fishery (Berkes et al., 2001; Charles, 2001; Pauly et al., 2003; Castilla and Defeo, 2005). As Defeo et al. (2007, p. 15) clearly state:

> The status of the world's fisheries is worrying and the factors that have led to the global decline are biological, social, political and cultural in nature. Marine fisheries are in trouble ... the trouble has occurred in the context of a well-developed fisheries science

that has largely focused on the resource and the biophysical aspects that control them but with less focus on the societal aspects of resource management.

Historically, people and their social systems have been at the periphery of fisheries management. However, particularly from a small-scale fisheries perspective, there is an increasing acknowledgement of revised approaches to governance that aim to achieve sustainability in the broader context, not just in terms of the fish stocks themselves (Berkes *et al.*, 2001; Charles, 2001; Kooiman *et al.*, 2005; Symes, 2006; McClanahan and Castilla, 2007b). Berkes *et al.* (2003, p. 2) explain that sustainability '... implies maintaining the capacity of ecological systems to support social and economic systems'. The aim, therefore, is to secure marine resources at the same time as securing the livelihoods of fishers. However, the ability to achieve goals that may conflict in reality is a complex exercise, and Hanna (2003, p. 309) argues that it is ultimately a '... search for balance in the distribution of authority and power'. The objectives of environmental sustainability, for example, are often prioritized in many countries, often at the expense of socio-economic considerations (Jentoft and McCay, 2003). Jentoft and Chuenpagdee (2009) make this point by defining fisheries and coastal governance as a 'wicked problem', emphasizing that biological, social and economic goals are often in conflict, but need to be understood in relation to their diversity, complexity, dynamics and scale. Further, they argue that how a problem is defined, and by whom, indicates a wider social and political construct, and probably reflects a problem that is further embedded in other problems.

A conceptual framework

Systems thinking has guided the conceptualization of more recent compliance research, recognizing that an understanding of fisheries compliance cannot focus on only one aspect of the system. In the past, for example, compliance was often understood and analysed from an economic perspective, exploring the costs and benefits of fishers' actions (Sutinen and Andersen, 1985; Anderson and Lee, 1986). Other studies began to explore compliance more widely, investigating social and institutional issues related to morality and legitimacy (Kuperan and Sutinen, 1998; Hatcher *et al.*, 2000; Hønneland, 2000; Gezelius, 2002, 2003, 2004; Raakjær Nielsen and Mathiesen, 2003). Further research has expanded the investigation further by seeking to understand and assess the different aspects of the fishery system as a means of understanding compliance behaviour (Hauck, 2008, 2009a). As McClanahan *et al.* (2009, p. 42) state:

> An integrated approach that addresses multiple needs, that at first appear peripheral to conservation and fisheries management, is essential in poor countries where there are multiple pressing priorities for action and a lack of infrastructure to deal with the costs of monitoring, control and surveillance of management measures.

This is reiterated by Berkes *et al.* (2001), who argue that many small-scale fishers are exposed to a myriad of socio-economic and political issues that are likely to influence their behaviour to comply. Thus, an integrated understanding of compliance behaviour arguably requires a more thorough analysis of the different components of the fishery system, and how they interrelate. Figure 11.1 outlines a proposed conceptual framework for understanding small-scale fisheries compliance.

This framework, which was originally conceptualized in relation to empirical studies in South Africa, is arguably applicable to small-scale fisheries more broadly. There are two key components of this framework. The first is the recognition that there is a diversity of factors that influence fisher behaviour. This draws on existing compliance theories that argue that a variety of variables need to be investigated to understand, and address, non-compliance. The framework therefore emphasizes the need to understand compliance within a fishery system, acknowledging the social, economic, institutional and biophysical factors that impact on whether or not fishers comply with rules and laws. Each of these factors will be discussed briefly below, but it is important to emphasize that they are themselves influenced by power and

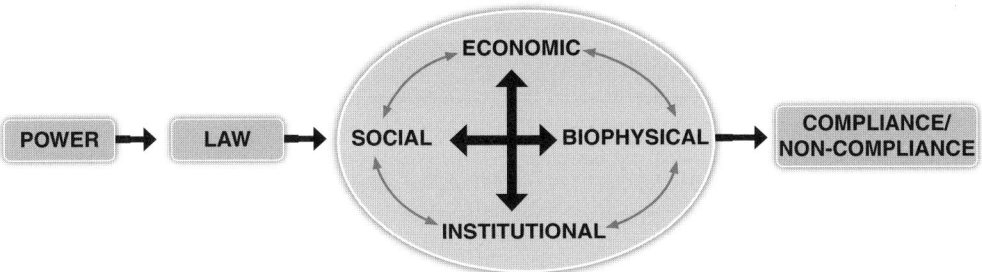

Fig. 11.1. A conceptual framework for understanding small-scale fisheries compliance.

law. The second key component of the framework, therefore, is that law cannot be taken as a given. Rather than conducting research only to enhance compliance of *existing* rules, it is necessary to take the analysis of fisheries compliance one step further. It is equally important to understand *how* law has evolved, its history and the power dynamics that have shaped it. Social and economic inequities found in laws and policies need to be identified as potential factors influencing compliance behaviour of small-scale fishers. Thus, it is argued that existing laws that marginalize small-scale fishers need to be understood, and challenged, in an attempt to enhance fisheries compliance.

Each of the factors highlighted in the framework will be briefly discussed below, but ongoing empirical research is required to explore these further, their linkages and whether or not they influence fisher behaviour in different fishery contexts. An understanding of power and law, which expand existing fisheries compliance theory, relates to the historical role of government, and other powerful elites, in managing fisheries, formulating laws and establishing socio-economic policies. It is recognized that 'crimes' and laws are socially constructed and are embedded in power differentials in which the criminalized are often those who are also socially, economically and politically oppressed (Scraton and Chadwick, 1991; Lynch and Stretsky, 2003). In the fisheries context, this relates to the allocation of rights, rules to harvest resources, and whether customary practices, food security and poverty relief are incorporated into fisheries laws and policies (Chuenpagdee *et al.*, 2005; Hernes *et al.*, 2005; Béné *et al.*, 2010). The marginalization and exclusion of small-scale fishers needs to be understood in this context. Furthermore, the role of customary law in influencing behaviour and potential conflicts that arise with state law need to be explored in relation to compliance (Bavinck, 2005; Rakotoson and Tanner, 2006; Ruddle and Hickey, 2008).

Institutional factors relate to the formal and informal rules and norms that govern resource use. Fisheries management is often made up of nested institutions that determine the allocation of rights to harvest a resource, the rules related to harvesting and the organizational structures established to manage these rules (Ostrom, 2000; Acheson, 2006). Many issues concerning institutional norms are political (Hoel and Kvalvik, 2006) and lead to questions of legitimacy (Jentoft, 1989). Management arrangements that explore the role of different stakeholders in managing the resource are critical to this understanding. As Chuenpagdee *et al.* (2005, p. 33) state: 'Despite the important impact on their livelihoods, coastal communities are often excluded from decision-making processes and debates on their livelihood options, such as access to the resources they depend on.'

In her discussion of collective action and the evolution of rules and norms, Ostrom (2000, p. 148) states that empirical field research indicates that: '… when the users of a common-pool resource organize themselves to devise and enforce some of their own basic rules, they tend to manage local resources more sustainably than when rules are externally imposed on them.'

Nevertheless, the need to monitor resource use is also recognized as a necessary element of governance in order to minimize 'free-riding' and to legitimize the system by reassuring those who are complying with rules that others are also conforming (Ostrom, 2000; Dietz et al., 2002; Schlager, 2002). It is argued, however, that costly monitoring and enforcement can be minimized if laws and rules are considered legitimate (Jentoft, 2000; FAO, 2005). Critical to this legitimacy is the need to ensure that institutions have moral force, thereby reflecting the norms, values and rights of fishers themselves (Jentoft, 2004b).

Related to this is the social context of the fishery, which particularly refers to the normative and social influences affecting fishers' behaviour. Social cohesion needs to be explored in relation to the development and support for rules and broader social norms (Ostrom, 2000; Gezelius, 2003; Acheson and Gardner, 2004; Acheson, 2006). A culture of compliance indicates that there is a *general* moral obligation to obey formal law (Zaelke et al., 2005; Gezelius, 2007). However, as noted by Gezelius (2002, p. 313), formal law may be: '… overruled by moral requirements which are perceived as more fundamental than the obligation to obey the law whenever the contents of specific laws conflict with certain moral norms of civil society.'

This is the case, for example, when *economic* needs are considered more legitimate than formal law, resulting in widespread acceptance of non-compliance as a means to support one's family. This is also the case when customary fishing practices are perceived as 'rights', protected through local rules or pre-existing (i.e. customary) systems of governance. The legitimacy of formal law, therefore, rests on it reflecting fishers' perceived rights and values (Jentoft, 2000; Berkes et al., 2001; Gezelius, 2002, 2004; FAO, 2005).

Economic aspects of the framework relate to poverty, livelihoods and the impact of the market on global and local trade. It is widely acknowledged that poverty is rife in small-scale fishing communities around the world (Berkes et al., 2001; Béné, 2003), that fishers live in relatively fragile livelihood conditions with few opportunities for alternative employment (Manning, 2001; Kooiman et al., 2005) and that they are vulnerable to food insecurity (Béné and Heck, 2005; Kooiman et al., 2005; Sowman and Cardoso, 2010). These conditions can lead to resource overexploitation, as few alternatives exist (Béné, 2003; Chuenpagdee et al., 2005). Furthermore, the role of globalization and the international market for fish products has had unprecedented impacts on fisheries governance at a local level (Dietz et al., 2003; Kooiman et al., 2005; Ahmed et al., 2006). Increased economic opportunities also provide incentives for free-riders to maximize their gains (Ostrom, 2000).

Finally, the biophysical aspects of fisheries resources and the ecosystem on which they depend are also important in understanding fishers' behaviour. The accessibility of inshore resources, for example, and biological characteristics all impact on harvesting strategies (Berkes et al., 2001). The dynamics of the natural system may be unpredictable (Charles, 2001; Clark, 2006), and even the weather can significantly influence the level, and intensity, of harvesting. Furthermore, certain ecosystem effects may impact on catch levels and changes to resource abundance, which in turn impact on the social and economic circumstances of fishers. The impacts of environmental variability, therefore, need to be understood in the context of management and how they impact on access to resources and other fisher decisions. The dynamics of the natural system, and the resource constraints, are important to understand in relation to the fishery system as a whole.

Research on fisheries compliance has only recently begun to explore the variety of relevant variables that influence fishers' behaviour (Kuperan and Sutinen, 1998; Hønneland, 2000; Gezelius, 2002; Raakjær Nielsen and Mathiesen, 2003; Roncin et al., 2004). These investigations have not only identified the complex interactions taking place in fisheries, but have begun to highlight a concern for traditional crime control strategies that are required to respond to, and address, resource over-exploitation and decline. Future empirical research needs to explore these linkages, and the various factors that influence fishers' decisions to comply, or not, with rules and laws.

Preconditions for Understanding and Enhancing Small-scale Fisheries Compliance

In attempting to understand the diversity of issues that affect compliance behaviour, three preconditions[2] have emerged that are considered necessary for understanding and enhancing small-scale fisheries compliance. Figure 11.2 highlights these preconditions, which have, as their foundation, equitable laws and policies. It is argued that the most critical precondition for achieving small-scale fisheries compliance is social justice. This encapsulates the concept of human rights, and more specifically the critical importance of acknowledging and protecting customary fishing practices and the livelihoods of fishers. Emanating from social justice is the concept of legitimacy, which in its broadest sense reinforces the importance of moral support for the institutional arrangements developed to govern a fishery. The last precondition is deterrence, which is considered key to reinforcing laws and rules, and in enhancing the legitimacy of the management system. Thus, both the normative action and rationalist approaches to compliance are highlighted. Fundamental to an understanding of Fig. 11.2, however, is that before deterrence can have effect, legitimacy needs to be in place, and before the concept of legitimacy can have effect, the underlying elements of social justice need to be embraced. Underpinning all of this is the acknowledgement that none of these preconditions can be adopted without a supporting legal and policy framework.

Social justice

Social justice is underpinned by a human rights perspective, recognizing the importance of equitable distribution and the minimization of social, economic and political harm (Scraton, 2002; Barton et al., 2007b). A key premise of social justice is the eradication of inequities that exist in human systems (Halsey, 1997; White, 2007). In their discussion of environmental crime, Halsey and White (1998, p. 356) argue that it is as important to: '... abolish not only those political and economic relations which lead to the domination and exploitation of ecosystems but also those relations of production which are premised upon the domination and exploitation of human beings'.

In a fisheries context, the importance of human rights is increasingly being recognized (see Charles, Chapter 4, this volume), with social justice referring to the importance of acknowledging and protecting customary rights and practices, enhancing food security, sustaining livelihoods and addressing the social, economic and political marginalization of fishers (Chuenpagdee et al., 2005; Hernes et al., 2005; Béné et al., 2010). An understanding of social justice, therefore, requires an understanding of the historical role of government, and other powerful elites, in formulating laws and establishing socio-economic policies. The two subsections below will highlight two particular examples of social justice issues in small-scale fisheries.

Protecting customary rights

The allocation of fishing rights is fundamentally a political issue (Hoel and Kvalvik, 2006). Jentoft and McCay (2003, p. 302) argue that it is therefore necessary to understand political influences by asking: '... who are the stakeholders and what are their political assets, how are public and private interests played out, how do unequal distribution of power among stakeholders and user-groups impact on decisions made, and who benefits and who loses from fisheries management'.

In South Africa, empirical research has highlighted the reality that allocation of rights to some fishers has deemed them 'legal', while others who have been excluded have been deemed 'illegal'. The latter group of

---

[2] 'Preconditions' in this context refer to the concepts that are argued to be key to influencing fisher behaviour. In other words, the discussion is kept at a conceptual level and does not necessarily differentiate these preconditions from the key motivators of compliance behaviour. Attempts to do the latter are explored in more detail by Gezelius and Hauck (in press).

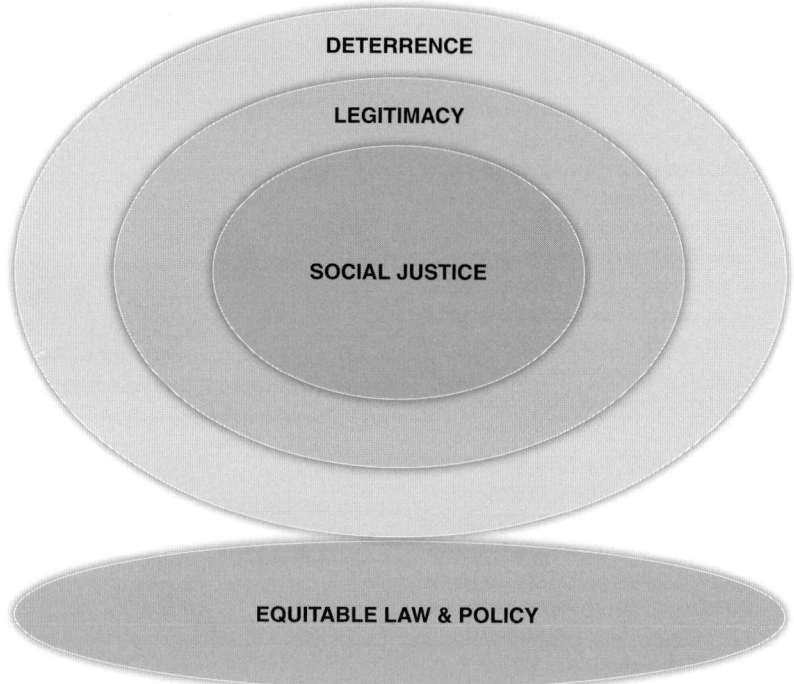

Fig. 11.2. The underlying preconditions required to guide a more integrated approach to small-scale fisheries compliance.

fishers are therefore considered 'poachers', and sanctioned through law enforcement measures (Hauck, 2009b). The process of criminalization is discussed in criminology literature, and it is widely recognized that power, and the interests that it protects, play a critical role in identifying that which is considered 'legal' or 'illegal' by the state (White, 1999, 2003; Lynch and Stretsky, 2003).

Power influences policies, which prioritize some objectives over others, often further marginalizing the powerless (Scraton and Chadwick, 1991; Scraton, 2002; Barton *et al.*, 2007a, b). For example, preferential policies that favour large-scale, industrialized fishers over small-scale, traditional ones have emerged in many fisheries around the world (McGoodwin, 1987; Ghee and Valencia, 1990; Mathew, 1990; Sunderlin and Gorospe, 1997; Fisheries Action Coalition Team, 2001; Silvestre *et al.*, 2003; Gupta and Sharma, 2004). This problem was identified as a key issue by the FAO Advisory Committee on Fisheries Research (2003, p. 9):

> One of the main policy thrusts in the past has been to promote economic growth at a national level, based on the assumption that all sectors of society (including small-scale fishers) will benefit. In fisheries, this has tended to favour the development of large-scale approaches over small-scale ones and the resources being concentrated in fewer and fewer hands.

As Platteau (1989) explains, many governments in developing countries in the 1950s and 1960s initiated modernization programmes, with resources being absorbed in the industrialized fisheries at the expense of the small-scale sector.

What is important to emphasize here is that political decision-making based on an unequal distribution of power is likely to lead fishers to break rules that are imposed on them (Jentoft and McCay, 2003). The economic and political power of large commercial companies and/or the state has been highlighted through various case studies, emphasizing how these interests have been entrenched to the detriment of small-scale fishers (McCay, 1984; van Sittert, 1994; Rakotoson and Tanner, 2006). As a result, traditional fishers who have been marginalized through specific laws that favour other groups (often more powerful elites) continue to harvest marine resources despite it being 'illegal' to do so. McCay (1984), in her ethnographic research of inshore 'piracy', clearly highlighted the cultural context of illegal fishing, describing how marginalized fishers used piracy as a tool for 'social protest' over discriminatory laws. Historical research by van Sittert (1993, 1994) has also described in detail the active marginalization of small-scale lobster fishers in South Africa by the joint economic interests of large-scale industry and the state. Although this led to increased rules and regulations, with an attempt by industry to entrench its exclusive access to the resource, customary fishing practices were driven 'underground', fuelling the emergence of a black market trade (van Sittert, 1993, 1994). As Meinzen-Dick and Pradhan (2001) explain, local community institutions often establish their own legal order, including their own property regimes. From the fishers' perspective, therefore, there is no moral obligation to obey formal law they perceive to be unjust, or contrary to their existing traditions and local laws. Customary practices are perceived as rights that should be protected and sustained through formal legal processes (Rakotoson and Tanner, 2006), or formally recognized under existing local institutions (Ruddle and Hickey, 2008). The FAO (2005, p. 41) argue that: 'Recognizing the existing rights of fishing communities is a fundamental element in building a successful fisheries management system. Doing so provides a basis of legitimacy, which can significantly enhance system compliance.'

Therefore, what is argued here is that without this formal recognition of fishers' rights and existing fishing practices, resistance to state-driven rules and regulations will persist (Hernes *et al.*, 2005).

Reference to legal pluralism is particularly relevant in this context, acknowledging that more than one legal order may be operating in the same society (Griffiths, 1986; Meinzen-Dick and Pradhan, 2001). This is particularly acute in post-colonial societies in which colonial law was imposed on indigenous institutions. In relation to fisheries, Bavinck (2003, 2005) refers to sea tenure (i.e. property rights in the sea) and the implications that emerge when different sea tenure systems operate simultaneously. In addition to the conflicts that result, Ruddle and Hickey (2008) argue that coastal communities are increasingly marginalized by being subjected to centralized management approaches (and laws) that are imposed on them without adequate recognition of local systems already in place. When state and customary laws collide, it is often the powerful interests of the state that are enforced, with marginalized groups being defined as 'illegal' or 'criminal' (Chambliss, 1975; Scraton, 2002). These inequities and power imbalances need to be understood and acknowledged in order to protect customary fishing practices and incorporate local institutions into new approaches to fisheries governance.

Sustainable livelihoods

Conflicting policy objectives, as discussed above, need to be emphasized in relation to the livelihoods of small-scale fishers and their right to meet basic needs. The powerful interests of big industry, to benefit from neoliberal policies and the export-orientated focus of fisheries policy, need to be understood in this context of social justice. Small-scale fishers, on the other hand, are increasingly relying on marine resources for food security and basic income as a result of vulnerable livelihoods and few economic alternatives (Berkes *et al.*, 2001; Béné, 2003; Béné and Heck, 2005; Pomeroy *et al.*, 2009; Sowman and Cardoso, 2010). This is emphasized by Chuenpagdee *et al.* (2005, p. 33), who explicitly state that: '… social justice is directly related to power and poverty …' and in order for fishers '… to make a living when no alternative sources of employment are available and one's bargaining position is weak, the only response ... is for fishers to increase their fishing efforts.'

Similarly, in his ethnographic study of two fishing villages in Norway and Canada, Gezelius (2002, 2003) explained that the perceived *economic need* to violate formal rules of the fishery resulted in no moral obligation within the collective to abide by that rule. Thus, he argued, the social and economic well-being of individual fishers, as well as the future of the community (in recognizing that fishing is an important social, cultural and economic activity), were recognized as 'rights', and the importance of securing these rights was 'regarded as a moral buffer against the obligation to obey the law' (Gezelius, 2002, p. 312). The same can be said for small-scale fishers in South Africa. Until their human rights are secured, and sustainable livelihoods are achieved, they will remain disenchanted with formal laws and policies, and they will 'vigorously defend perceived traditional rights', even if it is illegal to do so (Cardoso *et al.*, 2005, p. 35). The fact that there are significant resource constraints, however, needs to be recognized, and supplementary income-generating activities should be explored. In Vietnam, for example, Pomeroy *et al.* (2009, p. 427) explain that inshore resources are over-exploited, but: '... policies that reduce the number of fishers in small-scale fisheries without creating non-fishery livelihood opportunities will inevitably fail ... fishers will merely fish illegally ... to continue to make a living in order to feed their family.'

The identification of complementary livelihoods has been highlighted as a critical aspect of sound fisheries management, in order to sustain fishers' livelihoods, reduce pressure on diminishing resources and limit vulnerability during times of resource shortage or due to environmental variability (FAO, 2005; Allison and Horemans, 2006). The need to embed small-scale fisheries governance in broader poverty reduction strategies and economic development opportunities is therefore critical (Berkes *et al.*, 2001; Hara and Raakjær Nielsen, 2003; Kooiman *et al.*, 2005; McClanahan and Castilla, 2007b; Charles, Chapter 4, this volume; Allison *et al.*, Chapter 12, this volume).

Legitimacy

Legitimacy is directly linked to the principles underpinning social justice. As Hernes *et al.* (2005, p. 105) explain, if a management system is considered unjust, fishers are likely to resist it: 'in order to be legitimate, (fisheries) decisions must satisfy some basic criteria – or principles – of justice'. Thus, legitimacy is increasingly recognized as an important variable influencing fisheries compliance. There are many case studies around the world that have emphasized the importance of developing fisheries rules that reflect the norms, values and beliefs of fishers, in order to enhance legitimacy and, it is argued, compliance (Bavinck, 1996; Acheson, 1998; Kuperan and Sutinen, 1998; Ostrom, 2000; Berkes *et al.*, 2001; Gezelius, 2002, 2003, 2004; Dietz *et al.*, 2003; Raakjær Nielsen, 2003; Acheson and Gardner, 2004; van Ginkel, 2005). If rules are considered fair, social mechanisms will often develop to ensure adherence to these rules (Berkes, 1987; Ostrom, 2000). However, if formal laws are prioritized and conflict with customary practice and morality, as discussed above, an obligation to comply with these laws is eroded (Jentoft, 2000; Gezelius, 2002, 2003, 2004; Rakotoson and Tanner, 2006).

Fundamental to any fisheries system are institutional arrangements, which are the rules and structures in place to manage resource use (Ostrom, 2000). Key to an understanding of institutions, however, is the recognition that they are not developed in isolation from other factors, but are in fact embedded in social, economic, political and ecological realities (McCay, 2002; Jentoft, 2004a, b; Acheson, 2006). Thus, institutions are a fundamental component of the management system, and crucial to getting them 'right' is ensuring that they are '... ethically sound and socially just ...' (Jentoft, 2007, p. 361). Furthermore, Jentoft (2004a, p. 94) argues that '... fisheries management institutions must work from the realization that they are nested in social structures, moral norms and values that impinge on them ...'. Institutions fundamentally govern fisheries resources by creating the limits – and incentives – to ensure sustainability (ecologically, socially and economically) (Charles, 2001; McClanahan and Castilla, 2007b).

It is argued that institutions are legitimate insofar as they are created through participatory processes, ensuring that the fishers who are affected by them are also directly involved in creating them (Jentoft, 1989; Ostrom, 2000; Berkes et al., 2001). The assumption is that by moving away from conventional, centralized management, stakeholders other than government share decision-making power, thereby leading to a greater acceptance of rules and norms. As Hall-Arber (2005, p. 144) states: 'It is now almost a cliché to note that those who participate in the development of regulations are more apt to abide by them.'

Fisheries compliance research has also identified cases where law enforcement is weak (usually due to lack of capacity and resources), and where penalties are low, but the majority of fishers still comply with the regulations (Sutinen and Kuperan, 1999; Gezelius, 2003). Thus, other than the small number of chronic violators, other factors have been identified that help shape compliant behaviour. These factors, which are shaped by the theory of normative action, include moral and social norms, social pressure, perceived legitimacy of the management authority and fisher involvement in decision-making and management (Kuperan et al., 1997; Kuperan and Sutinen, 1998; Sutinen and Kuperan, 1999; Hatcher et al., 2000; Hønneland, 2000; McKinlay and Millington, 2000; Gezelius, 2002, 2003, 2004; Raakjær Nielsen, 2003; Raakjær Nielsen and Mathiesen, 2003). The co-management approach, therefore, has been implemented worldwide as one mechanism to enhance legitimacy in fisheries.

Significant experimentation, research and reviews have been undertaken on fisheries co-management around the world (Pomeroy et al., 2001; Viswanathan et al., 2003; Wilson et al., 2003; Pomeroy, Chapter 5, this volume), and some argue that it is *the* way forward for fisheries management (Gray, 2005). However, the results of co-management are highly variable. Although some have argued that co-management arrangements are still relatively government-driven through top-down processes (Hara and Raakjær Nielsen, 2003), others argue that co-management has been successful in strengthening participation, self-regulation and compliance, and has contributed to improving fishers' livelihoods and resource sustainability (Ahmed et al., 2006; Castilla et al., 2007; Morenzo et al., 2007; McClanahan et al., 2009).

Thus, although there is a great deal of international rhetoric to move towards participatory approaches to management, Symes (2006) argues that policy making in this regard is more apparent than real. He states (p. 116): 'There is common concern that institutional changes are not keeping pace with the needs of the new forms of governance and that until these transformations are complete, power will remain in the hands of the old oligarchy.'

As a result, the legitimacy of institutional arrangements requires power imbalances to be addressed, equitable sharing of benefits to be achieved and participatory institutional arrangements to be implemented.

Deterrence

The principle of deterrence ensures that the costs of prohibited actions are greater than the benefits of those actions. This emanates from an economic model of compliance, and rests on the assumption that rational individuals will comply with rules and laws if the perceived costs outweigh the benefits (Becker, 1968). In fisheries, this principle has led to enhanced law enforcement mechanisms as a means to increase the costs, through the probability of detection and conviction (Sutinen et al., 1990). Although crime control models have largely been relied upon by states to impose fisheries compliance, it is increasingly recognized in the literature, as already discussed, that there are other strategies that should be explored to address the underlying drivers of fisher behaviour (Kuperan and Sutinen, 1998; Sutinen and Kuperan, 1999; Hatcher et al., 2000; Hønneland, 2000; Gezelius, 2002, 2003, 2004; Raakjær Nielsen, 2003). Nevertheless, it is argued through the rationalist approach to compliance theory, that both formal and informal controls and sanctions are important for the small percentage of chronic violators in every fishery who are motivated by economic gain (Kuperan and Sutinen, 1998; Sutinen et al., 1990). Furthermore, the importance of enforcing rules has also been emphasized as a means to enhance the legitimacy of the management system (Tyler, 1990; Hønneland, 2000; Gezelius, 2002). In common property theory, Ostrom (2000) and others (Dietz et al., 2003) have argued that there will always be 'free-riders', and there is a need to ensure that rules are enforced. However, what is emphasized in other research is that the enforcement of rules must first be based on the assumption that rules themselves are considered fair and just, and are therefore accepted (Kuperan and Sutinen, 1998; Sutinen and Kuperan, 1999; Jentoft, 2000; Gezelius, 2002, 2003, 2004).

If law enforcement is expected to enforce laws that are not morally binding, costs will be high. Although it is recognized that command and control methods are not economically efficient (Dietz et al., 2003), they are increasingly relied upon to enforce rules (Raakjær Nielsen, 2003), particularly those that are devised through top-down management strategies (Hanna, 1995). However, as Levi argues, 'If an institution depends only on coercion for the successful implementation of its policies, the costs of enforcement will be insupportably high' (quoted in Jentoft, 2004b, p. 146). Nevertheless, Gezelius (2002) warns that if the authorities are not enforcing the rules, the incentive for fishers to obey rules that have no moral integrity is diminished. The result, as is the case in many small-scale fisheries in developing countries, is that both fisheries recognized by government (i.e. 'legal') and traditional or customary fisheries (sometimes perceived by the state as 'illegal') are operating simultaneously (van Sittert, 1993; Bavinck, 2003, 2005; Acheson and Gardner, 2004). The centralized establishment of rules, which hold little legitimacy on the ground, is coupled with weak (or non-existent) law enforcement, which translates to the state's inability to enforce the rules that it creates.

However, an important distinction has emerged in empirical research on small-scale fisheries compliance between those who are maximizing gains from illegal fishing (perceived to be making large profits) and those who are simply relying on resources for their livelihood ('just getting by' or 'putting food on the table'). This distinction, between fishing for 'need or greed' (Hauck, 1999), has also been highlighted by Gezelius (2002, 2003, 2004). Gezelius explained that illegal fishing as a result of economic need was generally tolerated by the community, whereas illegal fishing that was practised for profit was considered unacceptable by the community and sanctioned. Thus, 'The fear of social degradation or exclusion from the collectivity was usually enough to prevent opportunistic, utility-maximizing action beyond the law' (Gezelius, 2002, p. 310). In South Africa, although it was recognized that large economic gain had motivated fishers to violate rules, social pressure was weak largely due to the perceived illegitimacy of the rules. However, in the more organized illegal fisheries (such as the abalone fishery) for example, when outside opportunists emerged in the fishery, there was an increased acceptance of the need to enhance formal law enforcement controls (Hauck, 2009a, b).

The ineffectiveness of formal law enforcement is argued to be highly problematic, as it may lead to the 'domino effect' in which fishers violate rules because they see others getting away with it (Sutinen et al., 1990; Kuperan and Sutinen, 1994, 1998). As Sutinen et al. (1990, p. 246) explain, 'Non-violators stand to lose out on the resource if much of it is taken by the violators, thus pushing them to violate as well.' The importance of law enforcement to deter chronic violators, therefore, has been identified as important for reasserting the legitimacy of rules and regulations, and for enhancing the moral obligation of fishers to comply (Kuperan and Sutinen, 1994, 1998; Sutinen and Kuperan, 1999; Raakjær Nielsen and Mathiesen, 2003).

Research has, therefore, emphasized two critical points in relation to law enforcement and deterrence. The first relates to the theory of normative action. It is unrealistic to expect law enforcement to address fisheries non-compliance if it is implemented in isolation from broader strategies that address the social, political, economic and cultural factors that are driving fishers to behave in the way that they do. The second incorporates the rationalist approach to compliance, emphasizing that law enforcement, and other informal controls, are necessary for deterring chronic, profit-maximizing individuals, and for enforcing the legitimacy of the management system. Thus, while the monitoring and enforcement of rules have been identified as important for enhancing compliance, they need to be implemented through legitimate, socially valued institutions that are accepted and supported.

These three preconditions of small-scale fisheries compliance, which are argued to enhance our understanding of fisher behaviour, reflect the theoretical developments in fisheries compliance theory. Legitimacy is embraced by the theory of normative action, while deterrence is embraced by the rationalist perspective of compliance. The precondition of social justice, however, has been highlighted as an important concept that needs to broaden current compliance thinking. By challenging existing laws, and attempting to understand the history and power behind their development, an important analytical process is added to our understanding of fisher behaviour. While the concept of social justice has been emerging in fisheries discourse more broadly (Chuenpagdee et al., 2005; Hernes et al., 2005), it is argued here that it needs to be adopted more vigorously into compliance theory. As Jentoft (2000, p. 142) states '… a management system that cannot be defended on grounds of social justice is likely to be challenged, however solid its legal foundation.'

Conclusions

Traditionally, the key objective of fisheries management has been to sustain fish stocks and ensure the ecological integrity of the natural system (Raakjær Nielsen et al., 2004). The concept of sustainability, however, has significantly evolved over the years to acknowledge the inextricable link between the social and natural systems. In this context, a sustainable fisheries system is one that leads to social, economic, ecological and institutional sustainability of the fishery as a whole (Charles, 2001). Threats to any of these components of the fishery system need to be understood and addressed in order to achieve sustainability in this broad sense. Non-compliance has been identified as a threat to resource sustainability (Sutinen et al., 1990), and efforts to enhance compliance have largely been implemented in isolation of understanding the broader factors that may influence fisher behaviour. As a result, despite law enforcement efforts, many fisheries remain threatened, with both the resources and the fishers at risk. This has highlighted the need to gain a broader understanding of the factors that motivate fishers to comply (or not) with rules and laws.

This interconnectivity between human and natural systems is directly embraced by new approaches to fisheries governance (Bavinck et al., 2005; McClanahan and Castilla, 2007a; de Young et al., 2008; Garcia et al., 2008; Mahon et al., 2008). However, what underpins the small-scale fisheries compliance framework introduced in this chapter is that an understanding of compliance requires an understanding of power and law – how law is formulated and in whose interests. Furthermore, it has been argued that social justice, legitimacy and deterrence are preconditions for understanding and addressing compliance in a more integrated manner. By adopting strategies that aim to achieve these preconditions, fisheries policies will shift from a sole reliance on criminal justice approaches to achieve compliance, to a more integrated approach that aims to sustain the fishery system as a whole.

Acknowledgements

Financial assistance for conducting empirical research and writing this paper on small-scale fisheries compliance is gratefully acknowledged to the following: the Norwegian–South African regional cooperation agreement, the South Africa–Netherlands Research Programme on Alternatives to Development and the South African National Research Foundation. The opinions expressed are those of the author and are not necessarily to be attributed to the organizations above. Furthermore, I would like to acknowledge the constructive debates and comments from colleagues within the Environmental Evaluation Unit at UCT, as well as colleagues from abroad.

References

Acheson, J.M. (1998) Lobster trap limits: a solution to a communal action problem. *Human Organization* 57, 43–52.

Acheson, J.M. (2006) Institutional failure in resource management. *Annual Review of Anthropology* 35, 117–134.

Acheson, J.M. and Gardner, R.J. (2004) Strategies, conflict and the emergence of territoriality: the case of the Maine lobster industry. *American Anthropologist* 106, 296–307.

Ahmed, M., Siar, S.V., Wilson, D.C. and Muir, J. (2006) Governance and institutional changes in fisheries – impact on poverty reduction and environmental integrity in developing countries. In: Siar, S.V., Ahmed, M., Kanagaratnam, U. and Muir, J. (eds) *Governance and Institutional Changes in Fisheries: Issues and Priorities for Research*. WorldFish Center Discussion Series No. 3, WorldFish Center, Penang, Malaysia, pp. 1–24.

Allison, E.H. and Horemans, B. (2006) Putting the principles of the Sustainable Livelihoods Approach into fisheries development policy and practice. *Marine Policy* 30, 757–766.

Anderson, L.G. and Lee, D.R. (1986) Optimal governing instruments, operation level, and enforcement in natural resource regulation: the case of the fishery. *American Journal of Agricultural Economics* 68, 678–690.

Barton, A., Corteen, K., Scott, D. and Whyte, D. (2007a) Introduction: developing a criminological imagination. In: Barton, A., Corteen, K., Scott, D. and Whyte, D. (eds) *Expanding the Criminological Imagination: Critical Readings in Criminology*. Willan Publishing, Cullompton, UK, pp. 1–14.

Barton, A., Corteen, K., Scott, D. and Whyte, D. (2007b) Conclusion: expanding the criminological imagination. In: Barton, A., Corteen, K., Scott, D. and Whyte, D. (eds) *Expanding the Criminological Imagination: Critical Readings in Criminology*. Willan Publishing, Cullompton, UK, pp. 198–213.

Bavinck, M. (1996) Fisher regulations along the Coromandel coast: a case of collective control of common pool resources. *Marine Policy* 20, 475–482.

Bavinck, M. (2003) The spatially splintered state: myths and realities in the regulation of marine fisheries in Tamil Nadu, India. *Development and Change* 34, 633–657.

Bavinck, M. (2005) Understanding fisheries conflicts in the South – a legal pluralist perspective. *Society and Natural Resources* 18, 805–820.

Bavinck, M., Chuenpagdee, R., Diallo, M., van der Heijden, P., Kooiman, J., Mahon R. and Williams, S. (2005) *Interactive Fisheries Governance: a Guide to Better Practice*. Eburon Academic Publishers, Delft, Netherlands.

Becker, G.S. (1968) Crime and punishment: an economic approach. *Journal of Political Economy* 76, 169–212.

Béné, C. (2003) When fishery rhymes with poverty: a first step beyond the old paradigm on poverty in small-scale fisheries. *World Development* 31, 949–975.

Béné, C. and Heck, S. (2005) Fish and food security in Africa. *NAGA, WorldFish Centre Quarterly* 28, 8–12.

Béné, C., Hersoug, B. and Allison, E.H. (2010) Not by rent alone: analysing the pro-poor functions of small-scale fisheries in developing countries. *Development Policy Review* 28, 325–358.

Berkes, F. (1987) Common property resource management and Cree Indian fisheries in subarctic Canada. In: McCay, B.J. and Acheson, J. (eds) *The Question of the Commons: the Culture and Ecology of Communal Resources*. University of Arizona Press, Tucson, Arizona, pp. 66–91.

Berkes, F., Mahon, R., McConney, P., Pollnac, R. and Pomeroy, R. (2001) *Managing Small-scale Fisheries: Alternative Directions and Methods*. International Development Research Centre, Ottawa, Canada.

Berkes, F., Colding, J. and Folke, C. (2003) *Navigating Social-ecological Systems: Building Resilience for Complexity and Change*. Cambridge University Press, Cambridge, UK.

Cardoso, P., Fielding, P. and Sowman, M. (2005) Overview and analysis of social, economic and fisheries information to promote artisanal fisheries management in the BCLME region – South Africa. Final report and recommendations. Environmental Evaluation Unit, University of Cape Town, Cape Town, South Africa.

Castilla, J.C. and Defeo, O. (2005) Paradigm shift needed for world fisheries. *Science* 309, 1324–1325.

Castilla, J.C., Gelcich, S. and Defeo, O. (2007) Successes, lessons and projections from experience in marine benthic invertebrate artisanal fisheries in Chile. In: McClanahan, T.R. and Castilla, J.C. (eds) *Fisheries Management: Progress Towards Sustainability*. Blackwell Publishing, Oxford, UK, pp. 25-42.

CEC (Commission of the European Communities) (2007) *On a New Strategy for the Community to Prevent, Deter and Eliminate Illegal, Unreported and Unregulated Fishing*. Communication from the Commission to the European Parliament, the Council, the European Economic and Social Committee and the Committee of the Regions, 17 October 2007, Brussels.

Chambliss, W.J. (1975) Toward a political economy of crime. *Theory and Society* 2, 1949–1970.

Charles, A. (2001) *Sustainable Fishery Systems*. Blackwell Science, Oxford, UK.

Charles, A.T., Mazany, R.L. and Cross, M.L. (1999) The economics of illegal fishing: a behavioural model. *Marine Resource Economics* 14, 95–110.

Christie, P., Fluharty, D.L., White, A.T., Eisma-Osorio, L. and Jatulan, W. (2007) Assessing the feasibility of ecosystem-based fisheries management in tropical contexts. *Marine Policy* 31, 239–250.

Chuenpagdee, R., Degnbol, P., Bavinck, M., Jentoft, S., Johnson, D. and Pullin, R. (2005) Challenges and concerns in capture fisheries and aquaculture. In: Kooiman, J., Jentoft, S., Pullin, R. and Bavinck, M. (eds) *Fish for Life: Interactive Governance for Fisheries*. Amsterdam University Press, Amsterdam, Netherlands, pp. 25–37.

Clark, B.M. (2006) Climate change: a looming challenge for fisheries management in southern Africa. *Marine Policy* 30, 84–95.

de Young, C.D., Charles, A. and Hjort, A. (2008) Human dimensions of the ecosystem approach to fisheries: an overview of context, concepts, tools and methods. FAO Fisheries Technical Paper No. 489, FAO, Rome, Italy.

Defeo, O., McClanahan, T.R. and Castilla, J.C. (2007) A brief history of fisheries management with emphasis on societal participatory roles. In: McClanahan, T.R. and Castilla, J.C. (eds) *Fisheries Management: Progress Towards Sustainability*. Blackwell Publishing, Oxford, UK, pp. 3–21.

Dietz, T., Dolšak, N., Ostrom, E. and Stern P.C. (2002) The drama of the commons. In: Ostrom, E., Dietz, T., Dolšak, N., Stern, P.C., Stonich, S. and Weber, E.U. (eds) *The Drama of the Commons*. National Academy Press, Washington, DC, pp. 3–36.

Dietz, T., Ostrom, E. and Stern, P.C. (2003) The struggle to govern the commons. *Science* 302, 1907–1918.

Durrenberger, E.P. and King, T.D. (eds) (2000) *State and Community in Fisheries Management – Power, Policy, and Practice*. Bergin and Garvey, Westport, Connecticut.

EJF (Environmental Justice Foundation) (2005) *Pirates and Profiteers: How Pirate Fishing Fleets are Robbing People and Oceans*. Environmental Justice Foundation, London.

FAO (Food and Agriculture Organization) (2001) *International Plan of Action to Prevent, Deter and Eliminate Illegal, Unreported and Unregulated Fishing*. FAO, Rome, Italy.

FAO (Food and Agriculture Organization) (2003) A discussion paper for the ACFR Working Party on small-scale fisheries. Prepared for the *ACFR Working Party Meeting*, October 2003, Bangkok, Thailand.

FAO (Food and Agriculture Organization) (2005) *Increasing the Contribution of Small-scale Fisheries to Poverty Alleviation and Food Security*. Technical Guidelines for Responsible Fisheries vol. 10, FAO, Rome, Italy.

Fisheries Action Coalition Team (2001) *Feast or Famine? Solutions to Cambodia's Fisheries Conflicts*. Fisheries Action Coalition Team, Phnom Penh, Cambodia.

Flewelling, P., Cullinan, C., Balton, D., Sautter, R.P. and Reynolds, J.E. (2003) Recent trends in monitoring, control and surveillance systems for capture fisheries. FAO Fisheries Technical Paper No. 415, FAO, Rome, Italy.

Garcia, S.M., Allison, E.H., Andrew, N.L., Béné, C., Bianchi, G., de Graaf, G., Kalikoski, G.J., Mahon, R. and Orensanz, J.M. (2008) Towards integrated assessment and advice in small-scale fisheries: principles and processes. FAO Fisheries and Aquaculture Technical Paper No. 515, FAO, Rome, Italy.

Gezelius, S.S. (2002) Do norms count? State regulation and compliance in a Norwegian fishing community. *Acta Sociologica* 45, 305–314.

Gezelius, S.S. (2003) *Regulation and Compliance in the Atlantic Fisheries: State/Society Relations in the Management of Natural Resources*. Kluwer Academic Publishers, Dordrecht, Netherlands.

Gezelius, S.S. (2004) Food, money, morals: compliance among natural resource harvesters. *Human Ecology* 32, 615–635.

Gezelius, S.S. (2007) Three paths from law enforcement to compliance: cases from the fisheries. *Human Organisation* 66, 414–425.

Gezelius, S.S. and Hauck, M. (in press) Towards a theory of compliance in state regulated livelihoods: A comparative study of compliance motivations in developed and developing world fisheries.

Ghee, L.T. and Valencia, M.J. (1990) *Conflict over Natural Resources in South-east Asia and the Pacific*. United Nations University Press, Singapore.

Gray, T.S. (2005) Theorising about participatory fishery governance. In: Gray, T.S. (ed.) *Participation in Fisheries Governance*. Springer, Dordrecht, Netherlands, pp. 1–26.

Griffiths, J. (1986) What is legal pluralism? *Journal of Legal Pluralism and Unofficial Law* 24, 1–55.

Gupta, C. and Sharma, M. (2004) Blurred borders: coastal conflicts between India and Pakistan. *The Economic and Political Weekly*, 3 July.

Hall-Arber, M. (2005) Co-management at the eleventh hour? Participation in the governance of the New England groundfish fishery. In: Gray, T.S. (ed.) *Participation in Fisheries Governance*. Springer, Dordrecht, Netherlands, pp. 141–162.

Halsey, M. (1997) Environmental crime: towards an eco-human rights approach. *Current Issues in Criminal Justice* 8, 217–242.

Halsey, M. and White, R. (1998) Ecophilosophy and environmental harm. *Theoretical Criminology* 2, 345–371.

Hanna, S. (1995) User participation and fishery management performance within the Pacific Fishery Management Council. *Ocean and Coastal Management* 28, 23–44.

Hanna, S. (2003) The future of fisheries co-management. In: Wilson, D.C., Raakjær Nielsen, J. and Degnbol, P. (eds) *The Fisheries Co-management Experience: Accomplishments, Challenges and Prospects*. Kluwer, Dordrecht, Netherlands, pp. 309–319.

Hara, M. and Raakjaer Nielsen, J. (2003) Experiences with fisheries co-management in Africa. In: Wilson, D.C., Raakjaer Nielsen, J. and Degnbol, P. (eds) *The Fisheries Co-management Experience: Accomplishments, Challenges and Prospects*. Kluwer Academic Publishers, Dordrecht, Netherlands, pp. 81–98.

Hatcher, A., Jaffry, S., Thebaud, O. and Bennett, E. (2000) Normative and social influences affecting compliance with fishery regulations. *Land Economics* 76, 448–462.

Hauck, M. (1999) Regulating marine resources in South Africa: the case of the abalone fishery. In: Glazewski, J. and Bradfield, G. (eds) *Environmental Justice and the Legal Process*. Juta, Cape Town, South Africa, pp. 211–228.

Hauck, M. (2007) Non-compliance in small-scale fisheries: a threat to security? In: Beirne, P. and South, N. (eds) *Issues in Green Criminology: Confronting Harms Against Environments, Humanity and Other Animals*. Willan Publishing, London, UK, pp. 270–289.

Hauck, M. (2008) Rethinking small-scale fisheries compliance. *Marine Policy* 32, 635–642.

Hauck, M. (2009a) Rethinking small-scale fisheries compliance: from criminal justice to social justice. PhD thesis. University of Cape Town, Cape Town, South Africa.

Hauck, M. (2009b) Crime, environment and power: revisiting the abalone fishery. *South African Journal of Criminal Justice* 22, 229–245.

Hauck, M. and Kroese, M. (2006) Fisheries compliance in South Africa: a decade of challenges and reform 1994–2004. *Marine Policy* 30, 74–83.

Hauck, M. and Sowman, M. (eds) (2003) *Waves of Change: Coastal and Fisheries Co-management in South Africa*. University of Cape Town Press, Cape Town, South Africa.

Hauck, M. and Sowman, M. (2005) Coastal and Fisheries Co-management in South Africa: is there an enabling legal environment? *South African Journal of Environmental Law and Policy* 12, 1–21.

Hemming, B. and Pierce, B.E. (1997) Fisheries enforcement: our last fisheries management frontier. In: Hancock, D.A., Smith, D.C., Grant, A. and Beumer, J.P. (eds) *Proceedings of the Second World Fisheries Congress*. CSIRO Publishing, Brisbane, Australia, pp. 675–679.

Hernes, H.K., Jentoft, S. and Mikalsen, K.H. (2005) Fisheries governance, social justice and participatory decision-making. In: Gray, T.S. (ed.) *Participation in Fisheries Governance*. Springer, Dordrecht, Netherlands, pp. 103–118.

Hoel, A.H. and Kvalvik, I. (2006) The allocation of scarce natural resources: the case of fisheries. *Marine Policy* 30, 347–356.

Hønneland, G. (2000) Compliance in the Barents Sea fisheries: how fishermen account for conformity with rules. *Marine Policy* 24, 11–19.

Jentoft, S. (1989) Fisheries co-management: delegating government responsibility to fishermen's organisations. *Marine Policy* 13, 137–154.

Jentoft, S. (2000) Legitimacy and disappointment in fisheries management. *Marine Policy* 24, 141–148.

Jentoft, S. (2004a) The community in fisheries management: experiences, opportunities and risks. In: Hersoug, B., Jentoft, S. and Degnbol, P. (eds) *Fisheries Development: the Institutional Challenge*. Eburon, Delft, Netherlands, pp. 93–130.

Jentoft, S. (2004b) Institutions in fisheries: what they are, what they do, and how they change. *Marine Policy* 28, 137–149.

Jentoft, S. (2007) The limits of governability: institutional implications for fisheries and coastal governance. *Marine Policy* 31, 360-370.

Jentoft, S. and Chuenpagdee, R. (2009) Fisheries and coastal governance as a wicked problem. *Marine Policy* 33, 553–560.

Jentoft, S. and McCay, B.J. (2003) The place of civil society in fisheries management: a research agenda for fisheries co-management. In: Wilson, D.C., Raakjaer Nielsen, J. and Degnbol, P. (eds) *The Fisheries Co-management Experience: Accomplishments, Challenges and Prospects*. Kluwer, Dordrecht, Netherlands, pp. 293–307.

Kooiman, J., Bavinck, M., Jentoft, S. and Pullin, R. (eds) (2005) *Fish for Life: Interactive Governance for Fisheries*. Amsterdam University Press, Amsterdam.

Kuperan, K. and Sutinen, J.G. (1994) Compliance with zoning regulations in Malaysian fisheries. Paper presented at the *IIFET VII International Conference of International Institute of Fisheries Economics and Trade*, Taiwan, 18–21 July.

Kuperan, K. and Sutinen, J.G. (1998) Blue water crime: deterrence, legitimacy and compliance in fisheries. *Law and Society Review* 32, 309–337.

Kuperan, K., Abdullah, N.M.R., Susilowati, I., Siason, I.M. and Ticao, C. (1997) Enforcement and compliance with fisheries regulations in Malaysia, Indonesia and the Philippines. Research report, Department of Natural Resource Economics, University of Pertanian, Serdang, Malaysia.

Lynch, M.J. and Stretsky, P.B. (2003) The meaning of green: contrasting criminological perspectives. *Theoretical Criminology* 7, 217–238.

Mahon, R., McConney, P. and Roy, R.N. (2008) Governing fisheries as complex adaptive systems. *Marine Policy* 32, 102–112.

Manning, P. (2001) Small-scale fisheries management in sub-Saharan Africa. A background document for the *FAO Expert Consultation on Small-scale Fisheries Management in Sub-Saharan Africa*, December, Accra, Ghana.

Mathew, S. (1990) *Fishing Legislation and Gear Conflicts in Asian Countries*. SAMUDRA Monograph No. 1, International Collective in Support of Fishworkers (ICSF), Chennai, India.

McCay, B. (1984) The pirates of piscary: ethnohistory of illegal fishing in New Jersey. *Ethnohistory* 31, 17–37.

McCay, B. (2002) Emergence of institutions for the commons: contexts, situations and events. In: Ostrom, E., Dietz, T., Dolsak, N., Stern, P.C., Stonich, S. and Weber, E.U. (eds) *The Drama of the Commons*. National Academy Press, Washington, DC, pp. 361–402.

McClanahan, T.R. and Castilla, J.C. (eds) (2007a) *Fisheries Management: Progress Towards Sustainability*. Blackwell Publishing, Oxford, UK.

McClanahan, T.R. and Castilla, J.C. (2007b) Healing fisheries. In: McClanahan, T.R. and Castilla, J.C. (eds) *Fisheries Management: Progress Towards Sustainability*. Blackwell Publishing, Oxford, UK, pp. 305–326.

McClanahan, T.R., Castilla, J.C., White, A.T. and Defeo, O. (2009) Healing small-scale fisheries by facilitating complex socio-ecological systems. *Reviews in Fish Biology and Fisheries* 19, 33–47.

McGoodwin, J.R. (1987) Mexico's conflictual inshore Pacific fisheries: problem analysis and policy recommendations. *Human Organization* 46, 221–232.

McKinlay, J.P. and Millington, P.J. (2000) Fisher obligations in co-managed fisheries: the case for enforcement. In: Shotton, R. (ed.) *Use of Property Rights in Fisheries Management*. FAO, Rome, Italy.

Meinzen-Dick, R.S and Pradhan, R. (2001) Implications of legal pluralism for natural resource management. *IDS Bulletin* 32, 10–17.

Morenzo, C.A., Barahona, N., Molinet, C., Orensanz, J.M., Parma, A.M. and Zuleta, A. (2007) From crisis to institutional sustainability in the Chilean sea urchin fishery. In: McClanahan, T.R. and Castilla, J.C. (eds) *Fisheries Management: Progress Towards Sustainability*. Blackwell Publishing, Oxford, UK, pp. 43–67.

MRAG (Marine Resources Assessment Group) (2005) Review of impacts of illegal, unreported and unregulated fishing on developing countries. Final Report submitted to the UK's Department for International Development (DFID), London.

Ostrom, E. (2000) Collective action and the evolution of social norms. *The Journal of Economic Perspectives* 14, 137–158.

Pauly, D., Alder, J., Bennett, E., Christensen, V., Tyedmers, P. and Watson, R. (2003) The future for fisheries. *Science* 302, 1359–1361.

Platteau, J.P. (1989) The dynamics of fisheries development in developing countries: a general overview. *Development and Change* 20, 565–597.

Pomeroy, R.S., Katon, B.M. and Harkes, I. (2001) Conditions affecting the success of fisheries co-management: lessons from Asia. *Marine Policy* 25, 197–208.

Pomeroy, R.S., Anh Thi Nguyen, K. and Xuan Thong, H. (2009) Small-scale marine fisheries policy in Vietnam. *Marine Policy* 33, 419–428.

Raakjær Nielsen, J. (2003) An analytical framework for studying compliance and legitimacy in fisheries management. *Marine Policy* 27, 425–432.

Raakjær Nielsen, J. and Mathiesen, C. (2003) Important factors influencing rule compliance in fisheries lessons in Denmark. *Marine Policy* 27, 409–416.

Raakjær Nielsen, J., Degnbol, P., Viswanathan, K.K., Ahmed, M., Hara, M. and Abdullah, N.M.R. (2004) Fisheries co-management – an institutional innovation? Lessons from South East Asia and Southern Africa. *Marine Policy* 28, 151–160.

Rakotoson, L.R. and Tanner, K. (2006) Community-based governance of coastal zone and marine resources in Madagascar. *Ocean and Coastal Management* 49, 855–872.

Rigg, K., Parmentier, R. and Currie, D. (2003) Halting IUU fishing: enforcing international fishing agreements. Report prepared for Oceana by the Varda Group. Available at: www.vardagroup.org (accessed 6 September 2009).

Roncin, N., Bailly, D. and Raux, P. (2004) Determinants of fishermen's compliance – lessons from a coastal fishery in Biscay Bay. In: Matsuda, Y. and Yamamoto, T. (eds) What are responsible fisheries? *Proceedings of the 12th Biennial Conference of the International Institute of Fisheries Economics & Trade (IIFET)*, 20–30 July, Tokyo.

Ruddle, K. and Hickey, F.R. (2008) Accounting for the mismanagement of tropical nearshore fisheries. *Environment Development and Sustainability* 10, 565–589.

Schlager, E. (2002) Rationality, cooperation and common pool resources. *Science* 45, 801–819.

Scraton, P. (2002) Defining 'power' and challenging 'knowledge': critical analysis as resistance in the UK. In: Carrington, K. and Hogg, R. (eds) *Critical Criminology: Issues, Debates, Challenges*. Willan Publishing, Cullompton, UK, pp. 15–40.

Scraton, P. and Chadwick, K. (1991) The theoretical and political priorities of critical criminology. In: Stenson, K. and Cowell, D. (eds) *The Politics of Crime Control*. Sage, London, pp. 166–185.

SIF (Stop Illegal Fishing) (2008) *Stop Illegal Fishing in Southern Africa*. Stop Illegal Fishing, Gabarone, Botswana.

Silvestre, G.T., Garces, L.R., Stobutzki, I., Ahmed, M., Valmonte-Santos, R.A., Luna, C.Z. and Zhou, W. (2003) South and South-east Asian coastal fisheries: their status and directions for improved management: conference synopsis and recommendations. In: Silvestre, G., Garces, L.R., Stobutzki, I., Ahmed, M., Valmonte-Santos, R.A., Luna, C.Z., Lachica-Alino, L., Munro, P., Christensen, V. and Pauly, D. (eds) *Assessment, Management and Future Directions for Coastal Fisheries in Asian Countries*. WorldFish Center Conference Proceedings 67, The WorldFish Center, Penang, Malaysia, pp. 1–40.

Sowman, M. and Cardoso, P. (2010) Small-scale fisheries and food security strategies in countries in the Benguela Current Large Marine Ecosystem (BCLME) region: Angola, Namibia and South Africa. *Marine Policy* 34, 1163–1170.

Sunderlin, W.D. and Gorospe, M.G. (1997) Fishers' organizations and modes of co-management: the case of San Miguel Bay, Philippines. *Human Organization* 56, 333–343.

Sutinen, J. and Andersen, P. (1985) The economics of fisheries law enforcement. *Land Economics* 61, 387–397.

Sutinen, J. and Kuperan, K. (1999) A socio-economic theory of regulatory compliance. *International Journal of Social Economics* 26, 174–193.

Sutinen, J.G., Rieser, A. and Gauvin, J.R. (1990) Measuring and explaining non-compliance in federally managed fisheries. *Ocean Development and International Law* 21, 335–372.

Symes, D. (2006) Fisheries governance: a coming of age for fisheries social science? *Fisheries Research* 81, 113–117.

Tyler, T.R. (1990) *Why People Obey the Law*. Yale University Press, New Haven, Connecticut.

UN (United Nations) (2006) *UN General Assembly Resolution on Sustainable Fisheries*. A/RES/61/105, UN, New York.

van Ginkel, R. (2005) Between top-down and bottom-up governance: Dutch beam trawl fishermen's engagement with fisheries management. In: Gray, T.S. (ed.) *Participation in Fisheries Governance*. Springer, Dordrecht, Netherlands, pp. 119–140.

van Sittert, L. (1993) 'More in the breach than the observance': crayfish conservation and capitalism. *Environmental History Review* 17, 20–46.

van Sittert, L. (1994) *Red, Gold and Black markets: the Political Economy of the Illegal Crayfish Trade, c.1890 – c.1990*. Department of History, University of Cape Town, South Africa.

Viswanathan, K.K., Raakjaer Nielsen, J., Degnbol, P., Ahmed, M., Hara, M. and Abdullah, N.M.R. (2003) *Fisheries Co-Management Policy Brief: Findings from a Worldwide Study*. The WorldFish Center, Penang, Malaysia.

White, R. (1999) Criminality, risk and environmental harm. *Griffith Law Review* 8, 235–257.

White, R. (2003) Environmental issues and the criminological imagination. *Theoretical Criminology* 7, 483–506.

White, R. (2007) Green criminology and the pursuit of social and ecological justice. In: Beirne, P. and South, N. (eds) *Issues in Green Criminology: Confronting Harms Against Environments, Humanity and Other Animals*. Willan Publishing, Cullompton, UK, pp. 32–54.

Wilson, D.C., Raakjaer Nielsen, J. and Degnbol, P. (eds) (2003) *The Fisheries Co-management Experience: Accomplishments, Challenges and Prospects*. Kluwer Academic Publishers, Dordrecht, Netherlands.

Zaelke, D., Stilwell, M. and Young, O. (2005) What reason demands: making law work for sustainable development. In: Zaelke, D., Kaniaru, D. and Kruzikova, E. (eds) *Making Law Work: Environmental Compliance and Sustainable Development*. Cameron May, London, pp. 29–51.

12 Poverty Reduction as a Means to Enhance Resilience in Small-scale Fisheries

Edward H. Allison, Christophe Béné and Neil L. Andrew

Introduction

The vast majority (over 95%) of small-scale fisherfolk (fish farmers, fishers, traders and related occupations) are from low-income developing countries (FAO, 2009). A critical task for contemporary fisheries management is to find ways to ensure that fishery systems (including people) are resilient in the face of multiple drivers of change. If fisheries are to be sustained, adapted or transformed in ways that allow them to continue making useful contributions to society, then challenges to the sustainability of the ecological, social and economic elements of the system must be addressed (Jentoft, 2000; Charles, 2001). One such challenge is the high dependence of poor fishing communities on these systems in the developing world (Béné et al., 2007). Traditional, sectoral approaches to poverty reduction in fisheries have sought ways to reduce fishing costs, catch fish more effectively and make more profit. While these remain worthwhile objectives, if they involve increasing fishing capacity they are unlikely to be of long-term benefit, given the need to reduce fishing effort in many fisheries. Other approaches must be found.

Addressing poverty among fishing communities is complicated by the multidimensional nature of poverty. For example, cash incomes from fishing are often higher than earnings from agriculture or other wage-labour options, but vulnerability and insecurity may be higher (Allison, 2005; Béné et al., 2009a, b). Continued poverty is made more likely because fisherfolk are often marginalized, politically and economically (Béné, 2003), sometimes because they comprise a minority ethnic group, but more often because they are simply overlooked in development planning processes (Allison, 2005; Thorpe et al., 2007). Fisherfolk are often excluded from access to other employment opportunities, from equitable access to land, social services such as health and education, and may have weak political representation. They may also be poorly served by roads, markets and other infrastructure. These dimensions of livelihood insecurity and lack of social and human capital limit people's ability and motivation to participate in resource governance, and hinder their capacity to engage successfully with globalizing markets and other opportunities for economic development. This, in turn, leads to further marginalization and vulnerability among fishing-dependent communities, along with reduced income and assets, and increased pressure on resources that are already heavily exploited or at risk of degradation. All these processes undermine the resilience of small-scale fisheries (SSF) and provide the rationale for addressing poverty,

© CAB International 2011. *Small-scale Fisheries Management*
(eds R.S. Pomeroy and N.L. Andrew)

vulnerability and marginalization as integral parts of governing fisheries for sustainability.

To address the complex set of interacting causes and effects of poverty it is first necessary to identify its main dimensions and drivers; only then can practical measures to address or work around these constraints be identified and acted upon. Most useful are approaches that build on people's strengths, and address failures of basic entitlements, markets or policies, as required (Allison and Ellis, 2001). Most often, this will require partnerships with development actors outside the fishing sector, such as health and education service providers, or microfinance organizations, or local government, the judiciary and land courts. Understanding such connections between fisheries and broader economic, social and political processes is particularly important in the context of evolving development policy in a sector that seeks to use the potential of fish stocks to generate wealth for poverty reduction – a purpose for which there is wide consensus, albeit with debate about the appropriate mechanisms to achieve this (Cunningham et al., 2009; Béné et al., 2010).

In this chapter we argue that people who are poor, vulnerable or discriminated against are much less likely to be effective resource stewards. It is therefore a legitimate development objective to improve their material, relational and subjective well-being, as a step towards improving fisheries governance and the net benefits that flow from fisheries to wider poverty reduction. Of course, fisheries can be a vehicle to improve rights, wealth and well-being, but fisheries must be placed in the broader context of people's livelihoods and societal constraints and opportunities.

Following on from this rationale, development actions can address poverty in fishing-dependent communities in ways that do not put additional pressure on heavily exploited resources. None of the potential actions we identify should be taken as generalizable, as the 'binding constraints' to improved well-being and ecological status in any fishery system are locale-specific and, if unworkable or irrelevant blue-print solutions are to be avoided, will have to be identified through some diagnostic process (Rodrik, 2006; Andrew et al., 2007; Ostrom, 2007).

The chapter is organized first to draw on advances in ways in which poverty and livelihoods are understood and analysed. We describe the multiple dimensions of poverty and related 'state of being' such as vulnerability and social exclusion, with reference to several important aspects of vulnerability, including gender, climate change, HIV/AIDS and child labour. The Sustainable Livelihoods Approach (SLA) has proved useful in analysing poverty and people's livelihoods – we briefly review the approach and its key principles. With a broader development context in mind we then make the case that there is a natural order or sequencing of development interventions. We briefly review this perspective on development. Finally, we identify a set of principles to ensure that poverty reduction and resource governance efforts are synergistic, rather than antagonistic, and combine to strengthen the role that fishery social-ecological systems can play for the poor.

Poverty and Livelihoods

Before interventions to improve the livelihood of fish-dependent communities can be identified, it is necessary first to understand how people make a living, what constrains or enables them to do so, what options are available to them and in what ways they are succeeding or failing. It is also necessary to understand the nature of poverty in its multiple dimensions and the complex reasons that some people remain poor, become poor or escape poverty.

Understanding poverty

While livelihoods analysis has been effective at drawing attention to the factors influencing livelihood outcomes in fishing communities, it largely does so by emphasizing those outcomes in terms of incomes and assets. These material dimensions of wealth or poverty are obviously important in considering development, but there are other non-material dimensions of poverty that can also be important and are not necessarily directly linked to the material ones. The multiple dimensions of deprivation faced by many people involved in small-scale fisheries can be considered in relation to three main concepts, drawn from the wider understanding of poverty in development policy and practice:

Poverty is typically described in terms of low income and ownership of limited capital assets. Standardized measures of poverty of this type can be compared to standards (e.g. poverty lines) and are useful for assessing the impacts of economic policy and for targeting social protection measures. Increasingly, however, broader definitions of poverty are being used, for example:

> ... a human condition characterized by the sustained or chronic deprivation of the resources, capabilities, choices, security and power necessary for the enjoyment of an adequate standard of living and other civil, cultural, economic, political and social rights.
> (UN Committee on Social, Economic and Cultural Rights, 2001)

Vulnerability is understood in terms of people's exposure to risks, the sensitivity of their livelihood systems to these risks and their capacity to use their assets and capabilities to cope with and to adapt to these risks. Two commonly used applications of this concept are in the World Food Programme's famine vulnerability mapping (World Food Programme, 2007) and the IPCC's mapping of vulnerability to climate change (IPCC, undated).

While poverty and vulnerability are sometimes thought of as 'end results' of natural stresses combined with policy failures of various kinds, the third concept, that of *marginalization* or *social exclusion*, describes a process by which certain groups are systematically disadvantaged because they are discriminated against on the basis of their ethnicity, race, religion, sexual orientation, caste, gender, age, education, class disability, HIV status, migrant status or where they live (Atkinson, 1998; DFID, 2005).

Income and asset poverty, vulnerability and marginalization are interrelated and overlapping conditions (Fig. 12.1). For example, the poor tend to be more vulnerable to external 'shocks' because they lack assets to absorb and recover from the impacts of events such as destructive floods or tropical storms; fishers impacted by the December 2004 Asian tsunami are an example (Pomeroy *et al.*, 2006).

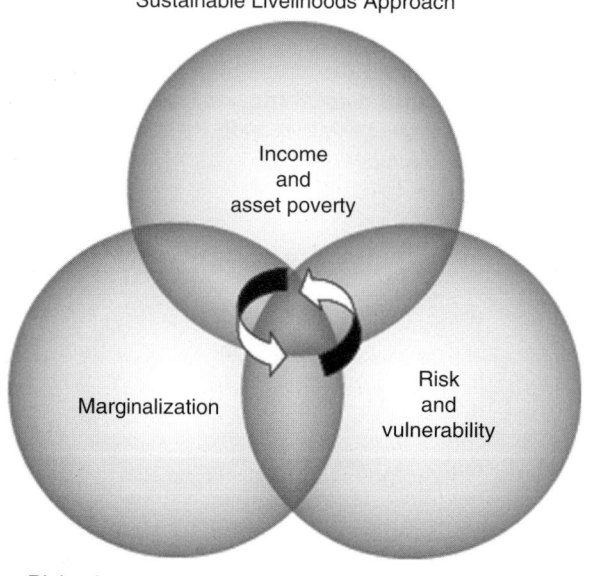

Fig. 12.1. Multiple overlapping and reinforcing dimensions of poverty, and three frameworks or approaches that can be used to address these (redrawn from Allison *et al.*, 2006).

Those who are vulnerable because their livelihoods are exposed to and sensitive to physical risks may need years fully to recover financially from one of these extreme events. Similarly, income-poor people can become impoverished because they are excluded from the rights and opportunities available to others, sometimes on grounds of ethnicity, citizenship or gender. The poor, lacking assets such as access to education and information, can become marginalized in political processes such as local development planning and are thus denied rights of participation.

Understanding livelihoods

The concept of a 'livelihood' brings together the critical factors that affect the vulnerability or strength of individual or family survival strategies (Ellis, 2000, p. 10): 'A livelihood comprises the assets (natural, physical, human, financial and social capital), the activities, and the access to these (mediated by institutions and social relations) that together determine the living gained by the individual or household.'

Livelihoods incorporating small-scale fishing are typically either occupationally diverse or geographically dispersed, and sometimes both (Ellis, 1998; Allison and Ellis, 2001). Such livelihood strategies may be composed of a portfolio of activities only some of which may be related to fishing. Mobility and migration is an important component of many fisherfolk's livelihood strategies (both men in the catching sector and women in the postharvest sector). Strategies can also relate to people's consumption choices (e.g. 'doing without' or the sale of assets). Short- and long-term measures to ensure survival are often distinguished as 'coping' and 'adapting', respectively (Ellis, 2000).

The fundamental social and economic unit considered in analyses of livelihoods is typically the household, conceived as being the social group that resides in the same place, shares the same meals and makes joint or coordinated decisions over resource allocation and income pooling. A sustainable livelihood is one in which people are able to maintain or improve their standard of living related to well-being and income or other human development goals, reduce their vulnerability to external shocks and trends, and ensure their activities are compatible with maintaining the natural resource base – in this case the fish stocks.

Analysing poverty and livelihoods: the Sustainable Livelihoods Approach

The Sustainable Livelihoods Approach (SLA) and its underlying 'capitals and capabilities' framework has proved useful in understanding livelihoods and the multiple dimensions of poverty (Allison and Ellis, 2001; Allison and Horemans, 2006). The SLA is typically set out in the form of a set of guiding principles and an analytical framework. The core principles that underlie SLA thinking can be summarized as:

- *Putting people's social and economic activities at the centre of the analysis.* This acknowledges that attempts to reduce fishing pressure or change the distribution of access to fisheries in support of the poor require us to understand people's circumstances.
- Taking a view of the options for management and development intervention that involve *transcending traditional sectoral boundaries* such as fisheries, agriculture, pastoralism, wage labour or small enterprise, and that incorporate overarching issues that affect all people, irrespective of occupation, such as access to social services (e.g. health, education, social security), financial services (savings, loans, insurance), political representation and judicial services.
- *Making micro–macro links.* Livelihood approaches encourage explicit consideration of links between local issues (such as resource allocation among different types of resource users in a fishery), meso-level processes (such as decentralization of planning and financial management of fishing ports or landing sites, from national to local authorities) and wider concerns, including national

policy and economic or social change (such as the adoption of a new fisheries policy or legislation, the liberalization of markets and the withdrawal of production-related subsidies).
- *Being responsive and participatory* in addressing management priorities. This normally involves working in partnership with fishers and other stakeholders in the public and private sectors and promotes a dynamic, adaptive and learning approach to management.
- *Building on strengths.* Although the focus of development is to address problems such as low incomes, poor health, lack of education, food insecurity, social exclusion or vulnerability, the livelihoods approach encourages ways of addressing these issues that make the most of people's existing strengths and abilities. In fishing communities, these may include extensive local or indigenous technical knowledge, strong vocational skills and diverse and flexible livelihood strategies that are highly responsive to change in potentially positive ways.
- *Taking a broad view of sustainability.* The four key dimensions to sustainability – economic, institutional, social and environmental (Charles, 2001) – are all important to sustainable fishery management. The livelihoods approach makes these dimensions explicit. The SLA also implicitly recognizes the dynamism of people's lives and does not view sustainability in static, equilibrial terms. Sustainability is viewed instead as the capacity of the elements of a livelihood system (people, institutions, environment and economy) to withstand shocks and adapt to change.

While none of these principles are new or unique to the livelihoods approach, taken together they represent a way of working in development that has yielded some positive results in other areas of rural and natural resource development (Chambers and Conway, 1992; Ellis, 2000). The reader is referred to Ellis (2000) and Allison and Ellis (2001) for more detailed discussions of the SLA.

The SLA has been criticized because it does not immediately draw the attention of analysts to the important dynamics of power that take place in all human interactions. Gender, class and ethnicity are all key elements in power relationships and can contribute to social exclusion (Kabeer, 2000). A focus on social exclusion leads to consideration of the broad framework of human rights law as a means of addressing neglect, discrimination and abuse. Later in this chapter, we illustrate the application of social exclusion theory and responses to two specific forms of inequity found within some fishing communities – gender inequity and the exploitation of children.

The definition of sustainability used within the SLA (above) is echoed in recent definitions of 'resilience' (see Andrew and Evans, Chapter 2, this volume for a review in this context). Resilience, as it is usually defined, is a value-free property of a social-ecological system (Walker *et al.*, 2004; Brand and Jax, 2007). As Walker *et al.* (2010) and others note, poverty is a resilient, but bad, 'state' for SSF, but the word 'resilient' is mostly referred to as a positive attribute of a system (and certainly of people). Andrew and Evans (Chapter 2, this volume, p. 16) define a resilient small-scale fishery as one that '. . . absorbs stress and reorganizes itself following disturbance, while still delivering benefits for poverty reduction'. By directly asserting a generic value, this definition makes resilience thinking more useful in policy development; it is the right of people within the fishery along with legitimate duty-bearers in government to define the specific mechanisms by which an SSF contributes to poverty reduction.

Livelihood sustainability is also affected by external factors, sometimes referred to as the *vulnerability context*, comprising cycles (e.g. seasonality), trends and shocks that are outside the control of the household. Trends might include decreasing catch rates, increasing prices for fish, and factors unrelated to fisheries that nevertheless impact on fishing households, such as rising costs of food staples or medicines. Shocks might include storm events that damage shore facilities, toxic algal blooms or sudden fuel-price hikes and currency devaluations that affect the costs of fishing inputs and market prices for fishery products. At a household level, illness or death of a family member and the theft or loss of a fishing net are obvious shocks.

Because the conceptualization of vulnerability is quite vague in the SLA, livelihoods-based analyses to inform poverty reduction actions are often supplemented with more explicit frameworks for vulnerability analysis. One such framework was developed by the Inter-governmental Panel on Climate Change (IPCC, 2001), and clearly distinguishes three elements of vulnerability – exposure to risks, sensitivity of the system to those risks and the capacity of actors or agents within the system to undertake planned adaptation to reduce either exposure or sensitivity – or both. The framework can be applied to any type of risk (e.g. economic risks, climate change risks) and the system under consideration can be adapted according to context – it can be applied to individuals, households, communities, sectors or national economies. Later in this chapter we illustrate its application to fishery systems facing risks from climate change and from exposure of fisherfolk to HIV and AIDS.

The concepts and methods of livelihoods analysis have been applied to understanding the role that fisheries play in the rural economy in coastal, lakeshore and floodplain areas in both developing and developed countries, in order to inform policy debates on fisheries management and development (e.g. Allison and Ellis, 2001; Allison, 2005; Allison and Horemans, 2006; Béné et al., 2007; Ahmed et al., 2008).

Understanding how people succeed or fail in sustaining their livelihoods in the face of shocks, trends and seasonality can help to design policies and interventions to assist peoples' existing coping and adaptive strategies. These interventions may include improving access to education and health care facilities, strengthening rights to land for settlement and agriculture (i.e. not just rights of access to fish stocks), reforming local tax and licence systems, providing financial and enterprise development services (and not just credit for purchase of fishing gear) and promotion of diversification – all issues seldom addressed in fisheries management and policy, and only recently addressed in fisheries development projects. We now turn to giving examples of the kinds of practical interventions of this type that have successfully been attempted under each of the three areas of poverty reduction.

Addressing Social Exclusion

With fishing often taking place on the neglected and poorly regulated margins of society, serious human development concerns can emerge around discrimination (including gender and social discrimination), labour market issues (including poor and insecure working conditions), power asymmetry and abusive labour relations or widespread incidence of child labour – as well as involvement of fishing boats and fisherfolk in international crime, including drug and arms smuggling and people-trafficking (Sharma, 2003; FAO, 2007; Allison and Kelling, 2009).

Emerging partnerships between fishery sector agencies and ILO, UNICEF, UNHCR, labour unions, international police forces and human-rights organizations are beginning to address some of these issues, but the links between these initiatives and improved fisheries governance have not yet been articulated (Allison et al., (submitted)). Strategies for reducing fishing-dependent peoples' marginalization are an integral part of any attempt to improve the governance of fisheries. This can be achieved by improving their access to regular health services, by governing fishing-related labour markets more effectively, supporting gender equity, addressing justice and security issues, and upholding basic human rights. A well-governed fishery sector will bring benefits to society that go well beyond securing a sustainable supply of fish, to include improved human security – and not just for fisherfolk.

Responses to gendered differentials and child labour issues in development outcomes are two examples of the results of processes of social exclusion that result in human-rights violations, and which can be addressed by recourse to international and national human rights law.

Women's rights and gender

Drawing on a range of sources, WorldFish Center (2010) provides some useful definitions of gender-related terms in fisheries and aquaculture:

Gender refers to the socially constructed roles and status of women and men, girls and boys. Culturally specific norms and beliefs define the social behaviour of women and men, and the relationship between them. Gender roles, status and relations vary according to place (countries, regions, villages), groups (class, ethnic, religious, caste), generations and stages of the life cycle of individuals. Gender is thus not about women but about how the categories of men and women are constructed and how those constructions affect the day-to-day lives of men and women.

Gender equity is the process of providing equal opportunities to women and men. To ensure fairness, measures are often needed to compensate for historical and social disadvantages that might have prevented women and men from having access to a level playing field. Equity is a precondition to equality.

Gender equality is the condition when women and men enjoy the same status. Gender equality means that women and men have equal conditions for realizing their full human rights and potential to contribute to national, political, economic, social and cultural development, and to benefit from the results. The concept of equality acknowledges that differential treatment of women and men may sometimes be required to achieve sameness of results, because of different life conditions or to compensate for past discrimination. Gender equality also signifies that society places equal value on the similarities and differences between women and men, and the varying roles that they play.

Gender analysis is a tool better to understand the different social, economic, cultural and political realities of women and men, girls and boys. At its core is an understanding of how culture (underlying values, norms and beliefs) defines gender identities and inequalities. It aims to uncover the dynamics of gender differences across a variety of issues. These include gender issues with respect to: (i) social relations (how 'male' and 'female' are defined in the given context, their normative roles, duties, responsibilities); (ii) economic/livelihood activities (gender division of labour in productive and reproductive work within the household and the community); (iii) access to and control over assets, resources, services, institutions of decision-making and networks of power and authority; and (iv) well-being in terms of needs and aspirations, the distinct indicators and perceptions of well-being and the needs and aspirations of men and women, both practical (i.e. given current roles, without challenging social norms) and strategic (i.e. needs which, if met, would change their position in society).

Gender disparities in fisheries and aquaculture can result in lower labour productivity within the sector and inefficient allocation of labour at household and national levels (reviewed by Weeratunge et al., 2010). Laws, customary beliefs, cultural norms and/or unfavourable regulatory structures of the state often reduce women's access to fisheries resources and assets (Okali and Holvoet, 2007). This implies that women are likely to constitute a larger proportion of the poor within this sector, as they do in agriculture, forestry and industry.

Addressing gender inequities by improving women's incomes and educational levels, as well as their access to information and decision-making processes, enhances human capabilities of the household, as well as society in general. Important for sustainable change are measures to improve governance, especially enhanced voice and accountability, and public sector capacity to be responsive to gender-specific needs. Cash earned by women contributes to the local economy, and in some areas is provided as capital to male producers to improve their productive assets. There is increasing evidence that those countries which have performed well towards achieving gender equity have also reached higher levels of economic growth and/or social well-being in general (World Economic Forum, 2006, 2007).

A range of directed activities can take place from sectoral entry points (e.g. health, education, enterprise development) and through local initiative (e.g. formation of women's marketing cooperatives and savings and insurance clubs), successful examples of which are reviewed in the special volume of World Development (Choo *et al.*, 2008) and in Williams *et al.* (2005), Okali and Holvoet (2007), Williams (2008) and Weeratunge *et al.* (2010). Overall, however, scaling-out efforts will require high-level legislative, political and economic commitment in the form of gender mainstreaming programmes.

Eliminating child labour

The International Labour Organization (ILO, undated) defines child labour as '. . . work that deprives children of their childhood, their potential and their dignity, and that is harmful to physical and mental development'. Fisheries is one of the most dangerous sectors for children to work in. If both ILO estimates and FAO statistics on total numbers of people employed in fisheries are accurate, children make up a significant part of the labour force. Estimates for four developing countries (Table 12.1) indicate that child labour in fisheries accounts for between 2.5% and 5.2% of the total number of child labourers in those countries, with boys estimated to account for between 86% and 91% of child labour in the sector. In these four countries, on average, every 10th person working in fisheries is a child. A study of Senegal's artisanal fishery by ILO found just under one third (29%) of the workforce were children (O'Riordan, 2006).

Appropriate interventions to ensure that children's work in the sector does not compromise their development or violate their rights are therefore vitally important. Not all work done by children is damaging to them, and interventions to remove children from work in fisheries must proceed with caution. Especially in fisheries, where sons follow their fathers, the work of children is often seen as professional training, preparing them for adulthood. And, it must be borne in mind, in setting expectations for the benefits of eradicating child labour, that both adults and children in fishing communities are often marginalized and deprived of access to education and health service support. From a fisheries governance perspective, the use of cheap child labour in fisheries signals probable economic over-exploitation, as children are usually paid less than adults to work. Arguably, children working for their own family can also play a critical role in contributing to maintaining a minimum living standard.

To date, there has been little policy response to this issue from the fisheries sector itself, despite the growing concern of other agencies around safety and health/working conditions in the fisheries sector and its widespread employment of children. Agencies involved in labour rights and children's rights have, however, worked with children and their communities to secure a better future for child situations of worst or hazardous forms of labour. The International Organization for Migration, for example, has taken a leading role in trying to free children working as bonded labour in Ghana's Lake Volta (ICSF, 2006).

Other prominent child labour campaigns have been associated with global commodity chains – for example, reports about abusive conditions and the use of labour, including children employed in shrimp farming, have raised consumer awareness, led to boycotts on particular companies' products and sometimes to overall bans on exports from entire countries (Silp, 2007; Solidarity Center, 2008). This type of action illustrates the power of

Table 12.1. Child labour in fisheries (selected countries, 2001–2002). Data from FAO (2008).

	Child labourers in fisheries (*n*)	Children in fisheries workforce (%)	Fisheries contribution to overall child labour (%)	Boys as child labourers in fisheries (%)
El Salvador	10,085		4.5	86.0
Ghana	49,185	9.4	2.5	87.2
Bangladesh	209,733	11.9	2.8	86.1
The Philippines	208,000	10.4	5.2	91.3

mobilizing consumer concern and choice, but does not directly lead to any improvement for the children, who depend on their jobs, unless consumer boycotts are followed up by legal action by exporter countries and practical action to comply with labour standards by companies.

Sustainable solutions to the problem of child labour lie beyond campaigns and legislation. Addressing the fundamental reasons of why children work, to the detriment of their health and education, means tackling the problems of poverty and inequality in the wider society as opposed to interventions targeting only the fishing-dependent communities. One example of an integrated solution was the ILO-IPEC Medan fishing programme aimed at eliminating child labour on jermals (fishing platforms) in North Sumatra, Indonesia. The programme built on a partnership with the Government of Indonesia and included surveys, dialogues, local capacity building with provincial government, withdrawal and rehabilitation activities (shelters, vocational training and provision of microfinance to children), but also aimed at prevention through awareness training and education programmes. The ILO's International Programme on the Elimination of Child Labour (IPEC) has a wealth of information on this programme and other child labour initiatives (ILO-IPEC, undated).

Reducing Vulnerability: Reducing Exposure to Risks and Building Adaptive Capacity

Environmental, political and economic hazards, adverse trends and shocks of various types can negate many years of hard work and careful accumulation of assets of those attempting to climb out of poverty. Fisherfolk face a range of such hazards – from extreme weather, theft or loss of fishing gear, to high susceptibility to ill health from exposure to water-borne diseases.

There are three basic ways in which fisherfolk's vulnerability to hazards can be reduced. One can either find a way to reduce exposure to risk, to reduce the sensitivity of their livelihood systems to the unavoidable risks to which they are exposed, or to build their capacity to adapt the consequences of a livelihood system that is exposed to some irreducible degree of risk and will always be sensitive to some risks. For example, most natural resource-based activities will always be sensitive to climate risks, and the only way to reduce sensitivity is to diversify out of natural resource-based activities – something not possible or desirable for many fisherfolk. We illustrate what can be done to reduce vulnerability in practice with reference to two major hazards confronting many fishing communities: climate change and HIV/AIDS.

Reducing vulnerability to climate change

Fisheries managers and fisherfolk have historically had to adapt to the vagaries of weather and climate (see, for instance, Glantz and Thompson, 1981; Cole, 1996; Gordon and Munro, 1996; Flaaten et al., 1998; Lauck et al., 1998; Allison et al., 2005; Rothschild et al., 2005). Uncertainty is inherent in fisheries, so there is an expectation of change and a stock of knowledge and experience of coping with it and adapting to it (Miller and Fluharty, 1992). However, current rates of change are historically unprecedented (MacKenzie and Schiedek, 2007; Dulvy et al., 2008). Furthermore, multiple stressors like over-exploitation and anthropogenic-driven degradation of marine and freshwater habitats are already threatening fisheries around the world and eroding their capacity to adapt to change in general.

Adaptation to climate change impacts will vary across scales (local, regional, national, global), by sector of activity (aquaculture, fisheries, agriculture) or by actors (individuals, communities, private sector, governments). They can be anticipatory strategies or reactive responses (Thompson and Adger, 2005) and should include: (i) management approaches and policies that build the livelihood asset base, reducing vulnerability to multiple stressors, including climate change; (ii) an understanding of current response mechanisms to climate variability and other shocks, in order to inform planned adaptation; (iii) a recognition of the opportunities that climate change could bring to the sector; (iv) adaptive strategies designed with a multi-sector perspective; and (v) recognition of fisheries' potential contribution to mitigation efforts. The last two aspects call for a greater inclusion of fisheries in ongoing climate policy discussions (Dulvy and Allison, 2009).

Reducing vulnerability to multiple stressors and helping build adaptive livelihoods is how institutions can foster socio-ecological systems that provide building blocks for the maintenance of livelihoods in the face of critical and pervasive threats, and resilient fisheries that can absorb disturbances and reorganize themselves following perturbation while still delivering benefits for poverty reduction. While some specific investment will be needed (i.e. risk reduction and transfer initiatives such as early warning systems, storm-shelters, managed retreat and insurance), adapting to climate change becomes a matter of addressing the fundamental problems of fisheries management and the underlying factors that cause vulnerability (Westlund *et al.*, 2007; Coulthard, 2008; Munday *et al.*, 2008; Pratchett *et al.*, 2008; Brander, 2010; Drinkwater *et al.*, 2010).

Understanding autonomous adaptation to past and current stresses, such as extreme events, can aid in designing measures that reduce the adverse impacts of future climate change and the implementation of inadequate measures (maladaptation). For instance current patterns of livelihood diversification in response to environmental changes, including geographical mobility, require specific policy and institutional changes that can inform planned adaptation. When analysing the adaptive responses of fishers to environmental change in the Inner Niger Delta in Mali, Morand *et al.* (submitted) stressed the degree to which these communities were highly sensitive to changes in the hydro-climatic conditions of the delta, but also that they were remarkably limited in the ways they could mitigate the impacts of these changes. For fish-dependent households that have adopted a diversified set of activities through involvement in farming, a close analysis reveals that the high seasonality and constraints characterizing their main activities (fishing and farming) does not offer any real possibility of switching between the activities. The situation is no better for specialized fishers that migrate to follow the fish and do not farm. The high density of the population in the delta has drastically reduced the possibility of finding new migration routes or new settlement sites within the delta. In sum, although migration and diversification are often presented in the literature as strategies adopted by households or individuals to reduce vulnerability, Morand and his colleagues demonstrate that, in the case of fish-dependent communities, these strategies alone will not be sufficient to help the local populations of the Inner Niger Delta to face the increasing constraints associated with a changing climate.

Negative impacts are extensively presented in the literature, while positive impacts of climate variability and change on the fisheries sector are seldom mentioned. The impacts of climate change will not be distributed equally. There will be relative 'winners' and 'losers', some communities may suffer significant losses due to physical damages or changes in fish distribution, while others will be less affected – or may even benefit by, for instance, positive changes in abundance of certain species. Successful identification of policies that enhance adaptation will occur only if the opportunities brought by climate change are identified.

Indirect impacts of climate change arising from adaptive strategies pursued by different sectors will require a more holistic planning perspective to ensure that adaptive strategies are designed with a multi-sector perspective to minimize net impacts. Pressure from other resources (e.g. water, agriculture coastal defence) might restrict the ability of the fishery sector to adapt to climate change in some cases, and enhance it in others (mangroves and reefs for coastal defence – enhance coastal fisheries; irrigation and flood control – disrupt inland fisheries).

Climate change will bring about new challenges to fisheries-based livelihoods in the coming decades. To tackle these, a diverse portfolio of responses is needed, where poverty and vulnerability reduction, fisheries governance and climate policies agenda are reconciled. The additional investment needed for local communities to adapt to climate change, if well targeted, can yield direct and ancillary benefits in the short and long term, resulting in positive returns on investment and 'win-win' situations.

Reducing people's vulnerability to HIV and AIDS

A synthesis of surveys conducted between 1992 and 2004 in ten low- or middle-income countries in Africa, Asia and Latin America for which data were available (Democratic Republic of Congo, Kenya, Uganda, Cambodia, Indonesia, Malaysia, Myanmar, Thailand, Brazil and Honduras) shows that, in all except one (Brazil), HIV prevalence in fishing communities is 4–14 times higher than the countries' national average prevalence rate for adults aged 15–49 (Kissling et al., 2005). These rates of HIV infection place fisherfolk among groups more usually identified as being at high risk; they are in fact higher than those for both truck drivers and the military in all countries (again except Brazil) for which comparative data are available. Because there are many fisherfolk compared with people in other sub-populations with high HIV prevalence, such as injecting drug users, the number of fisherfolk likely to be HIV positive is very large, making them a priority for support for prevention, treatment and care programmes for HIV and AIDS.

Many of the characteristics of small-scale fisherfolk in developing countries that make them vulnerable to HIV, such as mobility, lack of health care facilities, poverty and social exclusion, also make them likely to miss out on access to prevention, treatment and care, unless the scale of the problem is more widely known (Seeley and Allison, 2005).

Although there are a range of useful responses to high prevalence of HIV and incidence of AIDS in fishing communities (Box 12.1), there is still a need for greater

---

Box 12.1. Examples of actions in addressing HIV and AIDS in fishing communities (adapted from FAO, 2005).

Prevention
- workplace-based prevention measures in major seafood companies in Namibia (National AIDS Commission of Namibia);
- AIDS awareness-raising campaigns in Melanesia (Secretariat of the Pacific Community);
- behaviour change, e.g. through peer-to-peer education in Congo-Brazaville, Benin and Ghana (FAO/DFID Sustainable Fisheries Livelihoods Programme); and
- toolkit for HIV prevention among fishermen in Vietnam (Asian Development Bank and UNDP).

Care
- providing primary health services to mobile and migrant fishers in Tanzania and Congo (USAID, SIDA and National AIDS Commissions); and
- providing nutritional and positive living support for orphans and people living with HIV/AIDS in villages surrounding Lake Victoria, Uganda (The AIDS Support Agency – TASO).

Mitigation
- saving schemes developed for vulnerable women and girls in fishing communities in Congo (National AIDS Commission and FAO/DFID SFLP);
- training fishermen in alternative occupations to increase opportunities for livelihood diversification in Zambia (Médecins Sans Frontières);
- development of small-scale aquaculture for people living with HIV and AIDS in Malawi (WorldFish Center);
- junior farmer field and life schools for orphans and vulnerable children in fishing/farming communities in western Kenya (FAO and World Food Programme); and
- community-initiated safety nets – local fishing crew associations and beach management units donating a proportion of their daily catch to support orphans' education (Lakes George and Edward, Uganda; DFID/CARE Integrated Lake Management project).

visibility of the issues in both AIDS response planning and fisheries-sector investment strategies. Such visibility would ensure that responses are coordinated to include both prevention (e.g. behaviour change campaigns, improved availability of condoms and treatment for other sexually transmitted infections that raise probabilities of viral transmission) and treatment and care. Most importantly, such health-sector approaches should be integrated with attention to underlying causes of high HIV prevalence and high levels of morbidity and mortality from AIDS-related illness, such as poverty, social exclusion and poor health and nutrition, and to mitigation of the impacts of AIDS on affected households and communities. A major step forward in identifying and testing such integrated approaches was the investment of the Swedish International Development Agency (SIDA) in action-research to identify effective integrated solutions, scaled out to encompass many of Africa's most important HIV-affected fishing areas (WorldFish and FAO, 2010).

In some fishing communities, client–patron relationships can be responsible for the occurrence of transactional sex arrangements ('fish for sex'), whereby female fish-traders engage in sexual relationships with male fishers to secure a supply of fish (Béné and Merten, 2008). These researchers demonstrated that the economic impoverishment discourse, which is often put forward to explain fish-for-sex transactions, is in fact too simplistic to capture the complexity of the phenomenon. Béné and Merten's (2008) evidence suggests that women are neither entirely victims nor in total control of these relationships. Yet, the prevalence of HIV/AIDS makes fishers, traders and processors highly vulnerable to disease, which has a wide societal and economic impact (Allison and Seeley, 2004). Transactional sex is not, however, confined to the fish trade and is found in a variety of other contexts where women rely on men for access to resources and financial support (Chatterji et al., 2004).

These risk behaviours need to be changed by addressing both their proximate factors – men's behaviour – and their root causes – poverty, vulnerability and the 'risk environment'. Such efforts require novel partnerships between donors, fishery and health agencies, and within and between communities themselves.

The above examples illustrate how issues outside the normal domain of fishery managers impact on the livelihood and well-being of fishing-dependent communities (see also Hall, Chapter 8, this volume). These examples help highlight the fact that in situations where vulnerability to risks such as HIV and climate change are high, fishers may not perceive vulnerability related to fish stocks to be such a high priority, simply because the livelihood risks associated with fish stock collapse are minor compared with the risks of contracting HIV or the perceived risks of future damage from increased frequency of climate change-associated extreme events (Allison, 2005). More globally, recent empirical research suggests that, in some parts of the developing world, the main sources of vulnerability of fishing communities are rarely perceived to be directly related to the status of the resources. Instead, it is often a question of access to the most basic human needs such as food security, access to health or drinking water (Mills et al., 2009). Failure to address these and other sources of vulnerability in fishing communities thus undermines efforts to engage fisherfolk in participatory governance, including fisheries co-management.

Improving Income and Assets

A central element of most planned poverty reduction efforts is to improve the incomes and the asset base of the poor so that they are able to take advantage of a wider range of income-generating opportunities, or to cope with shocks. Poverty reduction in fishing-dependent people has tended to focus on ways to improve fishing incomes and access to fishing-related assets (including property rights related to fisheries). This sectoral focus is problematic in the context of widespread concern with biological overfishing, economic inefficiency and increasing conflict over access to resources, and more recent development advice has emphasized the need to raise incomes and strengthen assets in ways that do not necessarily put additional pressure on heavily exploited fish stocks.

Improved fishing incomes can be achieved, for example, by increasing fishing capacity, reducing fishing costs or increasing fishing efficiency, adding value to fishery products through local processing or improved handling and preservation. The logic of raising income and productivity is derived from comparison with agriculture, where such a goal can be pursued sustainably in many cases but, in the situation where access to fisheries is open to many, such an approach may lead to accelerated resource depletion unless accompanied by effective enforcement of limits to the overall fishing effort. With raised efficiency or effort of individuals as a means of increasing fishing income, there is likely to be a need to reduce the number of people able to access fishing as a source of income, in order to maintain overall exploitation rates at sustainable levels (Charles, 2001). Where the fisheries sector is not intricately embedded in the wider local economy, such a strategy makes economic sense from both a sectoral and regional or national perspective. It is less obviously beneficial to local and national poverty reduction if fishing activities and income are used to support other rural enterprises and to secure livelihoods in uncertain and highly seasonal production contexts (Allison and Ellis, 2001).

In the absence of an ability to control access to resources, or if there a perceived need to ensure access to as many people as possible, the wisdom of raising incomes from fishing can thus be called into question. Ian Smith (Smith, 1979) was among the first to point this out, arguing that sustainable development in small-scale fisheries needed to raise the opportunity costs of fishing, by raising opportunity income of alternative livelihoods. This would then result in fluid entry and exit from fishing according to the availability of better opportunities elsewhere. Thus, promoting alternative livelihood options would result in a diversified local or household economy in dynamic equilibrium, within which access to fishing opportunities was an integral part. Livelihood diversification has since become a stated policy goal of many fisheries management and development programmes – for example, the DFID-funded Sustainable Fisheries Livelihoods Programme in West and Central Africa, 2000–2006 (Westlund *et al.*, 2007) and the Spanish government-funded Asia-Pacific Coastal Livelihood Diversification Programme, implemented by FAO, which started in 2009.

Diversification refers here to the continual adaptive process whereby households add new activities, maintain existing ones and drop others, thereby maintaining diverse and constantly changing livelihood portfolios. It does not refer to individuals switching full-time occupations, nor to the diversity of sub-sector enterprise types in rural areas (Ellis and Allison, 2005). Nevertheless, when contemporary fishery development programmes talk about diversification, there is a strong imperative to use diversification out of fishing as a means to effect 'capacity reduction' (Ward *et al.*, 2004; see also Pomeroy *et al.*, Chapter 7, this volume).

Household incomes can be improved by diversification, if options that provide higher returns to investment (of labour and capital), or reduced risks, and improved well-being can be identified. The problem with many suggested diversification options is that they either increase risks, require assets not held by some fishing people (such as access to land) or deliver returns to investment inferior to fishing and fish-trading income. Moreover, many suggested diversification options are 'boutique solutions' accessible to only a few (e.g. handicrafts, ecotourism) or require reconfiguration of the household unit (e.g. long-distance labour migration). Among people who enjoy the material and non-material rewards of fishing (Pollnac *et al.*, 2001), diversification out of fishing may not be an attractive option, for any one of the above reasons.

Diversification is most likely to be attractive and sustainable where it is accompanied by appropriate asset strengthening – both through improved asset provision (e.g. through infrastructure development) and improving the conditions of access to assets (e.g. through community-based natural resource management, or means of ensuring that the poor and marginalized get equitable access to ways of strengthening their human and financial capital).

The difficult balance that development projects in fishing communities have to achieve is in how to foster improved livelihoods, while simultaneously sustaining the natural resource base. Once again, prescriptive approaches, technical fixes and institutional magic bullets are unlikely to succeed. Ground-up, contextualized, integrative approaches that enable people to choose successful and sustainable options are the ideal, although difficult to achieve in practice. Some examples of promising initiatives and approaches undertaken as part of the West and Central Africa Sustainable Fisheries Livelihoods Programme are synthesized in Table 12.2.

Table 12.2. Interventions that can support strengthening of the 'asset platform' upon which livelihoods are constructed, with examples from SFLP (Sustainable Fisheries Livelihoods Programme) community projects. Reproduced from Allison and Horemans (2006).

Asset category	Possible areas of intervention	Examples of SFLP community projects
Human	Training, education, awareness-raising, improved food security and access to better diet, improved access to health and education facilities	All 60 community projects had a capacity-building component in areas such as: adult literacy (Niger, Sao Tomé and Principe), household resources management (Mali), livelihoods diversification (Cameroon, Senegal), postharvest techniques (Sao Tomé and Principe), fishing techniques (Burkina Faso) and addressing HIV/AIDS (Congo, Benin)
Natural	Assisting communities to use resources more sustainably; improving postharvest use; improving access to sectoral service provision; supporting rehabilitation of degraded environments	Reliable sources of fuel wood to female fish processors through establishment of community woodlots (Ghana), fisheries enhancement initiatives (Niger, Burkina Faso), fisherfolk participation in surveillance (Guinea, Nigeria) and fisheries co-management (eight countries)
Financial	Social organization can improve access to credit and savings mechanisms; awareness-raising in formal institutions can increase access to credit and insurance schemes; improving natural assets can improve financial flows; business training can improve financial management	Improving access of fisheries communities to existing savings and credit schemes through training (Benin, The Gambia, Niger, Nigeria), small business management and saving schemes for female fish traders (Sao Tomé and Principe)
Physical	Helping to improve access to infrastructure; providing access to information on improved technologies; building capacity in communities to improve or develop their own physical assets	The development of human and social assets in community projects is intended to facilitate access to and management of physical assets; participation of fisherfolk organizations in the management of fisheries infrastructure (Benin, Cape Verde); well construction (Burkina Faso); restoration of community fish ponds (Niger)
Social	Strengthening community organization skills; building on existing institutions; raising awareness of social structures and functions; building trust; providing leadership training; encouraging inclusion of marginalized groups; supporting networks	There is a social capital-strengthening component in each project; some are more specifically focused on the strengthening of fisherfolk associations (Benin, Gabon, Mauritania); particular attention is paid to gender aspects (Sao Tomé and Principe) and the inclusion of migrant fishers (Congo) in the organizations

The overall principle is to build the key assets that people need in order to secure their lives and a means to a livelihood, and to find their own routes out of poverty. These are likely to include: (i) a secure natural resource asset base (particularly a well-managed fishery); (ii) financial capital (loans, insurance, savings schemes); (iii) human capital (access to health and education and skills-building opportunities); (iv) social capital (inclusion in community development initiatives); and (v) physical capital (a decent place to live and access to infrastructure such as roads, clean water supplies, safe harbours, information and communications technologies).

The key point, once again, is that such interventions should not be sectoral in nature, and should support both the development policies of local and national governments, in their wider responsibilities for poverty reduction, and the aspirations and needs of fishing-dependent communities. Development is rarely, if ever, an uncontested and apolitical process, and such solutions as are proposed here will not be implemented simply through good technical assistance and bureaucratic process. They will require deep political engagement and respect for local and national political process – however imperfect. Direct technical interventions can provide only limited and temporary solutions to development problems, so asset-strengthening and diversification projects and programmes can only be part of the development process. This should be recognized in both their conception and subsequent assessment.

Sequencing Development Interventions

By gaining an understanding of the livelihood strategies, institutional processes, power dynamics and social relations in fishing communities, and the risks faced by fisherfolk in both fishing and non-fishing activities, it should be possible to identify a potential set of actions (whether undertaken by fisherfolk themselves, in partnership with other stakeholders or with development actors) to address some of the multiple dimensions of poverty and to build on identified strengths and opportunities. What is missing in this approach, however, is a way of prioritizing and sequencing such interventions (Fig. 12.2). The previous sections of this chapter gave a small subset of examples of the many worthy and effective things that can be done. A rigorous *ex ante* impact assessment of returns to investment in different activities of this type would ideally be undertaken but, more commonly, some

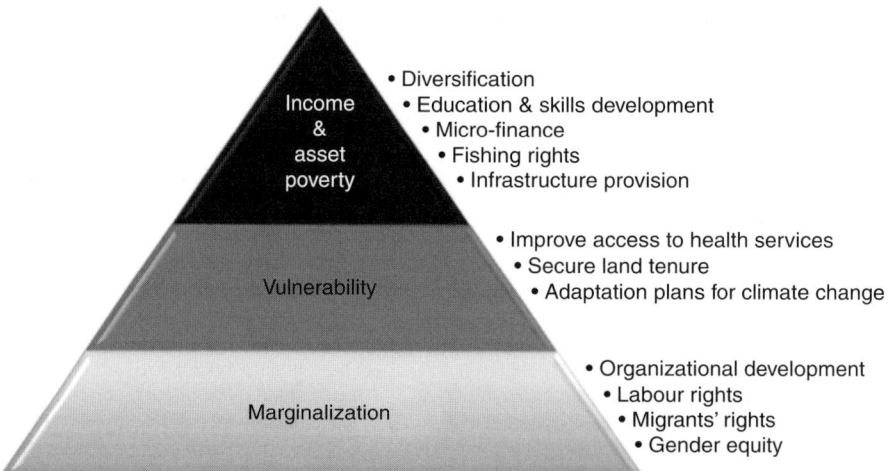

Fig. 12.2. A hierarchy of possible development actions to reduce poverty without necessarily increasing resource dependence. There will be many feedback loops and differences in staging depending on context and opportunity.

form of participatory ranking or prioritization exercise can be undertaken at the local level to decide what is feasible, and this will bring the greatest benefits (see Evans and Andrew, Chapter 3, this volume, for a review).

At the broadest level, interventions to strengthen governance need to foster the human security necessary for fisherfolk to feel they will be in a position to benefit in future. This will first require confidence that attempts to improve their lives will not be undermined by more powerful interests. Second, people will need to feel less vulnerable, both in terms of their lives and livelihoods. If they are expected to forego present rewards from (over)fishing and invest in resource stewardship, they must feel that they (or their children) will be around to reap the rewards of their restraint and acceptance of restrictions on their fishing activities. This confidence in the future is unlikely to apply in a community where people live in places where they can't send their children to school, or where they can't even secure the property right of the plot of land where they have built their houses, or where one in three people is HIV positive, as is the case in many fishing communities around Lake Victoria (Drimie et al., 2009).

In sum, any attempt to bring material benefits to people who are already disadvantaged or marginalized from the rest of the society is likely to fail – and it is easier to marginalize and exclude people who are poor and vulnerable, trapping them in a classic 'vicious cycle'. Indeed, 'elite capture' of benefits from community management – a result of failure of entitlements and lack of equitable and effective institutions for fair distribution of benefits – is a common complaint from community-based management of fisheries and forestry (e.g. Béné and Neiland, 2004; Resosudarmo, 2004; Wardell and Lund, 2006; Béné et al., 2010).

Once basic human rights are secure, and relevant local-level judicial and political systems reasonably equitable, the next priority should be accorded to reducing vulnerability to shocks and adverse trends. Reducing vulnerability by reducing risk exposure or sensitivity, or increasing adaptive capacity, allows people more time to accumulate assets.

Finally, where people's lives are sufficiently secure to encourage confidence in the future and longer time horizons for saving and investment, direct support to enable people to build their assets and improve their incomes can be transformative of the livelihoods of the poor – through support for improved access to infrastructure, markets and finance, strengthened rights of access to natural resources, and availability of new livelihood activities or improved production technologies.

In practice these actions may overlap and, within each category, the sequencing is not always as neat as implied above (e.g. investing in improved health care to build a human capital asset can reduce overall vulnerability) – the main point is that investing out of sequence risks, at best, disappointing returns on investment due to people's inability to capitalize on new opportunities due to social exclusion or failure to maintain and build their assets due to undiminished vulnerability. At worst, elite capture of development benefits will deepen existing inequality. Of course the ideal is to invest simultaneously in all processes supporting poverty reduction, but capacity to do so will usually be a major practical limitation. It is also important to remember that governmental or external support and intervention may only be required to resolve one or two key problems – this will then allow people to bring their institutions, networks of influence and their own resources to bear in addressing the remaining ones.

Sustaining Livelihoods and Communities as a Way to Sustain Fisheries

The variety and complexity of drivers and incentives that affect fisher livelihood and well-being at the local level are unlikely to be resolved by a homogeneous national policy response. As other commentators have noted (e.g. Carney, 1998; Allison and Ellis, 2001; Andrew et al., 2007), effective interventions require a more nuanced (local) knowledge of the particularities of poverty facing people who fish – or who are dependent on fishing for all or part of their livelihood. The nature of the risks and opportunities people face is a product of the ecological attributes of the fishery, markets, social dynamics that influence access to fishing and associated assets (e.g. boats, land and infrastructure), and their dependence on fishing. Just as the vulnerabilities and nature of poverty vary enormously, so too must the policies designed to meet the challenge. For example, policies to enhance fisheries management and reduce poverty in Cambodia's Tonle Sap Lake fisheries will be very different from those designed to improve the lives of households dependent on reef fisheries in the Solomon Islands (Andrew et al., 2007).

Within the diversity of policy responses required, however, there is a widespread consensus that the solution to fisheries decline in the small-scale sector lies in replacing or supplementing currently ineffective governance systems, where states manage inshore coastal and inland fisheries, with those based on better defined fishing rights, usually devolved to groups or communities. Granting fishing rights to communities is thought to make management more effective and efficient, as resource users are presumed to have a direct incentive to manage resources optimally so that they can derive the maximum future benefit from their property rights (Berkes, 1995; Pomeroy, 2001; Béné and Neiland, 2006; Pomeroy and Rivera-Guieb, 2006).

Strengthening of fishing rights can provide a route out of poverty if fisherfolk's poverty and vulnerability is mainly related to resource degradation brought about by insecure resource access and inadequate fishery management. Securing the right to fish (and to exclude too many others from fishing) doesn't, however, inure a fishing family against high incidence of malaria and HIV/AIDS, rent-seeking officials, theft of fishing gear, unsafe working conditions or forced eviction from their house. Yet this is the 'vulnerability context' faced by many small-scale fishers (Mills et al., 2009). They may also lack the power, education and cohesive social institutions to be aware of their rights, ability to self-organize and articulate their demands, negotiate with government officials and to carry out their responsibilities. In short, they are in need of social development in order to participate effectively as partners with government in fisheries management.

In the circumstances described above, the risk of resource degradation or stock collapse may be perceived as low or distant by many fishers in comparison with the exposure of their livelihood systems to the risks of ill health or death (particularly from malaria, HIV/AIDS, water-borne diseases and drowning and accidents), theft or loss of fishing gear, or lack of secure access to alternative productive assets, such as land.

The small-scale fisheries sector is also vulnerable to external factors beyond its control. These include floodplain modification and damming of rivers (e.g. Friend et al., 2009), displacement by aquaculture (e.g. Marshall, 2001), tourism and other coastal development, and pollution (see Hall, Chapter 8, this volume). Local systems allocating fishing rights can confront and prevent some of these threats, but not all – notably pollution and upstream modifications in river basins. Where fishing interests are historically overridden or unrepresented by competing claims, then people have no incentive to invest in managing their local fishery resources to optimize future yields.

The overall outcome of the set of circumstances described above is that, because of their continuing vulnerability and marginalization, many fishing people currently lack both the incentive and capacity to claim and defend systems of access rights that aim to conserve fish stocks for their exclusive use. Addressing small-scale fishing people's vulnerability and social exclusion should therefore be an important component of any programme that aims to define and strengthen rights of access as a means to improve the contribution of fisheries to poverty reduction and to rebuild fisheries to contribute to wealth creation and economic growth.

In order to interest and enable fisherfolk to participate in resource management, fisheries development programmes will need to address the factors that most immediately and directly threaten the sustainability of fisherfolk livelihoods. A strategy to bring together responsible fisheries with social development to strengthen capacity and incentives of fisherfolk to invest in defending their fishing rights should consider the following principles:

- Address over-exploitation that threatens resource sustainability and the flow of benefits from fisheries to the wider economy as the priority objective of a shift towards rights-based fishing. As well as defining rights to fish, the rights of present and future generations to benefit from the resources should be included. Building the value of the resources should be an explicit objective of fisheries management in the small-scale subsector.
- Integrate broader human rights of fishers to an adequate livelihood as part of an expanded rights-based approach to fisheries management. This means including poverty-reduction criteria as a key component of decisions over equitable allocation of rights, including decisions over inclusion and exclusion, and the protection of small-scale fishworkers' access to resources and markets. It also means addressing deficiencies in fishing people's rights of equitable access to health care, education, justice and the rule of law.
- Support empowerment of fishing communities, both through their social inclusion and building their capabilities. Transition to rights-based fishing requires relationships between fishing rights-holders and duty-bearers (such as governments) to be transparent and based on mutual trust and accountability of the duty-bearers. Social inclusion of fishing communities, together with improved fishery governance, would also help address many of the conditions that currently link the fishing sector with illegal activities – related to both fishing and other maritime and trans-national crime.
- Integrate responsible fisheries policies with wider poverty reduction policies, such as Poverty Reduction Strategy Papers, in countries where fisheries are economically important. This is a necessary condition to achieve inter-sectoral policy coherence and maximize the contribution of fisheries to meeting poverty targets such as the Millennium Development Goals. It is also important for ensuring that fisheries agencies receive an appropriate allocation of central and local government budgets.

If these principles are used to guide attempts to improve the governance of fisheries and the well-being of fishing communities, then there is a good chance that we will continue to enjoy the benefits that small-scale fisheries contribute to human development.

References

Ahmed, N., Allison, E.H. and Muir, J.F. (2008) Using the sustainable livelihoods framework to identify constraints and opportunities to the development of freshwater prawn farming in southwest Bangladesh. *Journal of the World Aquaculture Society* 39, 598–611.

Allison, E.H. (2005) The fisheries sector, livelihoods and poverty reduction in Eastern and Southern Africa. In: Ellis, F. and Freeman, H.A. (eds) *Rural Livelihoods and Poverty Reduction Policies*. Routledge, London, pp. 256–273.

Allison, E.H. and Ellis, F. (2001) The livelihoods approach and management of small-scale fisheries. *Marine Policy* 25, 377–388.

Allison, E.H. and Horemans, B. (2006) Putting the principles of the Sustainable Livelihoods Approach into fisheries policy and practice. *Marine Policy* 30, 757–766.

Allison, E.H. and Kelling, I. (2009) Fishy crimes: the societal costs of poorly governed marine fisheries. *Third Annual Convention of the Consortium of Non-Traditional Security Studies in Asia*, Singapore, 3–4 November. Available at: http://www.rsis-ntsasia.org/activities/conventions/2009-singapore/Edward%20Allison.pdf (accessed 28 July 2010).

Allison, E.H. and Seeley, J.A. (2004) HIV and AIDS among fisherfolk: a threat to 'responsible fisheries'? *Fish and Fisheries* 5, 215–234.

Allison, E.H., Adger, W.N., Badjeck, M.C., Brown, K., Conway, D. and Dulvy, N.K. (2005) *Effects of Climate Change on the Sustainability of Capture and Enhancement Fisheries Important to the Poor: Analysis of the Vulnerability and Adaptability of Fisherfolk Living in Poverty*. Fisheries Management Science Programme Project No. R4778J, Marine Resources Assessment Group, London.

Allison, E.H., Horemans, B. and Béné, C. (2006) Vulnerability reduction and social inclusion: strategies for reducing poverty among small-scale fisherfolk. Paper presented at the *Wetlands, Water and Livelihoods* Workshops, Wetland International, St. Lucia, South Africa, 30 January–2 February.

Allison, E.H., Kurien, J., Pomeroy, R., Ratner, B.D., Willmann, R. and Asgard, B. (submitted) Rights-based-fishing: should it include human rights? *Fish and Fisheries*.

Andrew, N.L., Béné, C., Hall, S.J., Allison, E.H., Heck, S. and Ratner, B.D. (2007) Diagnosis and management of small-scale fisheries in developing countries. *Fish and Fisheries* 8, 227–240.

Atkinson, T. (1998) *Exclusion, Employment and Opportunity*. Centre for Analysis of Social Exclusion Working Paper No.4, London School of Economics, London.

Béné, C. (2003) When fishery rhymes with poverty, a first step beyond the old paradigm on poverty in small-scale fisheries. *World Development* 31, 949–975.

Béné, C. and Merten, S. (2008) Women and fish-for-sex: transactional sex, HIV/AIDS and gender in African fisheries. *World Development* 36, 875–899.

Béné, C. and Neiland, A.E. (2004) Empowerment reform, yes . . . but empowerment of whom? Fisheries decentralization reforms in developing countries: a critical assessment with specific reference to poverty reduction. *Aquatic Resources, Development and Culture* 1, 35–49.

Béné, C. and Neiland, A.E. (2006) *From Participation to Governance: a Critical Review of the Concepts of Governance, Co-management, and Participation and their Implementation in Small-scale Inland Fisheries in Developing Countries*. WorldFish Center Studies and Reviews 29, The WorldFish Center, Penang, Malaysia and the CGIAR Challenge Program on Water and Food, Colombo, Sri Lanka.

Béné, C., Macfadyen, G. and Allison, E.H. (2007) Increasing the contribution of small-scale fisheries to poverty alleviation and food security. FAO Fisheries Technical paper No. 481, FAO, Rome, Italy. Available at: www.fao.org/docrep/009/a0965e/a0965e00.htm (accessed 3 February 2010).

Béné, C., Steel, E., Kambala, L.B. and Gordon, A. (2009a) Fish as the 'bank in the water': evidence from chronic-poor communities in Congo. *Food Policy* 34, 104–118.

Béné, C., Belal, E., Baba, M.O., Ovie, S., Raji, A., Malasha, I., Njaya, F., Na Andi, M., Russell, A. and Neiland, A. (2009b) Power struggle, dispute and alliance over local resources: analyzing 'democratic' decentralization of natural resource through the lenses of African inland fisheries. *World Development* 37, 1935–1950.

Béné, C., Hersoug, B. and Allison, E.H. (2010) Not by rent alone: analysing the pro-poor functions of small-scale fisheries in developing countries. *Development Policy Review* 28, 325–358.

Berkes, F. (1995) Community-based management and co-management as a tool for empowerment. In: Singh, N. and Titi, V. (eds) *Empowerment, Towards Sustainable Development*. Zed Books, London, pp.138–146.

Brand, F.S. and Jax, K. (2007) Focusing the meaning of resilience: resilience as a descriptive concept and a boundary object. *Ecology and Society* 12(1), art. 23. Available at: http://www.ecologyandsociety.org/vol12/iss1/art23/ (accessed 3 February 2010).

Brander, K. (2010) Impacts of climate change on fisheries. *Journal of Marine Systems* 79, 389–402.

Carney, D. (1998) *Approaches to Sustainable Livelihoods for the Rural Poor*. Poverty Briefing No.2, Overseas Development Institute, London.

Chambers, R. and Conway, G.R. (1992) *Sustainable Rural Livelihoods: Practical Concepts for the 21st Century*. Discussion Paper 296, Institute of Development Studies, Brighton, UK.

Charles, A.T. (2001) *Sustainable Fishery Systems*. Blackwell Science Ltd, London.

Chatterji, M., Murray, N., London, D. and Anglewicz, P. (2004) *The Factors Influencing Transactional Sex Among Young Men and Women in 12 Sub-Saharan African Countries*. POLICY programme funded by the US Agency for International Development, Washington, DC.

Choo, P.S., Nowak, B.S., Kusakabe, K. and Williams, M.J. (eds) (2008) Gender and fisheries. *Development* 51(2), 176–179.

Cole, C.F. (1996) Can we resolve uncertainty in marine fisheries management? In: Lemons, J. (ed.) *Scientific Uncertainty and Environmental Problem Solving*. Blackwell Science, Oxford, UK, pp. 233–263.

Coulthard, S. (2008) Adapting to environmental change in artisanal fisheries – insights from a South Indian lagoon. *Global Environmental Change* 18, 479–489.

Cunningham, S., Neiland, A.E., Arbuckle, M.A. and Bostock, T. (2009) Wealth-based fisheries management: using fisheries wealth to orchestrate sound fisheries policy in practice. *Marine Resource Economics* 24, 271–287.

DFID (2005) *Reducing Poverty by Tackling Social Exclusion*. Department for International Development, London. Available at: http://www.dfid.gov.uk/pubs/files/social-exclusion.pdf (accessed November 2006).

Drimie, S., Weinand, J., Gillespie, S. and Wagah, M. (2009) *HIV and Mobility in the Lake Victoria Basin Agricultural Sector: a Literature Review*. IFPRI Discussion Paper 00905, IFPRI, Washington, DC. Available at: http://www.ifpri.org/sites/default/files/publications/ifpridp00905.pdf (accessed 29 July 2010).

Drinkwater, K.F., Beaugrand, G., Kaeriyama, M., Kim, S., Ottersen, G., Perry, R.I., Portner, H.O., Polovina, J.J. and Takasuka, A. (2010) On the processes linking climate to ecosystem changes. *Journal of Marine Systems* 79, 374–388.

Dulvy, N. and Allison, E. (2009) A place at the table? *Nature Reports Climate Change* 0906, 68–70.

Dulvy, N.K., Rogers, S.I., Jennings, S., Stelzenmüller, V., Dye, S.R. and Skjoldal, H.R. (2008) Climate change and deepening of the North Sea fish assemblage: a biotic indicator of warming seas. *Journal of Applied Ecology* 45, 1029–1039.

Ellis, F. (1998) Household strategies and rural livelihood diversification. *Journal of Development Studies* 35, 1–38.

Ellis, F. (2000) *Rural Livelihoods and Diversity in Developing Countries*. Oxford University Press, Oxford, UK.

Ellis, F. and Allison, E.H. (2005) Linking livelihood diversification to natural resources in a poverty reduction context. Livelihoods Support Programme Briefing Notes, FAO, Rome, Italy.

FAO (2005) *Reducing Fisherfolk's Vulnerability Leads to Responsible Fishing. New Directions in Fisheries*. A series of policy briefs on development issues No. 1, Sustainable Fisheries Livelihoods Programme, FAO, Rome, Italy.

FAO (2007) *Social Issues in Small-scale Fisheries: How a Human Rights Perspective can Contribute to Responsible Fisheries*. Presentation to FAO Committee on Fisheries No. 27, Rome, Italy.

FAO (2008) *Country Profiles*. Available at: http://www.fao.org/fishery/countryprofiles/search (accessed 6 May 2008).

FAO (2009) *The State of World Fisheries and Aquaculture – 2008 (SOFIA)*. FAO, Rome, Italy.

Flaaten, O., Salvanes, A.G.V., Schweder, T. and Ulltang, O. (1998) Fisheries management under uncertainty – an overview. *Fisheries Research* 37, 1–6.

Friend, R.M., Arthur, R.I. and Keskinen, M. (2009) Songs of the doomed: the continuing neglect of fisheries in hydropower debate in the Mekong. In: Molle, F., Foran, T. and Käkönen, M. (eds) *Contested Waterscapes in the Mekong Region: Hydropower, Livelihoods and Governance*. Earthscan, London, pp. 23–54.

Glantz, M.H. and Thompson, J.D. (eds) (1981) *Resource Management and Environmental Uncertainty: Lessons from Coastal Upwelling Fisheries*. John Wiley and Sons, Chichester, UK and New York.

Gordon, D.V. and Munro, G.R. (eds) (1996) *Fisheries and Uncertainty. A Precautionary Approach to Resource Management*. University of Calgary Press, Calgary, Canada.

ICSF (International Committee in Support of Fishworkers) (2006) *Report of Workshop on Emerging Concerns of Fishing Communities: Issues of Labour, Trade, Gender, Disaster Preparedness Biodiversity and Responsible Fisheries*, Fortaleza, Brazil, 4–6 July. International Committee in Support of Fishworkers, Chennai, India. Available at: http://www.icsf.net/icsf2006/uploads/publications/reports/pdf/english/issue_7/ALL.pdf (accessed 5 August 2008).

ILO (International Labour Organization) (undated) *About Child Labour*. Available at: http://www.ilo.org/ipec/facts/lang-en/index.htm (accessed 27 July 2010).

ILO-IPEC (undated) *International Programme for the Elimination of Child Labour*. Available at: http://www.ilo.org/ipec/programme/lang–en/index.htm (accessed 28 July 2010).

IPCC (Intergovernmental Panel on Climate Change) (2001) *Climate Change 2001: Impacts, Adaptation and Vulnerability, Contribution of Working Group II to the Third Assessment Report of the Intergovernmental Panel on Climate Change*. Cambridge University Press, Cambridge, UK.

IPCC (undated) *Working Group II: Impacts, Adaptation and Vulnerability*. Available at: http://www.ipcc-wg2.org/ (accessed 29 July 2010).

Jentoft, S. (2000) Legitimacy and disappointment in fisheries management. *Marine Policy* 24, 141–148.

Kabeer, N. (2000) Social exclusion, poverty and discrimination: towards an analytical framework. *IDS Bulletin* 31, 83–97.

Kissling, E., Allison, E.H., Seeley, J.A., Russell, S., Bachmann, M., Musgrave, S.D. and Heck, S. (2005) Fisherfolk are among groups most at risk of HIV: cross-country analysis of prevalence and numbers infected. *AIDS* 19, 1939–1946.

Lauck, T., Clark, C.W., Mangel, M., Gordon, R. and Munro, G.R. (1998) Implementing the precautionary principle in fisheries management through marine reserves. *Ecological Applications* 8, S72–S78.

MacKenzie, B.R. and Schiedek, D. (2007) Daily ocean monitoring since the 1860s shows record warming of northern European seas. *Global Change Biology* 13, 1335–1347.

Marshall, J. (2001) Landlords, leaseholders, and sweat equity: changing property regimes in aquaculture. *Marine Policy* 25, 335–352.

Miller, K.A. and Fluharty, D.L. (1992) El Niño and variability in the north-eastern Pacific salmon fishery: implications for coping with climate change. In: Glantz, M.H. (ed.) *Climate Variability, Climate Change and Fisheries*. Cambridge University Press, Cambridge, UK, pp. 49–88.

Mills, D., Béné, C., Ovie, S., Tafida, A., Sinaba, F., Kodio, A., Russell, A., Andrew, N.L., Morand, P. and Lemoalle, J. (2009) Vulnerability in African small-scale fishing communities. *Journal of International Development* 26. Available at: http://dx.doi.org/10.1002/jid.1638 (accessed 31 May 2010).

Morand, P., Kodio, A., Béné, C., Sinaba, F., Andrew, N. and Lemoalle, J. (submitted) Vulnerability and adaptation of African rural populations to climate change: the case of fishers' communities in the Central Delta of Niger. *Climate Change*.

Munday, P.L., Jones, G.P., Pratchett, M.S. and Williams, A.J. (2008) Climate change and the future for coral reef fishes. *Fish and Fisheries* 9, 261–285.

Okali, C. and Holvoet, K. (2007) *Negotiating Gender Changes within Fisheries Development*. Sustainable Fisheries Livelihoods Programme, FAO, Rome, Italy.

O'Riordan, B. (2006) Child labour – growing pains. *Samudra Report* 44, 8–13.

Ostrom, E. (2007) A diagnostic approach for going beyond panaceas. *Proceedings of the National Academy of Sciences* 104, 15181–15187.

Pauly, D., Christensen, V., Dalsgaard, J., Froese, R. and Torres, F. (1998) Fishing down marine food webs. *Science* 279, 860–863.

Pollnac, R.B., Pomeroy, R.S. and Harkes, I. (2001) Fishery policy and job satisfaction in three Southeast Asian fisheries. *Ocean and Coastal Management* 44, 531–544.

Pomeroy, R. (2001) Devolution and fisheries co-management. In: Meinzen-Dick, R., Knox, A., and Di Gregorio, M. (eds) *Collective Action, Property Rights and Devolution of Natural Resource Management: Exchange of Knowledge and Implications for Policy*. Deutsche Stiftung für Internationale Entwicklung/Zentralstelle für Ernährung und Landwirtschaft (DSE/ZEL), Feldafing, Germany.

Pomeroy, R.S. and Rivera-Guieb, R. (2006) *Fishery Co-management: a Practical Handbook*. CABI Publishing, Wallingford, UK and International Development Research Centre, Ottawa, Canada.

Pomeroy, R., Ratner, B., Hall, S., Pimoljinda, J. and Vivekanandan, V. (2006) Coping with disaster: rehabilitating coastal livelihoods and communities. *Marine Policy* 30, 786–793.

Pratchett, M., Munday, P., Wilson, S., Graham, N., Cinner, J. and Bellwood, D. (2008) Effects of climate-induced coral bleaching on coral-reef fishes – ecological and economic consequences. *Oceanography and Marine Biology: an Annual Review* 46, 251–296.

Resosudarmo, I.A. (2004) Closer to people and trees: will decentralization work for the people and the forests of Indonesia? *European Journal of Development Research* 16, 110–132.

Rodrik, D. (2006) Goodbye Washington consensus, hello Washington confusion? A review of the World Bank's economic growth in the 1990s: learning from a decade of reform. *Journal of Economic Literature* 44, 973–987.

Rothschild, B.J., Chen, C. and Lough, R.G. (2005) Managing fish stocks under climate uncertainty. *ICES Journal of Marine Science* 62, 1531–1541.

Seeley, J.A. and Allison, E.H. (2005) HIV and AIDS in fishing communities: challenges in delivering antiretroviral therapies to vulnerable groups. *AIDS Care* 17, 688–697.

Sharma, C. (2003) The impact of fisheries development and globalization processes on women of fishing communities in the Asian region. *APRN Journal* 8, 1–12.

Silp, S. (2007) *US and Thailand Spar over Child Labor in Shrimp Industry*. Available at: http://www.irrawaddy.org/article.php?art_id=7693 (accessed 5 July 2008).

Smith, I.R. (1979) *A Research Framework for Traditional Fisheries*. ICLARM studies and reviews No. 2, International Center for Living Aquatic Resources Management, Manila, Philippines.

Solidarity Center (2008) *The Degradation of Work: the True Cost of Shrimp*. Solidarity Center, Washington, DC.

Thompson, E.L. and Adger, N.W. (2005) Defining response capacity to enhance climate change policy. *Environmental Science and Policy* 8, 562–571.

Thorpe, A., Andrew, N.L. and Allison, E.H. (2007) Fisheries and poverty reduction. *CAB Reviews: Perspectives in Agriculture, Veterinary Science, Nutrition and Natural Resources* 2, 1–12.

UN Committee on Social, Economic and Cultural Rights (2001) *Poverty and the International Covenant on Economic, Social and Cultural Rights*. Available at: http://www.unhchr.ch/tbs/doc.nsf/%28Symbol%29/E.C.12.2001.10.En (accessed 28 July 2010).

Walker, B., Holling, C.S., Carpenter, S.R. and Kinzig, A. (2004) Resilience, adaptability and transformability in social-ecological systems. *Ecology and Society* 9(2), art. 5. Available at: http://www.ecologyandsociety.org/vol9/iss2/art5/ (accessed 3 February 2010).

Walker, B.D., Sayer, J., Andrew, N.L. and Campbell, B.D. (2010) Resilience in practice: challenges and opportunities for natural resource management in the developing world. *Crop Science* 50, 10–19.

Ward, J.M., Kirkley, J.E., Metzner, R. and Pascoe, S. (2004) Measuring and assessing capacity in fisheries: basic concepts and management options. FAO Fisheries Technical Paper No. 433/1, FAO, Rome, Italy.

Wardell, A. and Lund, C. (2006) Governing access to forest in Northern Ghana: micro-politics and the rents of non-enforcement. *World Development* 34, 1887–1906.

Weeratunge, N., Snyder, K. and Choo, P.Z. (2010) Gleaner, fisher, trader, processor: understanding gendered employment in fisheries and aquaculture. *Fish and Fisheries*. Available at: http://onlinelibrary.wiley.com/doi/10.1111/j.1467-2979.2010.00368.x/abstract (accessed 16 September 2010).

Westlund, L., Poulain, F., Bage, H. and van Anrooy, R. (2007) *Disaster Response and Risk Management in the Fisheries Sector*. FAO, Rome, Italy.

Williams, M.J. (2008) Why look at fisheries through a gender lens? *Development* 51, 180–185.

Williams, S.B., Hochet-Kibongui, A.-M. and Nauen, C.E. (eds) (2005) Gender, fisheries and aquaculture: social capital and knowledge for the transition towards sustainable use of aquatic ecosystems. ACP-EU Fisheries Research Report No. 16, EU, Brussels.

World Economic Forum (2006) *The Global Gender Gap Report 2006*. World Economic Forum, Geneva, Switzerland. Available at: http://www.weforum.org/pdf/gendergap/report2006.pdf (accessed 28 July 2010).

World Economic Forum (2007) *The Global Gender Gap Report 2007*. World Economic Forum, Geneva, Switzerland. Available at: http://www.weforum.org/pdf/gendergap/report2007.pdf (accessed 28 July 2010).

World Food Programme (2007) *Vulnerability Analysis and Mapping (VAM)*. Available at: http://www.wfp.org/operations/vam/ (accessed 24 December 2007).

WorldFish Center (2010) *Analytical Framework, Matrix and Tools for Gender and Fisheries*. WorldFish Center, Penang, Malaysia.

WorldFish/FAO (2010) *Fisheries and HIV/AIDS in Africa: Investing in Sustainable Solutions*. Available at: http://www.worldfishcenter.org/wfcms/SF0959SID/ (accessed 29 July 2010).

Index

Page numbers in **bold** refer to illustrations and tables

access
    constraints 169
    control 48–49, 87, 88, 228
    rights 20, 36, 63–65, 68, 69, 72, 232
accountability 119, **120**, 190
active versus passive approaches 96
adaptation
    autonomous 225
    barriers 142
    capacity 21, 42, 143, 224–227
    differentiated from coping 134–135
    management 109
    strategies 219, 221, 225
    see also adaptive management; capacity, building
Adaptive Environmental Assessment and Management 95
adaptive management (AM)
    components 50–51
    conceptual development 94–95
    current practice 96–97
    defined 94
    diagnosis and management role 37
    key features 127
    targets 79, 99
    theory 95–96
    uncertainty mitigation tool 93–110
    uncertainty and sustainability vehicle 36
    use 97–110
Adaptive Management of Renewable Resources 95
Africa 61, 115, 117, 122
agencies 143, 223

AIDS (Acquired immune deficiency syndrome) 155–156, **226**–227
alliances 122
    see also co-management; partnerships
allocations 19–20, 67–68, 70, 124–125, 203–205
Analytical Hierarchy Process (AHP) procedure 43
aquaculture **133**, **134**, 151–152
Asia 77–79, 115, 117, 121
Asia-Pacific Coastal Livelihood Diversification Programme 228
assessment
    local adaptation aid 140–**141**
    research approach reference material sources 27
    targets and methods defining guidance 99
    techniques 24–**25**, 37–39, 43, 47, 80–83
    see also evaluation
assets 227–**230**
    see also capital
assumptions 100–101, 107, 108
audiences 183–185, **191**

Bangkok Statement 61
Beach Management Units (BMUs) 119, **120**, 121
Becker's neoclassical model of rational criminality 197
behaviour 124, 169–170, 197, 200–201, 202
benefits 19, 22, 126–127, 161, 162
bioeconomic model 82
biological status 82

239

biophysical aspects 135, **136**, **137–138**, 199–200, 202
boundaries 22, 39, 40, 116–117, **120**, 121
brainstorming 100
business service providers (BSPs) 165
buy-back programmes 84, 85

CANARI 117, 123–124, 125, 126
capacity
    adaptive 21, 42, 143, 224–227
    assessment 80–83
    building 22, 117–118, 121, 123–124, 128, 173
        see also resilience
    data lack 78
    defined 75, 80
    management 76, 79, 84
    measurement 79–83
    perceptions 84
    production 5
    reduction programmes 79
    target level identification 79
    see also overcapacity; resilience
capital 42, 135, **136**, 172–173, 216
    see also assets
capture methods 78
capture-fisheries 161, 166
Caribbean area 115, 117, 122
catch per unit of effort (CPUE) 78, 83, 161
catches
    data, countries included 4–5
    estimation 4–5, 8
    history 70
    limitation **84**, 85
    production and utilization **6**
    quota (harvest rights) 66–67, 68, 201
    rights 65
    total allowable 66, 82, 85, 86
    underestimation 8
    see also harvesting; productivity
change 108, 109, 122, 143, 151
    see also drivers
children 221, **223**–224
climate change
    change driver 132, **133**, **134**, 135–148
    effects **136**, **137–138**, **139**
    mitigation 148
    policies 136, 140
    response options 141
    vulnerability 218, 221, 224–226
co-management
    adaptive management key conditions 127, 128
    approach 207
    community-based framework 89
    definition 116
    design principle approach 116–121
    individual/household incentive 126–127
    interpretation differences 20
    key condition 116
    levels 121–126, 128
    success factor 118
    see also collaboration; participation; partnerships
Coastal Zones and Marine Ecosystems projects 148
Code of Conduct for Responsible Fisheries (CCRF) 19, 63, 72, 161
collaboration 10–11, 20, 37, 69–70, 127
    see also co-management; partnerships
collectives 67, **120**
Committee of Fisheries Management (CAP) 119
commons 19, 116, 160, 164
communication
    audiences 183–185, **187**, **191**
    barriers 180–181
    conversations 145
    data use 179
    dialogue 145, 156–157
    facilitation 188–189
    flows 181–182
    governance 190
    importance 178–179
    internet resources **195**
    leadership 189
    mass media **187–188**, 190–193
    messages 105, **179**–183, 185–186
    organizational chart **181**
    outputs and outcomes monitoring 186
    pathways 186, **187–188**
    planning 105, 189
    policy 189
    processes 179–180, 186–190
    products 186
    stages 180
    stakeholders 183–185
    strategies 182–183
    threats awareness 140
community-based approaches 20, 68–69, 88–89, 116, 121–126, **229**
    see also co-management; participation
community-based coastal resources management (CBCRM) 161, 162
Compass and Gyroscope 95
competition hypothesis 170–171
compliance 196–197, **201**, 203, **204**, 207–209
concepts development 94–95, 116
conceptual model (situation analysis) 99–100, 102, 107
conflicts
    causes 61, 62, 164–165
    legal pluralism problem 205
    management 118, 119, **120**, 186, 188
    overcapacity indicator 82–83

congruence **120**
conservation 84, 88, 123, 161, 184
constraints 35, 36, 117, 160, 169, 172–173
consumption figures **6**, 9, 151
contracts 172, 174
cooperatives 162–165
    see also co-management; collectives
coping strategies 134–135, 143, 219, 221
    see also adaptation
costs 19, 104, 126–127, **139**, 148, 152
    see also expenses; prices
countries 3–**7**, 4–5, 115, 148
credit 173
crime 197, 202, 205
    see also illegal fishing
CRiSTAL 42
customs 123, 142, 201, 205
cyclical feedback loop 96

data
    collection methods 104–105
    communication use 179
    countries included 4–5
    deficiencies 150
    envelopment analysis 81
    lack, fishing capacity 78
    management 104–105, 106–107
    pairing 105
    requirement 151
    secondary 99
    sources 42–43
    systems 2–3, 9–11
databases, online 42–43
decentralization 153, 154
degradation 88, 148, 232
demand 77–78, 151
democratization 154
demography 132, **133**, **134**, **139**, 150–151
deterrence 197, 203, **204**, 207–209
development **133**, **134**, 151, 153–154
DFID (Department for International Development), UK 50
diagnosis approaches 35–52, 145, 217
discrimination 218, 221
    see also exclusion; gender issues; marginalization
diseases 132, **133**, **134**, 154–156, 232
diversification 143, 145, 225, 228
documentation 109
drivers
    analysis 132–133
    change 140, 142, 148–150
    compliance, /non-compliance 199, 202
    concepts and definitions 132–133
    defined 133
    economic **139**

external 25, **36**, **42**, 122, 132–157, 220, 232
governance **133**, **134**
identification 39, 217, **218**
list **133**
market 151–153
need or greed 208
participation 153
social 105
tree 133, **134**
    see also change; governance; incentives; markets; poverty; subsidies
drought **136**, **138**

EAF (ecosystem approach to Fisheries) 18–19, 20, 22, 23–**24**, 29, 30
    see also ecosystem-based approaches
ecology 18, 42, 43, 94
economics 42, 82, 88–89, **139**, 166, 167–168
    see also markets; socio-economics; value chain approach
ecosystem-based approaches 17, 18–19, **27**, **42**
    see also EAF
efficiency 175
effort 19, 65–66, 78, 83, **84**, 161
EJF (Environmental Justice Foundation) 198
El Niño **137**, 142
Elimination of Child Labour (IPEC) Programme 224
empiricism 95
employment 3, 5, **6**–7, 9, 88–89, 160
    see also fishers; livelihoods
empowerment 123–124, 128, 233
    see also participation
enablement 121–122, 173–175
enforcement
    contracts 172, 174
    law 197, 199, 207, 208
    management rules 126
    methods 118
entitlements 88
    see also allocations; rights
entry, limiting 84
environment 88, 95, **133**, **134**, 140–141, 202
equity issues 83, 86, 203, **204**–205, 222
estimation methods 81–83
    see also assessment; measures
ethnography 206
evaluation 102–103, 105–107, 108, 110
    see also assessment; monitoring
evidence 109
exclusion 217, 219, 220, 221–224, 232
    see also marginalization
exit strategies **84**
expenses 104
    see also costs
experimentation 23, 50–51, 95

exploitation 23, 233
  see also overfishing
exports 8
extension services 175
externalities **36**, **42**, 122, 132–157, 220, 232
  see also drivers; opportunities; threats

facilitation 188–189
failures 142–143
FAO (Food and Agriculture Organization) 2, 18, 23, 154
FCAs (Fishery Cooperative Associations), Japan 164, 165
Fees, royalty 86
  see also incomes
finances 125, 148, 172–173
  see also economics; funding
fish-farming methods 164–165
fisheries *see* fishery systems
fishers
  active management engagement 98
  categories 80
  limitation **84**
  numbers 5, **6**, 78, 150, 160
Fishery Policy and Management Cycle, FAO 23
fishery systems
  classical dimension 24
  context 41–43
  current situation description 98–100
  defined 39–40, 133
  dynamics 51
  identity 39, 40–41
  perspectives 51
  restructuring 152
  status 82, 199–200
  user categories 88
flexibility 94
food security 8, 9, 61, 146
FOS (Foundations of Success) 100
fragmentation 161, 162
frameworks 17, 23–26, 48, 52, 119, 128
fuel 150, 152
funding 103, 104, 125, 146–148
  see also economics; income

gear restrictions **84**, 85, 88
GELOSE (Geston Locale Sécurisée) 119, **120**, 121
gender issues **7**, 141–142, 145–146, 221, 222–223
  see also women
Global Conference on Small-Scale Fisheries (2008) 71
goals 100, 102
  see also objectives
Gordon-Schaefer model 19

governance
  communication improvement role 190
  data 9–11
  defined 153
  development policy 153–154
  good, principles 20
  management perspectives 17
  new approaches 209
  reform 1–2
  research approaches **27**
  resource use monitoring 202
  value chain upgrading role 168
  see also institutions; legislation; management; policies
governments 122, 124, 146, 176, 199
groups 86, 87, 122–123, 162
guides and guidelines
  assessment targets and methods defining 99
  BMU setting up 119
  budget creation 104
  data management 107
  EAF into practice 20
  ecosystem approach to Fisheries 20
  goal and strategies defining 100
  institutional analysis 50
  management effectiveness evaluation 106
  monitoring plans development 103
  objectives and activities development 102
  threats identification and prioritization 99
  visioning understanding 44

habitat **139**
harvesting 66–67, 68, 82, 94, 201, 202
  see also catches
health 132, **133**, **134**, 154–156
health and safety standards 153
HIV (human immunodeficiency virus) 155–156, **226**–227
Holling, C. S. 94, 95
households 9–10, 116, 126–127
hunger focus 154
hygiene standards 153
hypotheses, testable 101

IBEFish 48
ICSF (International Collective in Support of Fishworkers) 60–61
illegal fishing 154, 161, 198–199, 205, 208
ILO (International Labour Organization) 154, 223, 224
implementation frameworks 17, 23–26, 52, 128
incentives 83–84, 86, 118, 126, 231–232
  see also drivers; subsidies
incomes 227–**230**

indicators
   baseline data 104–105
   food security 9
   management performance 42, 44, 45–46
   measurement 103
   overcapacity 82–83
   results chain process 101
   selection 95
   sustainability 45
   see also monitoring
individuals 67, 68, 85, 86, 116, 126–127
inequities 145–146, 204–205, 220
   see also gender issues; women
information
   acquisition 36
   exchange 179
   flow 117, 171–172
   gathering 98–99
   participation importance 123
   sources 42–43, **138**, **195**
   systems 2–3
   see also communication; data; knowledge systems
infrastructure 41, 174
inland sectors **6**, 7–8, 9, **138**
innovations 16, 17, 18, 30, 35, 43, 49
   see also adaptive management; co-management; EAF; ecosystem-based approaches; resilience
institutional analysis and development (IAD) framework 29, 50
institutions
   approaches 29
   arrangements 206–207
   assessment **27**
   building 171–175
   design 49
   framework inclusion 25, **36**
   functionality 49
   nested 118–119, **120**
   performance 116
   structures 46, 49–50
integrated approach 18, 30, 42, 200, 224, 233
Integrated Assessment and Advisory framework 24–**25**
interactions 46, 47–49, 59–72
International Institute for Environment and Development 50
International Organization for Migration 223
The International Plan of Action for the Management of Fishing Capacity 76
internationalism **133**, **134**, 202
internet **38**, **195**
interventions 156, 221, **229**, 230–231
interviews 98
investments 17, 174, 225
   see also funding

IPCC (Inter-Governmental Panel on Climate Change) 135, 218, 221
issues, identification **42**, 43

justice, social 203, **204**, 205–206, 209

Kearney, J. 60, 61, 67
Kenya 119–121
knowledge systems 50, 110, 115, 126, 127, 174
   see also information; learning

labour 151, 154, 221, **223**, 224
   see also fishers; women; workers
Lake Chilwa fishery, southern Malawi 40, 148
laws
   customary 201
   enforcement 197, 199, 207, 208
   evolved 201
   formal 202
   legal pluralism 205
   role 197
   understanding, compliance factor 209
LDCs (Least Developed Countries) 148
leadership 123, 189
learning
   adaptive **36**, 50, 51, 110, 128
   by doing 94, 96, 127
   capturing 109
   co-management factor 115
   diagnosis phase **36**, 39
   enhancement strategies 50
   experimentation 51
   from history 142–143
   opportunities 37
   see also knowledge
Lee, Kai N. 95
legislation 19, 119, 121–122, 161, 205
   see also laws
legitimacy 19, 202, 203, **204**, 206–207, 209
licenses 65, 70, 84, 85
limits 64–65, **84**, 85, 88
   see also gear restrictions; quotas
livelihoods
   adaptive 225
   alternative **84**, 88–89, 206
   analysis 221
   approaches 61
   complementary 206
   concept 219
   diversification 143, 145, 225, 228
   fragility 202
   insecurity 216
   metrics 3
   options 89

livelihoods *continued*
    strategies 142, 150–151, 219, 225
    sustainability **42**, 205–206, 220, 221, 231–233
    see also opportunities; SLA; threats

Madagascar 119–121
maladaptation 225
management
    adaptive phase importance 28
    approaches 18–23, 156–157
    assessment 42, 44, 45–46, 50
    change implications 108–109
    choices 27
    constituency 46–51
    evaluation 102
    failures 17
    framework **26**, **36**
    perspectives 16–17
    strategic activities 102
    structures 46, 49–51
    targets 99
    see also adaptive management (AM)
Management Planning and Implementation Cycle, FAO 23–**24**
managers 146
mangrove loss 140
mapping 47, 140–141, 145, 167–168, 218
marginalization
    causes 150–151
    compliance discourse inclusion 197
    compliance/non-compliance factor 201
    effects 198
    factors 164, 205, 216, **230**, 231
    power-driven policies effect 204
    reduction strategies 221
    systematically disadvantaged **218**
    see also exclusion
Marine Living Resources Act 62
marine protected areas (MPAs) 88, 123, 161, 184
marine sectors 7–8, 9
markets
    analysis 170–171
    development 160–176
    drivers 151–153
    globalization **133**, **134**, 202
    identification and examination 165
    information flow 171–172, 174
    integration 152
maximum sustainable yield (MSY) 23
measures 79–83, 84, 103, 153, 154
    see also metrics
Medan fishing programme 224
media **187–188**, 190–193
membership 117, **120**, 121
metrics 4, 5, 10
    see also measures

migration 150–151, 219, 225
Millennium Development Goals 20, 154
mobility 151, 164, 219
monitoring
    communication outputs and outcomes 186
    control and surveillance 198, 199
    implementation 104–105
    methods 103
    of monitors 121
    resources 119, **120**, 202
    see also evaluation; indicators
multi-criteria approaches 43

NAPA (National Adaptation Programme of Action) projects 148, **149**
natural resource management 16–17, 94, 96, 199, 229
needs 202, 206, 208
networks 46, 47, 48–49, 122, **181**
NGOs (non-governmental organizations) 165
normative action model 197, 203, 208–209

objectives 43–45, 101–102, 109, 125–126, 128
observations 94, 108
opportunities
    awareness 40
    external sources **36**
    identification 43
    management perspectives 16–17, 25
    non-sectoral drivers effect 25
    ranking and prioritizing 41–43
    sources 29, 30, 35, 42, 169
organizations 123, 124, 165, 166, **181**
overcapacity 75, 82–90
overfishing 41, 76, 150, 156, 161, 162
    see also overcapacity

Pacific area 115, 228
participation
    achievement 123–124
    approaches 18, 39, 87
    assessment techniques 37–39, 47
    barriers 145–146
    communication policy and planning 189
    compliance encouragement factor 207
    constraints 38–39
    design principles inclusion **120**
    diagnostic process 41
    governance 89
    meaningful 145
    problems 47–48
    resource management 232–233
    tools 44–45
    see also co-management

Participatory Diagnosis and Adaptive
    Management (PDAM) framework 35–52
Participatory Rural Assessments (PRAs) 37–39
partnerships
    development, value chain 171
    extension services provision 175
    financial 173
    poverty reduction role 217
    power sharing role 146
    public-private 171, 175
    sense of ownership 125
    social exclusion issues strategies 221
    success keys **147**
    trust building 128
    see also co-management
people 21, 28, **42**, 46–47, 145
    see also stakeholders
permissions, securing 103–104
phytosanitation 153
planning 97–103, 105–106, 189
policies 23, 136, 141, 152, 153–154, 233
    see also Code of Conduct for Responsible
        Fisheries; EAF; frameworks
port state measures (PSM) 154
post-harvest activities 160
    see also employment
post-harvest uses 4–5, **6**, 9
poverty
    compliance/non-compliance factor 199, 202
    data 9–11
    described 218
    dimensions and drivers identification 29,
        217, **218**
    illegal fishing driver 198, 199
    levels 77
    livelihoods link 217–221
    markets access constraint 165
    reduction 30, 146, 216–233
    sustainability connection 77
    understanding 217–221
    see also marginalization; vulnerability
Poverty Reduction Strategy Papers 233
power
    compliance factor 201, 209
    imbalances 204–205
    relationships 220
    relinquishing reluctance 122
    sharing 127, 146
presentations 193
press releases 190–193
prices 151–152, 165–166
    see also costs
privatization 86
processes *see* institutions; policies; shocks
productivity 5, **6**, 8–9, 10, 135, **139**
    see also catches
projections 43–45

projects 148, **149**, **229**
protection 62, **84**, 88, 123, 161, 184
public services 89

quotas 66–67, 68, 85, 86, 201

rapid-appraisal techniques 37–39, 82
rationalist model 197, 203, 207–209
recruitment **139**
Reduced Emissions from Deforestation scheme
    148
regionalization 153–154
regulations 118
    see also laws; rules
replicates 105
research-based approaches 17–18, 26–29, 30,
    37–39, 43, 50–51
resilience
    approaches 27–28, 30
    building 94, 150, 156
    collaborative management need 20
    definition 22, 135, 220
    enhancement 216–233
    maintaining 94
    management 16–17, 21–23, 30
    promotion factors 45
    theory testing bar 23
    see also capacity
The Resilience Alliance 28
resources management 87–88
resources securing 103–104
responses 16, 17, 40, 132–157
responsible fishing 154
results chain 101, 102, 107
revenue **139**
rights
    awareness 20
    categories
        children 223–224
        community 61, 66, 67–69, 70
        customary, protection 202, 203–205
        group fishing 86, 87
        harvest 66–67, 68, 201
        human 20, 23, 154, 203, 233
        individual 67, 68
        ownership 62, 124–125
        participation 20
        property 19–20, 119, 124–125, 160, 174
        resources 124–125
        tenure 64, 165, 174, 205
        use 64, 86, 87
        women 222–223
    defence principles 232–233
    forms 62–68
    impacts 87

rights *continued*
    operational-level 67
    recognition 205
    strengthening 232
    to organize **120**
    transferability 66, 71
    see also allocations
rights-based approaches
    human/fishery interaction 59–72
    implementation 69–71
    institutional factors 201
    limited entry approach 64–65
    management approaches 17, 67
    participation principles 30
    perspectives 29
    poverty reduction aspect **218**
    Web-based reference material **27**
    see also rights
risk 42, 43, 169, 224–227
Root cause analysis 100
rules 118, 119, 126, 160, 201

sanctions 118, **120**, 121, 207–208
sanitary measures 153
scale
    analysis 140
    attributes 39–40
    catch data disaggregation 4
    concept 116–117
    defined by fishing methods 1, 79–80
    description difficulty 1
    industrial fisheries 171, 198
    large fisheries **6**, 8–9
    scaling up 156
scenarios 44–45
sea cucumber fishery 40
sea levels **136**, 137
seasonality **84**, 164
sectors comparison 2, **6**, 7–8
Seri fishing community 118
sex risk 155, 227
sharing 107, 110, 127
shocks 218–219, 220
    see also threats
situation analysis (conceptual model) 99–100, 102, 107
size
    enterprises 153
    see also scale
SLA (Sustainable Livelihoods Approach) 28–29, 42, 217, **218**, 219–221
SLFP (Sustainable Fisheries livelihoods Programme) 228, **229**
social systems
    co-management preparation 123–124

    cohesion 202
    context 202
    dimensions 42
    governance approach inclusion 200
    justice 203, 205–206, 209
    services provision 174
social-ecological systems (SESs) 21, 133, 225
socio-economics 10, **38**, 82, 83, 161, 200
Sofala Bay, Mozambique 118–119
speaking, public 193
stakeholders
    adaptive management participation encouragement 103
    analysis 47, 98, 184–185
    engaging 98
    identification 46–47, 183–185
    matrix **185**
    objectives development 43–44
    resource management role 201
    targeting 183–185
    typologies 47
    see also people
standards 119, 153, 172, 174–175
stewardship role 165, 217
stocks 78, 150, 161–162, 164, 232
    see also overcapacity; overfishing
Stop Illegal Fishing report 198
storms **136**, **138**
storylines 44
strategies 67, 100, 136, 140–142
stressors 225
    see also shocks
structures 46, 49–50, 124, **136**, 148
subsidies 150, 152, 174, 175
    see also incentives
Subsistence Fisheries Task Group (SFTG), South Africa 61
suki relationship 169
support 124, **136**, 148
surveys 10, 76, 98
sustainability
    challenges 216
    co-management factor 207
    concept evolution 209
    definition 220
    explained 200
    focus 16
    holistic advice 27
    indicators 45
    management 16, 17, 36
    methods 231–233
    poverty connection 77
    programme 228, **229**
    resilience metaphor 21
    see also capacity; resilience
systems thinking 200

target catch ratio 82
target resource-orientated management (TROM) approach 27
taxes 86
team formation 104, 106
technology **133**, **134**
temperature 135, **136**, **137**, **138**
tenure 64, 165, 174, 205
territorial use rights in fishing (TURFs) 64, 86, 87
test-learn-adapt 94
Thomson, D. 2
threats
    awareness 40
    classification 99
    identification 43, 99
    management perspectives 16–17, 25
    prioritizing 41–43
    reduction result 101
    sources 29, 30
    understanding 136–142
    see also climate change
total allowable catch (TAC) 66, 82, 85, 86
trade **133**, **134**, 140, 151–153
    see also markets
traders 165–166, 168–170
transparency 190
    see also accountability
trial-and-error approach 96, 127

uncertainty 36, 93–110
UNCLOS (United Nations Convention on the Law of the Sea) 161
UNFCCC (United Nations Framework Convention on Climate Change) 143, 148
Unified Classification of Direct Threats 99
Universal Declaration of Human Rights principles 20

value 21–22, 146, 175
value of catch per unit of effort (VCPUE) 83
value chain approach 160, 165–168, 175
vessels 78, 79, 84, 85
visioning 44–45
vulnerability
    assessment 28, 140–142
    conceptualization 221
    explained 218
    external factors 220, 232
    fisheries fragmentation factor 160
    key factors 134
    mapping 140–141, 218
    poverty factor 29, 202, **230**
    reduction **218**, 224–227

Walters, Carl J. 94, 95
water availability **136**, **138**
wealth-based approaches **27**, 62
web-based reference material **38**, 195
welfare functions 62
well-being approaches 28
WFFP (World Forum of Fisher People) 60–61
WHAT (World Humanity Action Trust) Commission 86
women
    barriers 145–146, 169
    change adaptation agents 143
    fisheries employment 3, **7**, 160
    marketing role 165, 169
    rights 61, 222–223
    vulnerability 141–142, 155
    see also gender issues
work plan monitoring 104–105
workers **6**, 9
    see also employment; fishers; labour; women
World Bank 2
World Development Report (2008) 152
Worldfish 2